Sustainable Materials and Technology

Sustainable Materials and Technology (SMT) book series publishes research monographs (both edited and authored volumes) showcasing the latest developments in the field and comprehensively covering topics such as:

- Recycling of waste into useful material and their energy applications
- Catalytic action of Nano oxides for efficient carbon reforming process
- Sustainable technologies for plastic transformation
- Bifunctional nanoparticles for sustainable water splitting applications
- Sustainable dying and printing
- New materials from waste
- Sustainable Manure Management and Technology: Potentials, Uses and limitations
- Sustainable Mechanical Engineering Approach
- Sustainable biochemistry for the improvement of health
- Sustainable development of Mechanical recycling of automotive components
- Sustainable-waste recycling and conversion in useful materials for different applications
- Sustainable development of inexpensive Nano-photo catalysts
- Sustainable development of recycling of discarded lithium ion batteries
- Modern sustainable cement and concrete
- Sustainable adsorbent for hazardous removal
- Sustainable superior electromagnetic shielding materials
- Excellent sustainable nanostructured materials for energy storage device
- Sustainable development of heavy metal detoxification from water
- Carbon dioxide utilization for sustainable energy
- Sustainable development in green syntheses of materials
- Environment friendly and sustainable cloth for garments application
- Sustainable design and application of eco-materials
- Nanoparticles for sustainable environment applications
- Sustainable remediation of industrial contaminated water towards potential industrial applications
- Biomaterials for sustainable bioremediations

Ramzi Khiari · Mohammad Jawaid
Editors

Current Status and Opportunity in Fibre and Composites

Posidonia oceanica

Springer

Editors
Ramzi Khiari
Department of Textile
Higher Institute of Technological Studies
of Ksar Hellal
Ksar Hellal, Tunisia

CNRS, Grenoble INP, LGP2
Université Grenoble Alpes
Grenoble, France

Mohammad Jawaid
Department of Chemical and Petroleum
Engineering
College of Engineering
United Arab Emirates University
Abu Dhabi, United Arab Emirates

ISSN 2731-0426 ISSN 2731-0434 (electronic)
Sustainable Materials and Technology
ISBN 978-981-96-4545-9 ISBN 978-981-96-4546-6 (eBook)
https://doi.org/10.1007/978-981-96-4546-6

This Springer imprint is published by the registered company Springer Nature Singapore Pte Ltd.
The registered company address is: 152 Beach Road, #21-01/04 Gateway East, Singapore 189721, Singapore

The editors are honoured to present this book.
Ramzi Khiari dedicates his work to his father.

Dedicated to my father
"Mohamed Lazhar Khiari",

In memory of the man who shaped me, guided me, and filled my life with love. Although you left us on 24 May 2024, your spirit lives on in my heart and in everything I do. On this sorrowful occasion, we pray to Allah, asking that the Almighty covers the deceased with His immense mercy, grants him a place in the

highest of paradises in the company of the prophets and martyrs, and bestows upon our family and loved ones great patience and comfort. We belong to God, and to Him we shall return.

May Allah shower him with mercy!

This book is a tribute to your life, your wisdom, and your unfailing support. May your soul find peace and may you always be remembered.

With all my love,
Ramzi Khiari

Preface

Posidonia oceanica (*P. Oceanica*) is a major Mediterranean Sea grass as leaves or balls. It lives exclusively in the sea. It belongs to the marine phanerogam family and more specifically to the *Posidoniaceaes* family. *P. oceanica* is widespread, and its geographical distribution has been studied by numerous researchers. Aspects of its ecology and physiology have also been reported. *P. oceanica* was first accurately described in 1623 by G. Bauhin under the name Alga marina. It was not until the nineteenth century that Delile, in his work entitled "La flore d'Egypte" published in 1824, gave it its current name of *Posidonia oceanica* (Linnaeus) Delile. The structure of *P. oceanica* shoots has been extensively described in numerous works. They have the same cell structure and morphology as other marine phanerogams. It occurs in shallow and costs, and it plays an important role in the ecology of the Mediterranean basin. *P. oceanica* has been recognized for its capacity to stabilize marine environments, support biodiversity, and act as a carbon sink, but recent research has uncovered another promising avenue for its utilization: its fibres' potential in sustainable composites and other materials. It was investigated as source of lignocellulosic fibres for pulp and paper production and as biosorbent for pollutants removal. *P. oceanica* can be used in different forms, as raw material, chemically modified or transformed in activated carbon, cellulose, microcrystalline cellulose, and nanocellulose (CNC and CNF).

This book, *Current Status and Opportunity in Fibre and Composites—Posidonia oceanica*, provides an in-depth exploration of the current knowledge surrounding *P. oceanica*, with a focus on its application in fibre technology and composite materials. Published by Springer Nature, the book brings together leading experts from across disciplines, combining insights from marine biology, environmental science, materials engineering, and sustainable development. This book presents an extensive survey on recent improvements in the research and development of *P. oceanica* that are used to make fibre and/or raw material in various applications, because of its outstanding features like lightweight, low cost, environmentally friendly, and sustainability. The unique feature of this book is it presents a unified knowledge of *P. oceanica* on the basis of synthesis, characterization, properties, and applications. This book also focusses the recent advances in *P. oceanica* fibres from renewable resources and introduces potential applications of this materials. This book gives a sound knowledge on *P. oceanica* to the readers with numerous example illustrations, methods, and results for graduate students, researchers, and industrialists. Readers will have a quick reference by exploring the research literature on the subject with commercial value-added research applications of bamboo composites.

We hope that this volume will not only provide a better understanding of the ecological importance of *Posidonia oceanica*, but also spark new ideas and collaborations that will advance the use of natural fibres in high-value materials. The synergy between preserving our natural heritage and advancing technological innovation is at the heart of this work, and we are delighted to share it with the global scientific community. We are thankful to all authors contributed book chapters from different part of world in this edited book and make our imaginary thought into reality. Lastly, we are highly thankful to Springer Singapore team for continuous support at every stage to make it possible to publish on time.

Ksar Hellal, Tunisia/Grenoble, France — Ramzi Khiari
Abu Dhabi, United Arab Emirates — Mohammad Jawaid

Contents

About the Editors

Dr. Ramzi Khiari is a senior lecturer at the Higher Institute of Technological Studies in Ksar-Hellal (Monastir, Tunisia) in the department of Textile Engineering. He was graduated in 2005 at the National Engineering School of Monastir in the specialty of Textile Chemistry, before getting a Master degree (2007) from the same institution. Then, he performed a sandwich Ph.D. thesis (2010) between ENIM from University of Monastir in Tunisia and Grenoble INP in France. Finally, in 2017, he has got the diploma of "Habilitation Universitaire" from University of Monastir, and in 2020, he has got the diploma of "Habilitation à diriger des Recherches" in Grenoble INP. His research interests focus on the valorization of biomass at multi-scale levels, namely fibres, nanocellulose, lignin, hemicelluloses, and their uses as potential raw material in several industrial applications (textile, papermaking, polymeric materials, composites, and nanocomposites). A particular focus is given for vegetal biomass from annual plants and particularly agricultural residues and industrial wastes. During his career, he coordinated several research projects, mainly projects with industry, and supervised various postgraduation students. He has been a member of scientific committees and organizing committees and participated in national/international evaluation boards.

Dr. Mohammad Jawaid is currently working as a distinguished professor at Chemical and Petroleum Engineering Department, United Arab Emirates University, UAE. Earlier he worked as a senior fellow (professor) at Biocomposite Technology Laboratory, Institute of Tropical Forestry and Forest Products (INTROP), Universiti Putra Malaysia (UPM), Malaysia, and also has been a visiting professor at the Department of Chemical Engineering, College of Engineering, King Saud University, Riyadh, Saudi Arabia, since June 2013. He also worked as a distinguished visiting professor at Malaysia Japan International Institute of Technology (MJIIT)-Universiti Teknologi Malaysia (UTM), Kuala Lumpur, Malaysia. He has more than 24 years of experience in teaching, research, and industries. His area of research interests includes hybrid composites, lignocellulosic reinforced/filled polymer composites, advance materials: graphene/nanoclay/fire retardant, modification and treatment of lignocellulosic fibres and solid wood, biopolymers and biopolymers for packaging applications, nanocomposites and nanocellulose fibres, and polymer blends. So far, he has published 70 books, 90 book chapters, more than 500 peer-reviewed international journal papers, and several published review papers under top 25 hot articles in science direct during 2013–2020. He also obtained 6 Patents and 6 Copyrights/Industrial design. H-index and citation in Scopus are 92 and 36,560, and in Google Scholar, H-index and citation are 110 and 49,425; H-index = 79 (Web of Science). He is a founding series editor of Springer Three Book Series. He worked as a guest editor of special issues of several peer-reviewed journals. He is also professional membership of IOM, ACS, SPE, APA, MySET, etc. Besides that, he is also a reviewer of several high-impact international peer-reviewed journals of Elsevier, Springer, Wiley, Saga, ACS, RSC, Frontiers, etc. Forty Ph.D. and 17 Master's students graduated under his supervision in 2014–2023 in the fields of hybrid composites, green composites, nanocomposites, natural fibre-reinforced composites, nanocellulose, etc. He secured various research grants at university, national, and international levels on polymer composites of around 6 million Malaysian ringgits (USD 1.36 Million). He also delivered plenary and invited talks in international conferences related to composites in

India, Turkey, Malaysia, Thailand, the United Kingdom, France, Saudi Arabia, Egypt, and China. Besides that, he is also a member of technical committees of several national and international conferences on composites and material science. Dr. Mohammad Jawaid also received several awards at National and International Level. He is also the winner of Newton-Ungku Omar Coordination Fund: UK-Malaysia Research and Innovation Bridges Competition 2015.

Chemical Composition and Applications of *Posidonia oceanica* Fibers—A Comparison with Other Cellulosic Fiber Sources

Medhat E. Owda, Ragab Abouzeid, Saad Melhi, Eid H. Alosaimi, and Mahmoud A. Abdelaziz

Abstract Many tons of *Posidonia oceanica* are removed from the shores of the Mediterranean every year in order to maintain the quality and enjoyment of the beaches. Most of this waste is disposed of in landfills, where it is expensive to remove. Therefore, in this chapter, the chemical composition of *P. oceanica* and its potential applications in many fields will be examined and compared with other non-wood raw materials. These wastes can be blended with polymers to form composites consisting of natural fibers and polymers with outstanding properties. Compared to synthetic fibers, Posidonia fibers are an inexpensive, renewable, environmentally friendly, and suitable source for reinforcing polymer compounds.

Keywords Pulp and paper · *Posidonia oceanica* · Grass fibers · Chemical composition

1 Introduction

The rapid growth in population, industrialization and urbanization, and the demand and consumption of paper have increased significantly. The use of pulp and paper has affected by the development of many countries and markedly by the levels of civilization [74]. The best renewable natural materials for papermaking are wood,

M. E. Owda
Chemistry Department, Faculty of Science, Al-Azhar University, Nasr City, Cairo, Egypt

R. Abouzeid (✉)
Cellulose and Paper Department, National Research Centre, Dokki, Giza, Egypt
e-mail: r_abouzeid2002@yahoo.com

S. Melhi · E. H. Alosaimi
Department of Chemistry, College of Science, University of Bisha, P.O. Box 511, Bisha 61922, Saudi Arabia

M. A. Abdelaziz
Department of Chemistry, Faculty of Science, University of Tabuk, Tabuk, Saudi Arabia

R. Khiari and M. Jawaid (eds.), *Current Status and Opportunity in Fibre and Composites*, Sustainable Materials and Technology,
https://doi.org/10.1007/978-981-96-4546-6_1

non-wood materials, and recycled fibers [12]. Over the past century, wood is the major raw material in the pulp and paper industry and the dominant source for cellulose, providing for 90–95% of all produced cellulosic pulp [68]. Increased demand and cost of wood, as well as shortage supply, cause the fiber-based industry to find alternatives. As a result, alternative cellulose sources have been evaluated further in recent years to reduce the demand for wood-based biomass and increase process sustainability. There are a variety of non-wood plant fibers that contribute as fiber supplies in regions where forest resources are few, such as India, China, and Egypt, and are highly importance in these areas [5]. Environmental problems in the papermaking industries have led to development of cleaner technology based on the reduction of wood consumption. Cleaner technology is used to generate more output with little impact on the environment, as well as lower disposal costs, stability hazards, and resource costs, resulting in a lower environmental threat and increased profits in pulp and paper-based industries [1]. Furthermore, it consumes a large amounts of wood, which must be harvested and delivered on a regular basis [46]. This is a major contributor to deforestation and its associated climatic effects on society's livelihood. In this regard, it is critical to incorporate non-wood resources in their entirety or in combination with natural wood to address environmental, social, and ecosystem management issues. Also, non-wood potential alternatives are numerous in some regions, especially in the developing countries [11]. Agricultural residues have recently become one of the most important alternative resources, owing to an increase in worldwide demand for fibrous raw materials, a lack of trees in many countries, and increased sustainability consciousness [17, 44]. These features for all kinds of fibrous materials such as fruits, vegetables, bark residues, and agricultural residues. Naturally growing plants, agricultural by-products, and industrial crops are possible sources of non-wood fibers [38]. There are around 1000 different types of fiber plants that can be used to produce fibers [96]. Various studies have shown that these agricultural waste products can be used to make pulp and paper [45, 52, 60, 70, 76]. In the pulp and paper industry, fiber morphology is an important characteristic to consider. Interfiber bonding could require a specific fiber length [70, 111]. Non-woody and recycled fibers, therefore, offer a fantastic opportunity to substitute woody fibers [81, 98]. Non-wood pulp and paper have a number of advantages over wood pulp and paper, as excellent sources for special papers, ease of pulping, and high-quality bleached pulp [10, 84].

2 Chemical Composition of Biomass Resources

The major characteristics of biomass resources are cellulose, hemicellulose, and lignin. As the major component of plant biomass, lignocellulose is composed of these three polymers. Unlike cellulose and hemicellulose, lignin is made up of aromatic polymers that are formed from phenylpropanoid precursors. Cellulose is a type of linear polymer that is made up of D-glucose subunits that are connected together via ß-1,4 glycosidic linkages. Fibrils of cellobiose molecules are held together by van der Waals forces as well as hydrogen bonds. As a result of these linkages, the monomers have a crystalline structure and cannot be degraded by enzymes or swelled by water. High temperatures and pressure, however, can rupture the hydrogen-bound crystalline structure and hydrolyze the ß-1,4-glycosidic link to produce glucose monomers. In contrast, hemicellulose has a more random structure and fewer repeating ß-1,4-glycosidic linkages than cellulose. As a consequence, hemicellulose's structure is less crystalline than cellulose's structure. Hemicellulose is a polysaccharide that is made up of D-xylose, D-mannose, D-galactose, D-glucose, L-arabinose, 4-O-methyl-glucuronic acid, and D-galacturonic acid, all of which are linked not only by ß-1,4-glycosidic bonds but also by ß-1,3-glycosidic bonds. L-arabinose is also Because it is composed of branches, hemicellulose is more susceptible to being hydrolyzed or extracted via hydrothermal processes than cellulose. Lignin is an amorphous heteropolymer that is made up of phenylpropane units that are linked to one another by a variety of various types of linkages. Although it is made up of the monomers coniferyl alcohol, sinapyl alcohol, and p-coumaryl alcohol, the relative proportions of these three alcohols can be very different depending on the source. Grass lignin is mainly composed of guaiacyl and syringyl units, as well as p-coumaric acid and ferulic acid. A lignin-rich cell wall is a structural support, a permeability barrier, and a resistance to microbial attack. Lignin also provides resistance against oxidative stress. Figure 1 illustrates the architectures of cellulose, a common form of hemicellulose found in grass, and lignin monomers.

There is a significantly higher ratio of cellulose to lignin in herbaceous materials such as grass than in woody materials such as branches. Lignin is found in around 5–25% of non-woody biomass. Depending on a variety of factors, including the maturity of the plant, the amount may differ even within species. The cellulose content of lawn

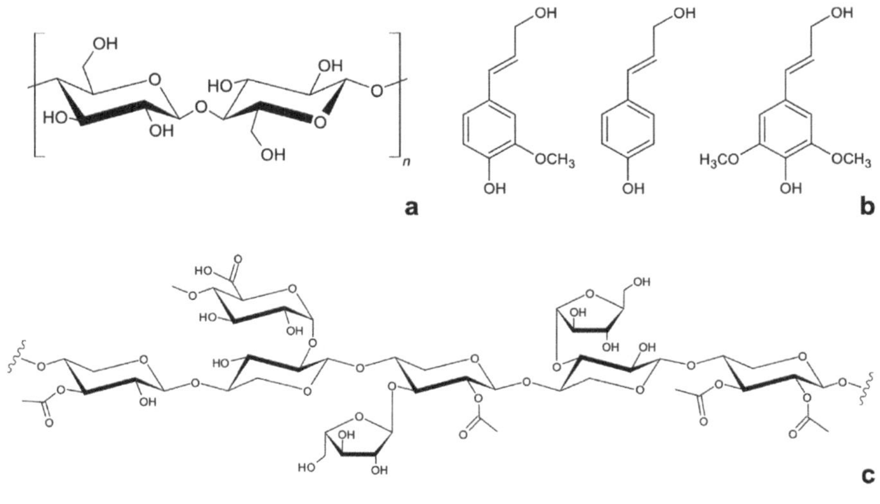

Fig. 1 Chemical structure of cellulose (**a**), the building blocks of lignin (**b**), and hemicellulose (**c**)

grass is 41.7%, the hemicellulose content is 35.8%, and the lignin content is only 8.0%.

3 Pulps and Papers from Grass Fibers: Sources and Compositions

The earliest paper was made from non-wood materials, primarily textile waste, in many early societies including the United States of America, China, and Egypt. So far, non-wood plant materials such as cereal straws, sugarcane bagasse, bamboo, reeds, esparto, kenaf, corn stalks, sorghum stalks, and *P. oceanica* have been tried for pulp production. In the last two decades, there has been intense research on alternative sources of pulp for the paper industry all over the world.

3.1 Papyrus

The papyrus plant is a perennial, non-woody sedge that can be abundantly found along the Middle East, Nile River, and Africa. The stem has a base diameter of 3–5 cm. Its height ranges from 2 to 5 m. The stalk is mainly composed of a husk or cortex that surrounds a pith segment. The pith is constituted of fiber packs that are surrounded by thin wall cells (parenchyma). There are additionally vessels, epidermal cells, and sclerenchyma present. The cortex is densely packed with fiber bundles. Papyrus plants grow at a rate of 45 tonnes per hectare every year [101]. 80–90% of the volume and 30–80% of the dry weight of papyrus stems are made up of the pith component. The inflorescence constitutes around 20% of the overall plant, less roots. The pith section of the fiber constitutes 10% of the total fiber weight. As a result, the amount of fibers in the stem can vary by 20–75%. Papyrus plant fibers are found in a wide variety of lengths and diameters, depending on the species, place of origin, and placement in the stem. Papyrus fibers typically measure 1.8 mm and 0.012 mm in length and width, respectively [101]. The harvesting of annual plants are used to make agro-based fiber. Wood, whether softwood or hardwood, contributes for 92% of worldwide paper and board production, while agricultural wastes account for about 8% [40]. Using these residues now is not a good idea because they are frequently burned in open fields under uncontrolled conditions, causing environmental harm due to the heavy smog produced [124]. In different countries, agricultural residues might be employed as active alternative materials to forest resources [97].

3.2 Esparto

Esparto grass grows naturally in southern Spain's Mediterranean grassland areas and North Africa. The grass is greyish green in appearance, and it has long rolled up leaves that can reach a length of 1 m. It grows in 3 m diameter bunches or clumps and is coarse and strong. When there is a lot of growth, the clumps tend to merge into one big mass that covers many square miles. Esparto takes about 12 years to grow from seed to harvestable form [14].

3.3 Reeds Grass

Reeds are long perennial grasses that grow in marshes and are widely distributed in the delta regions, river bottoms, and swamplands of Egypt, Russia, Romania, northern China, North Korea, and Spain. Based on soil characteristics, hydrologic conditions, nutrients, and pH, the diameter will range from 9 to 22 mm and the height from 2.5 to 5 m [2]. In October or November, the plant will be fully grown, but food will take another four to six weeks to accumulate in the root stock, ensuring its reproduction in the following season. In April, the moisture content of young stalks can reach 80%. By the end of July, the moisture content falls to between 60 and 65%, and between 26 and 27% by the end of December. The reduced moisture content lowers shipping costs and prevents reed damage during storage. Due to its low cellulose content and up to 10% silica, non-wood fiber covers are difficult to harvest. The average stalk fiber length is 0.35–3.35 mm, with a mean of 1.8 mm, and the average fiber diameter is 0.006–0.022 mm [2]. In the fusion of multiple sections, it is utilized to produce wrapping paper, duplex and triplex paper, corrugating medium, and linerboard. Green Sabai Grass. It is a tufted grass with stems that are 60–90 cm tall and long leaves that are primarily located close to the base. In India, Sabai grass is an important source of fiber for the paper industry, as well as Nepal and Pakistan. The species was a fibrous staple raw material that was known for its durability, strength, and stiffness [116]. The pulp has a similar quality to esparto and is hence used for printing paper and fine book. Sabai grass accounted for 22% of all fibrous material pulped in India in 1952 [1]. The fiber length ranged from 0.5 to 4.9 mm on average. The widths range from 0.01 to 0.016 mm. Ilvessalo-Pfäffli [51] describes fibers as a rather thick wall with a thin lumen and sharp, never branched ends.

3.4 Bamboo

Particularly in warm, tropical areas, bamboo grows. Bamboos have a widespread geographic range and are commonly found growing wild throughout the world. One of the plants that grows the fastest and is appropriate for pulp production is bamboo.

The stem typically has a diameter of 5–15 cm [18]. Bamboo fiber varies in length depending on the species. It changes according on the inter-nodal length in some species while varying from bottom to top in others. The average diameter of bamboo fibers, which have lengths between 2.7 and 4.0 mm, is 0.015 mm [107]. It is used to create newsprint, Bristol board, duplex, triplex paper, wrapping, bag paper, multiwall sacks, printing as well as writing paper.

3.5 *P. oceanica*

Due to its extensive distribution and productivity, *P. oceanica*, a member of the Posidoniaceae family, is the most significant species of seagrass for the Mediterranean Sea's ecologically coastal balancing [69]. *P. oceanica* is a slow-growing (>10 cm year^{-1}) marine phanerogam with roots, leaves, and rhizomes [69, 117]. *P. oceanica* colonizes vast underwater areas under suitable environmental conditions, developing extensive meadows from the surface to 40 m depth (Figs. 2 and 3) (although it has been documented in extremely clear waters till 48 m depth), comprising around 1.5% of the Mediterranean Sea's total surface area [20, 21, 86]. *P. oceanica* banquettes can grow up to 2.5 m thick [22, 108]. *P. oceanica* is an aquatic plant whose decaying leaves accumulated along the coastlines, giving a residue [13]. According to Dural et al. [31], total *P. oceanica* output in the Mediterranean basin ranges from 5 to 50 m per year. On the Tunisian coast, large amounts of *P. oceanica* fragments collect, causing beach cleaning every summer. The lignocellulosic fibrous leaves are residual biomass that is nonhazardous, light, renewable, and readily available without processing/preparation at a minimal cost [16]. The Mediterranean Coast of Egypt is the southern point of *P. oceanica*'s range in the Mediterranean Sea, where it grows in dense meadows ranging in depth from 5 to 30 m along the west coast [108]. Litter from *P. oceanica* washes up on Egyptian beaches, generating stranded deposits and egagropili beds [108]. In terms of marine coastal environments, a thick banquette of procambia leaves on a beach indicates the presence of healthy *P. oceanica* meadows, according to Boudouresque et al. [22]. As a result of its special qualities such as fire and moisture resistance, seagrass leaves are considered an environmentally friendly alternative for wood composite production. During the growth process, spherical aggregates of seagrass grow in underwater hollows, and the waves' dynamic movement intensifies the entanglement process [77]. *P. oceanica* is important to the Mediterranean basin's marine ecosystem because: (a) it captures large amounts of CO_2 and releases O_2 into the water. (b) It stabilizes the seafloor with its root system. (c) It provides food and shelter to a wide range of marine species. *P. oceanica*'s leaves are regularly renewed. During the autumn, the plant loses its leaves. Between 300 and 2000 g m^{-1} 2 year^{-1} of leaf production is predicted. As a result, vast quantities of dead *P. oceanica* leaves accumulate along the coast, and their clearance is costly. Their degradation produces hydrogen sulfide (H_2S). Typically, biomass is collected and disposed in landfills [22]. The leaves of *P. oceanica* are high in cellulose, hemicellulose, and lignin. *P. oceanica* lignin contains

(a)

(b)

Fig. 2 *P. oceanica* plant (**a**) and balls (**b**)

a lot of p-hydroxybenzoic acid, which is esterified to lignin by 2-methoxyphenol and 2,6-dimethoxyphenol groups [53, 54]. It also contains phenolic compounds, phenylpropanoic acid, and phenylmethane derivatives [27, 50].

One of the many biological functions provided by seagrass sediments is that they help to minimize beach erosion [110]. As well as contribute to the dissolved organic carbon (DOC) reserves in pelagic habitats along the coast and the microbiome [73], and providing habitat for a variety of invertebrates. Seagrass deposits on beaches are a nuisance, but they also have important ecological functions that must be studied in depth. By recovering the energy content of their lignocellulosic components, these deposits can be utilized as a renewable source of bi-product. For example, from *P. oceanica* egagropili, [28] developed high grade cellulose and a strategic industrial cellulose material [24] and showed that investigating the trace element (TE) composition of *P. oceanica* deposits could help to be used in industrial and agricultural applications. Trace elements (TE) come into the marine ecosystem from the air and the land.

Most of these sources are related to industrial, urban, and agricultural activities along the coast. Some TE, like Pb, Cd, Cr, As, and Hg, are highly poisonous non-essential elements, while others like Cu, Zn, Mn, Fe, and Ni are important to plant life and act as essential micronutrients [109]. Because trace elements stay in the environment and can be used by living things (Table 1), they move up the food chain to higher trophic levels and pose a threat to human health if they reach levels above

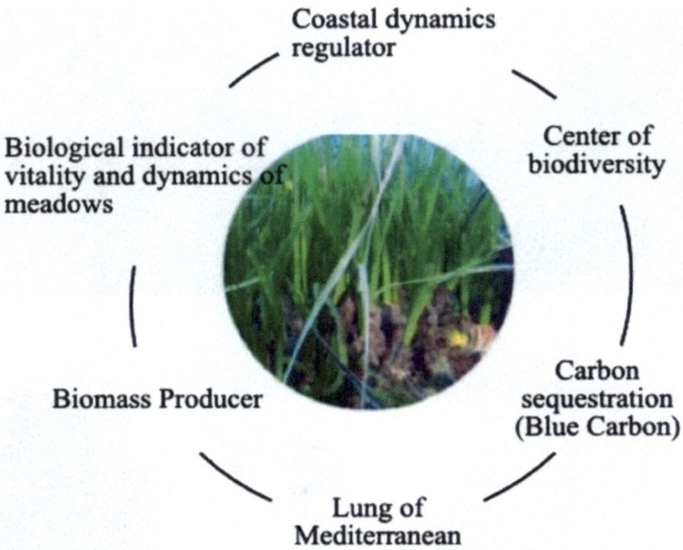

Fig. 3 Summary list of *P. oceanica* seagrass's ecological significance

Table 1 Chemical composition of *P. oceanica* [121]

Standards methods	Posidonia balls	Posidonia leaves
Extractives (ethanol/toluene)	10.7	19.2
Lignin (Klason lignin)	29.8	29.3
Hollocellulose	61.8	57.1
α-Cellulose	40.0	31.4
Ash	12.0	10.5

the toxicity threshold [19]. In Egypt, for example, *P. oceanica* has been used to measure pollution from trace elements [32, 57].

4 Chemical Composition of *P. oceanica*

The chemical composition of leaves was determined according to Tappi standard methods (Table 1), which indicated a high holocellulose content 57.1%, cellulose 31.4%, lignin 29.3%, extractives 19.2%, and ash 10.5%, while the cellulose content in balls is around 40.0% (ashes 12%, and extractives around 10.7%) [15].

Table 2 shows the chemical composition of *P. oceanica* and fiber grasses, which confirms that these raw materials contain comparatively high amounts of extractives, particularly ethanol–toluene extractives (10.7% for Posidonia and 13% for reeds).

Posidonia is higher in Posidonia (12%) than other grasses). The cellulose content of the raw materials studied is the same (40% for Posidonia and 40.8% for esparto) [115].

It can be concluded that the grasses contain higher cellulose content, as shown in Table 2. This suggests that using of these grasses would be more profitable for papermaking applications. In countries with insufficient forest resources, non-wood fibrous materials and recycled fibers can reduce or replace the use of wood fiber in pulp and paper production. In order to evaluate a raw material's pulping and paper-making potential, its proximate chemical composition must be determined, a pulping process and bleaching sequence must be identified, bleached and unbleached pulps must be evaluated, pulp fibers must be examined morphologically, and handmade paper sheets must be tested for their physical properties. Oceanica balls and other grasses. The ash compositions of different cellulosic grasses are compiled in Table 3 from published sources.

Table 2 Comparative analysis of the chemical of *P. oceanica* balls and other grasses

Amounts in % (w/w with respect to oven dried raw material) (%)	*P. oceanica* balls	Papyrus	Esparto	Reeds	Bamboo
	Khiari et al. [60]	Rosado et al. [92]	Nabais et al. [80]	Oleszek et al. [85]	Scurlock et al. [105]
Ethanol–toluene extractives	10.7	4.1	1.3	13	4.6
Holocellulose	61.8	40.1	93.7	59.7	87.8
Cellulose	40	18.4	40.8	35.6	49.1
Lignin	29.8	26.8	5.3	14.8	26.1
Ash	12	5	1.36	9.5	1.3

Table 3 Ash composition for *P. oceanica* and other grasses

%	*P. oceanica* balls	Papyrus	Esparto	Reeds	Bamboo
	Khiari et al. [60]	Rosado et al. [92]	Abdelhak [3]	Oleszek et al. [85]	Scurlock et al. [105]
Si	17.7	33.7	32.5	9.74	24.32
Ca	9.12	2.7	7.25	0.466	3.99
Mg	3.89	1.2	2.4	0.292	6.69
S	1.92	0.3	n.d.	0.185	14.05
K	2.04	28.2	1.32	1.84	34.23
Na	2.49	1.4	0.4	0.55	1.14
Fe	3.78	0.3	2.6	n.d.	2.38
Cl	0.72	4.1	n.d.	0.613	n.d.
P	0.12	2.1	0.6	0.247	2.58

(a) (b)

Fig. 4 *P. oceanica* (PO) leaves as collected (**a**) and after drying (**b**)

Due to the problems caused by the accumulation of dead leaves (especially during the tourist season) of *P. oceanica* (Fig. 4) on the coast various methods of utilizing this biomass have been proposed. Leaf litter has been suggested for the following applications: (i) Removal of contaminants from aqueous solutions by sorption processes. Fibers and particles derived from *P. oceanica* leaves have been used to remove pigments found in textile industry waste. (ii) As a fertilizer and composting material. (iii) As a source of cellulose for paper making and other applications. (iv) As a biofuel, and for the preparation of fiber-reinforced composites [26].

The data that was supplied regarding the chemistry of non-wood fibers, agricultural wastes, and annual plants, as well as the comparison of wood, non-wood, and fiber grass resources for the manufacturing of pulp and paper are shown in Tables 4 and 5, respectively. Figure 5 shows lignin, cellulose, and hemicellulose composition of wood, non-wood, and annual plant fibers. Overall, it seems that non-wood and annual plant fibers have the potential to be used for pulp and papermaking, though more research is needed to determine the feasibility of large-scale production.

The pulping of *P. oceanica* and date palm rachis was carried out using soda-anthraquinone. Table 6 lists the experimental conditions as well as the content of the resulting pulps. Cooking yield, lignin content, extractives, kappa number, and viscosity all decreased as the cooking temperature increased from 150 to 170 °C, regardless of the raw material. In holocellulose, on the other hand, there is an increase in relative content. The ash content is unchanged by the pulping process, which is surprising. Because the percentage of silicon in *P. oceanica* is so high, this behavior should be taken into account. In further analysis, it was found that date palm rachis displayed greater delignification with a mean cooking yield of 44% compared to 63% in Posidonia, a lignin content of 4% compared to 9% in Posidonia, and a Kappa number of 47 compared to 63 in Posidonia. Comparing lignin and holocellulose amounts is challenging because *P. oceanica* pulp contains 93% lignin, ash, and holocellulose while date palm rachis pulp contains only 80% [4]. In previous work, pulping the *P. oceanica* collected along Tunisia's coasts. 500 g of cleaned and crushed raw material were delignified in two steps. Initially, 10% (w/v) NaOH

Table 4 Summary of properties for non-wood plant fibers

Categories	Raw materials		Chemical properties	
		Lignin (w/w %)	Cellulose (w/w %)	Hemicellulose (w/w %)
Agricultural residue	Wheat straw [36]	15.30	38.20	36.30
	Banana stem [47]	18.21	59.18	17.50
	Rice straw [37]	21.90	41.20	19.50
	Bagasse [100]	21.70	42.34	28.60
Annual plant	Elephant grass [75]	17.70	45.60	–
	Bamboo [120]	31.00	43.00	39.00
	Switch grass [90]	23.89	41.20	–
Non-wood crops	Kenaf bast [65]	12.50	55.50	17.70
	Plant fruit [79]	18.54	37.01	31.51
	Date palm rachis [60]	27.20	45.00	29.80
	Date palm leaves [60]	31.20	30.30	–

Table 5 Comparison of wood, non-wood, and fiber grass resources for pulp and paper production [63, 95]

Description	Fiber resources	
	Wood	Non-wood and fiber grass
Growth cycle	Long	Short
Chemical used	Large amount of chemical	Small amount of the chemical
Lignin content	Higher	Lower
Cellulose content	High	Low
Time for pulping	Need long time	Short time
Environmental problem	Increase environmental problem	Reduce environmental impact
Washing after pulping and bleaching	Easy	Difficult
Speed of paper machine	High	Low
Operation cost	Expensive	Inexpensive
Chemical recovery	Possible	Difficult
Handling and storage	Easy	Difficult

was impregnated with alkali solution overnight at room temperature. A mixture of sodium hypochlorite solution and buffer solution at pH 4.7 was used to wash the sample before bleaching. It was necessary to change the bleaching bath every two hours. Continuing this procedure resulted in a completely bleached sample. To produce carboxy methylcellulose, the pulp was washed and lyophilized [60].

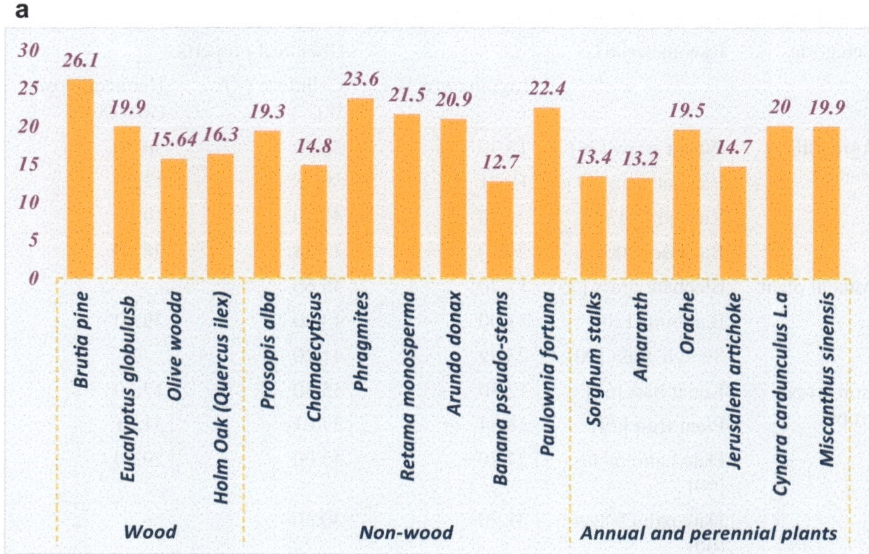

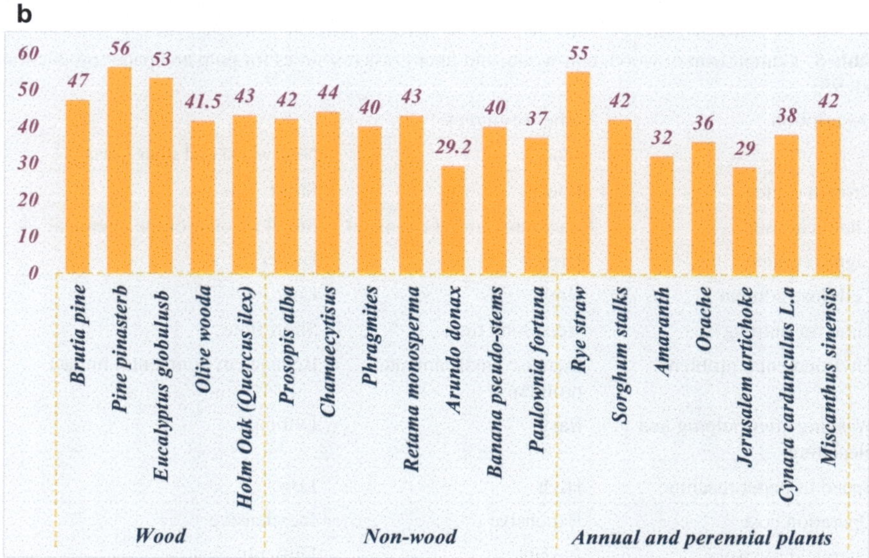

Fig. 5 **a** Lignin % in different types of wood, non-wood, and annual plants [61]. **b** Cellulose % in different types of wood, non-wood, and annual plants [61]. **c** Hemicellulose % in different types of wood, non-wood, and annual plants [61]

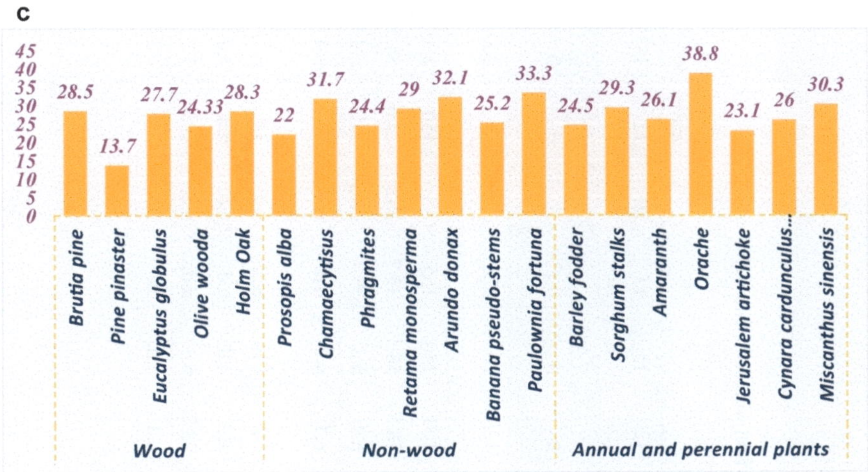

Fig. 5 (continued)

5 *P. Oceanica* as Reinforcement Fibers in Composites

Recently, a lot of research has focused on revalorization of *P. oceanica* waste. In addition to being a source of cellulose, it is also used as a means of reinforcing conventional polymers, as well as biodegradable packaging. This would create new natural fiber-reinforced plastics, which are becoming more popular because of recent concerns about the environment. In the last few decades, more and more natural fiber-reinforced plastics and wood plastic composites (WPCs) have been made and used. WPCs have a lot of advantages when it comes to money and the environment. They also have a lot of advantages from a technical point of view, since they can be shaped using standard manufacturing methods like hot-press molding, extrusion, and injection. It was proven that the species, chemical composition, geometry, harvesting method, hygroscopicity, processing, content, as well as their dispersion, and orientation inside the matrix were essential for fiber interaction in terms of strength development, compatibility, and hydration [55]. On the other hand, Due to stranding incidents on the Mediterranean coast of Spain, between Valencia and Alicante, the *P. oceanica* has attracted attention in its application as a reinforcement of polymeric composites [41, 89, 93]. Composites reinforced with fibers from *P. oceanica* grass in a wheat gluten binding as a matrix were developed. Fibers from plants ranging in size from 2 to 8 mm were collected in Valencia, Spain. The composites utilized wheat gluten, which has a high protein content (80% by weight), in concentrations ranging from 10 to 40%. SEM analysis revealed that the matrix and fibers are homogenous and had good interaction and that mechanical properties including modulus of rupture and elasticity improved with increasing the wheat gluten content in the composites [94]. This marine species waste can be used as a binding matrix, together with gluten with

Table 6 Chemical composition and pulping conditions of unbleached pulps from *P. oceanica* in comparison with date palm rachis or leaves

	Date palm rachis [61]			*P. oceanica* [35, 62]			Date palm rachis [35]						Date palm leaves [36]	
Total alkali charge expressed in NaOH (%)			20				20	23	12	15	13	20	12	18
Anthraquinone concentration			0.1				0.1		0	0	0.1		0	0
Time at constant temperature, min			120				120		120	20	120		20	300
Temperature (°C)	150	160	170	150	160	170	165–170		100	150	165		150	150
Cooking yield (%)	46.8	44.8	41.8	66.0	63.6	59.9	44.2	2.3	78.5	58.5	39.7	29.5	22.6	38.5
Screening yield (%)		94			96		43.1	43	n.d.	n.d.	39.7	28.2	n.d.	n.d.
Ethanol–toluene extractives (%)	n.d.	1.81	0.91	n.d.	0.57	0.26	n.d.	n.d.	n.d.	n.d.	n.d.	n.d.	n.d.	n.d.
Ash (%)		4			12		n.d.	n.d.	2.2	1.8	n.d.	n.d.	3.6	6.15
Holocellulose (%)	n.d.	69.3	75.2	n.d.	71.4	72.4	n.d.	n.d.	n.d.	n.d.	n.d.	n.d.	n.d.	n.d.
Lignin (%)	n.d.	5.2	3.4	n.d.	9.1	8.5	n.d.	n.d.	22.2	6.3	n.d.	n.d.	10.8	2.2
Kappa number	59	54	47	75	65	63	25.5	20.7	n.d.	n.d.	50	20.9	n.d.	n.d.

a high protein content, to generate new sustainable materials, the authors conclude. This reinforcement also included *P. oceanica* fibers. The viability of biocomposites in a biodegradable polyethylene matrix derived from sugarcane was investigated [48]. Plant fiber contents (250 m) ranging from 5 to 40% by weight were used to produce the biocomposites. A composite containing 40% marine phanerogam fibers has a greater modulus of rupture and modulus of elasticity than a composite without fibers by around 30 and 140%, respectively. In addition, composites with a high fiber content of *P. oceanica* are shown to be dimensionally stable forate. It includes items such as decks, furniture, flooring, handrails, and car interiors. In composites with polyethylene as the matrix, *P. oceanica* fibers from Alicante, Spain, were used [89]. Sodium hydroxide was applied to the fibers (2 and 10 mm). High-density polyethylene (HDPE) composites reinforced with *P. oceanica* fibers and compatibilized with PE-g-MA show improved mechanical properties, according to a new study. It was found that composites made with untreated *P. oceanica* fibers and PE-g-MA compatibilizer had the best tensile strength and stiffness. Alkalized fibers also improved the mechanical properties of HDPE composites. "*P. oceanica*" fibers are an abundant, sustainable, and biodegradable resource that can be used to reinforce HDPE composites. The use of these fibers can help reduce the environmental impact of composite materials [99]. Biocomposites were produced from untreated and treated *P. oceanica* plants using epoxy resin and 10% by weight fibers. Results demonstrate that adding fibers increases modulus of elasticity significantly. NaOH solution-treated fibers produce the best results, improving fiber compatibility with epoxy matrix, water resistance, and mechanical strength [30, 62]. It was demonstrated that fibers from the red alga Gelidium [71] could be used in composite polymer materials. In the characterization of the lignocellulosic composition of the fibers from the Gelidium, obtained mean values of 17% of α-cellulose, 4.5% of lignin, and 29.5% of hemicellulose, below those found in the lignocellulosic composition of Sargassum [9]. The synthesis conditions and the composition of natural fiber-reinforced cement matrices are shown in Table 7.

Hamdaoui et al. [50] studied the influence of fiber size modification on thermal and mechanical properties of *P. oceanica* fibers in cement composites. It has been observed that increasing the size of the fibers enhances the insulating qualities of the material: thermal conductivity is reduced by around 22% when about 20% of the fiber volume is employed. The compressive strength and compressibility increased by around 27% between 5 and 10% and then declined but remained greater than the samples without fibers. However, it appears that when the fiber fraction increases, so does the fracture toughness. It was also discovered that adding fibers to this cement matrix boosts the material's courseness and ability to seal fissures. Seagrass was studied because it was considered to be a viable building insulation material [40, 105]. Seagrass balls have also been used as a reinforcing material in the production of polyethylene composites [90]. Bettaieb et al. [15, 16] examined at the leaves of *P. oceanica* and concluded that they had promising potential as nanofillers for polymer matrices. González-García et al. [46] investigated the physical and mechanical properties of fiberboards derived from *P. oceanica* fibers and concluded that they can be used as a substitute for traditional fiberboards. Similar findings were reported by

Table 7 Synthesis of the composition of natural fiber-reinforced cement matrices

Matrix	Fibers	Water/ Cement	Volume fraction (%)	Fibers length (mm)	Chemical treatment	Reference
Concrete	Sisal	0.57	1–5% (mass fraction)	50	None	Swift and Smith [113]
Mortar	Sisal coconut	0.4	3%	37,525	NaOH/ $Ca(OH)_2$	Tolêdo Filho et al. [115]
Cement paste	Hemp	0.5	0–20%	20	None	Hamdaoui et al. [50]
Cement paste	Sisal eucalyptus	N/A	1–5% (mass fraction)	1.660.66	None	Roma Jr et al. [92]
Cement paste	*P. oceanica*	0.5–1	0–20%	N/A	None	Allègue et al., [7]
Cement paste	Flax	1	2%	25	Rhéomac treated fibers boiled fibers	Sawsen et al. [105]
Mortar	FlaxHemp	0.5	1–3%(mass fraction)	16	Cement coated fibers Plasma	Fantilli et al. [40]
Mortar	Hemp	0.5	0–3%(mass fraction)	6–12–18	None	Çomak et al. [30]
Geopolymer mortar	Sisal coconut	N/A	0, 0.5, 0.75, and 1.0%	3.5–4	N/A	Wongsa et al. [123]

others [6]. Despite extensive research on *P. oceanica* fibers, the leaves have received far less attention. Saval et al. [104] developed cement-bonded particleboards made from Posidonia oceanica (seagrass) leaves offer a sustainable building material, suitable for interior and potentially non-structural applications, especially where fire resistance and thermal insulation are crucial. The physico-mechanical properties of particleboard produced from seaweed glue and tapioca starch flour were investigated by Liew et al. [73]. They discovered that adding more tapioca starch flour to the seaweed glue increased its mechanical characteristics. Kuqo et al. [67] employed isocyanate resin to bind particleboards produced from *P. oceanica* leaves. Seagrass leaves are ideal for use in the construction and furniture industries. The availability of low-cost raw lignocellulosic materials, which may be used to replace slow-growing trees and reduce production costs, is a major concern in the composites industry. A variety of plants were studied and tested worldwide in composites manufacturing, including Tupinamba stalks [65], vine stalks [124], and cotton stalks [44, 83]. The mechanical properties of the board were not significantly affected by the addition of up to 10% PO leaves. More of the mechanical properties were affected by the internal bond strength than the external bond strength. A maximum of 25% PO leaves were added to the boards and did not improve their dimensional stability significantly. It was found that boards made up of a 50/50 mixture of wood particles, and PO leaves had the best swelling values for thickness. It is possible to improve the properties of the panel by increasing the resin dosage and using an alternative resin system, such as isocyanates. The researchers believe this might make it possible to incorporate more PO leaves into the panel. Panels made from cereal straw have been produced commercially using such strategies. It was investigated how seagrass fibers (*P. oceanica*) and pine wood particles (Pinus sylvestris) affect the mechanical properties of cement-bonded boards, as well as their fire and water resistance. Because seagrass fibers have more chemical and physical compatibility than wood particles, seagrass-based cement boards have greater mechanical strength and durability than wood-based cement boards. For standardized compliance, a seagrass fiber with a weight of up to 37 weight percent is acceptable. Increases in fiber proportions (wt%) reduce internal bond strength while having no effect on bending strength. Furthermore, seagrass-based cement boards demonstrated superior water resistance and flame resistance. Up to 37 weight percent of seagrass fibers are acceptable for standard compliance. Increases in the weight percentage of the fibers cause a reduction in the internal bond strength but have no effect on the bending strength. Additionally, cement boards made from seagrass showed superior qualities in terms of water resistance and flame resistance. Instead of delaminating and shattering like wood-based cement boards, seagrass-based cement boards transfer applied tensile stresses into the cement matrix. Given the current scarcity of wood, seagrass-based cement boards are a superior and sustainable substitute for wood-based cement boards that can be used in the construction industry. In order to support the shift to renewable resources and to explore new industrial applications, natural PO fibers might be utilized to reinforce lime matrix. This would have significant positive effects on the economy and the environment. Although some attempts have been made to use PO fibers in the production of energy, cellulose products, food packaging, the manufacture of pulp and chemicals, and other industries [13, 60].

6 *P. Oceanica* Fibers for Wastewater Treatment

In the past few years, some researchers have investigated how *P. oceanica* could be used as a renewable dye adsorbent or as a source of lignocellulosic fibers for making pulp and paper [4]. It has been demonstrated that *P. oceanica* fibers are low-cost and renewable adsorbents for removing dyes from aqueous solution and removing heavy metals and organic compounds as well [8, 25, 33, 67, 83]. Thermally treated *P. oceanica* residues could be utilized as a low-cost biowaste for phosphate adsorption, not only from aqueous synthetic solutions (90%), but also from real wastewater. *P. oceanica* fibers successfully adsorb orthophosphate, which contributes to water quality degradation in lakes, rivers, and coastal areas [119]. Using *P. oceanica* biomass, as a low-cost biosorbent, vanadium(III) and molybdenum(V) were removed from aqueous solution from dried leaves (from a beach). A single-metal system with

P. oceanica non-living biomass was developed to determine the performance of the biomass [87]. It was reported by Boulaiche et al. that *P. oceanica* was used to remove metals such as Pb(II), Cd(II), Ni(II), Cu(II), and Zn(II). Several factors were considered, including pH, contact time, ion concentrations, and temperature. The biosorption of metals was verified by kinetics and isotherm models [23].

7 *P. oceanica* for Cellulose Derivatives

Cellulose derivatives are a very important and promising area of products with added value that are made by changing low-cost, naturally occurring raw materials in a chemical way. Chemically modification of cellulose is a common method that uses well-known esterification, etherification, and oxidation reactions of its hydroxyl groups [28]. Using an environmentally friendly process that does not use chlorine and use H_2O_2 and organic peracids, high-grade cellulose was made from *P. oceanica*. Cellulose was easier to get out of *P. oceanica* after it was transformed into cellulose acetate (CA) [58]. Carboxymethyl cellulose and sodium cellulose carboxymethylate have been produced from cellulose extracted from *P. oceanica*. A large amount of *P. oceanica* fragments accumulate along coastal beaches in the form of ball-shaped (oval or spherical) dry materials of different sizes, called "egagropili", which are tightly packed and play an important ecological role. The fiber-like leaves at the base of *P. oceanica* rhizomes and stems cluster together very tightly, generating these characteristic brown free-floating balls. In flat beaches, long-wavelength surface waves provide the structure which forms the sea balls, also known as sea rissoles or sea potatoes. Their numbers accumulate along the sandy shores when they dry and become wind- and tide-driven (Fig. 6). Due to the negative visual impact of egagropili, municipalities are forced to remove and dispose of them in landfills as municipal waste at significant expense [78].

Microcrystalline cellulose (MCC) was produced by chemically delignifying and alkali-treating *P. oceanica* brown algae as seen in Fig. 7 [113]. Sanyang et al. [102] describe the entire extraction process as illustrated in Fig. 8. Nanocrystalline cellulose

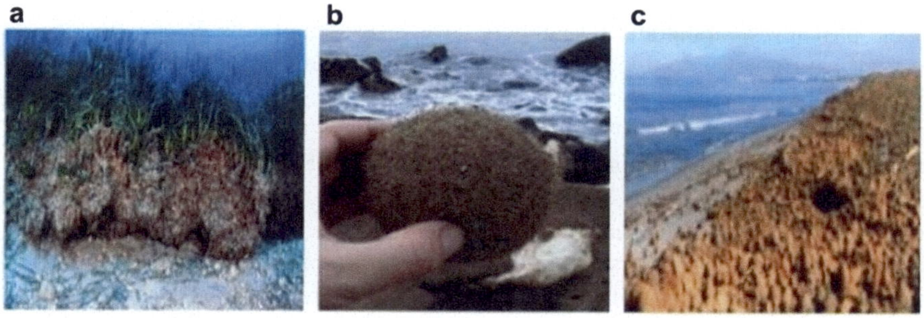

Fig. 6 Rhizomes and stems of *P. oceanica* (**a**) egagropili (**b**) that accumulate along the coast (**c**) [78]

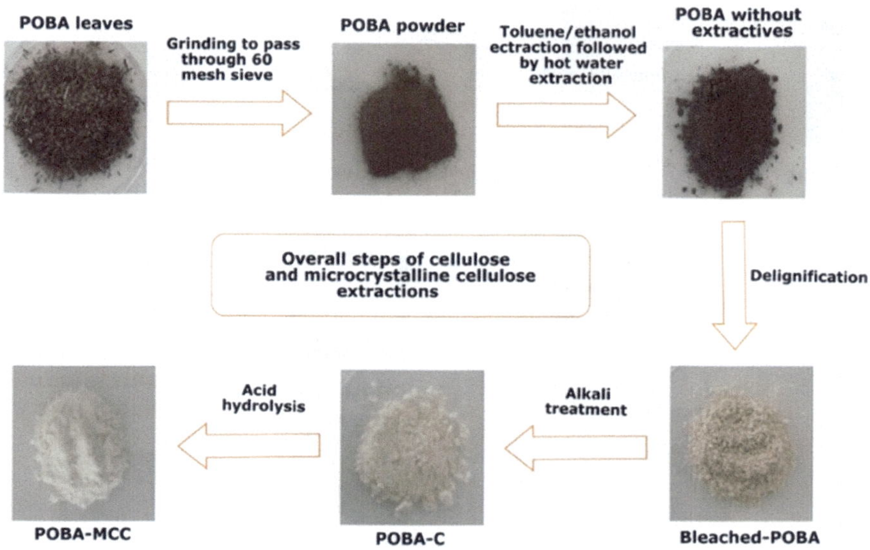

Fig. 7 Diagram showing the different steps required to isolate cellulose and micro-crystalline cellulose from *P. oceanica* [113]

(NC) is obtained by hydrolyzing egagropili cellulose with sulfuric acid. Converting a primary hydroxyl group to a carboxylate group of *P. oceanica* fibers. TEMPO (2,2,6,6-tetramethylpiperidine-1-oxide radical)-oxidation was used to modify *P. oceanica* fibers, followed by bleaching and refining at a concentration of 1% (w/w). Filters and deionized water were used to wash and filter the TEMPO-oxidized cellulose fibers until their conductivity was below 5 μS cm^{-1} [56].

To produce cellulose nanofibrils with a high lignin content, *P. oceanica* balls were used as raw materials. There is an abundance of low-cost raw material in Posidonia waste, which accumulates along the Tunisian coast of the Mediterranean Sea. However, its potential for value-added products and applications is relatively untapped. A large portion of it is currently used for energy production after beach cleaning for tourism purposes [15, 102]. To achieve microfibrillation, unbleached Posidonia fibers were mechanically treated (by twin-screw extrusion or steam explosion) before ultrafine grinding. Several alternative processes for replacing refining (fibrillation step) have been proposed for unbleached Posidonia pulp, including twin-screw extrusion and steam explosion [42].

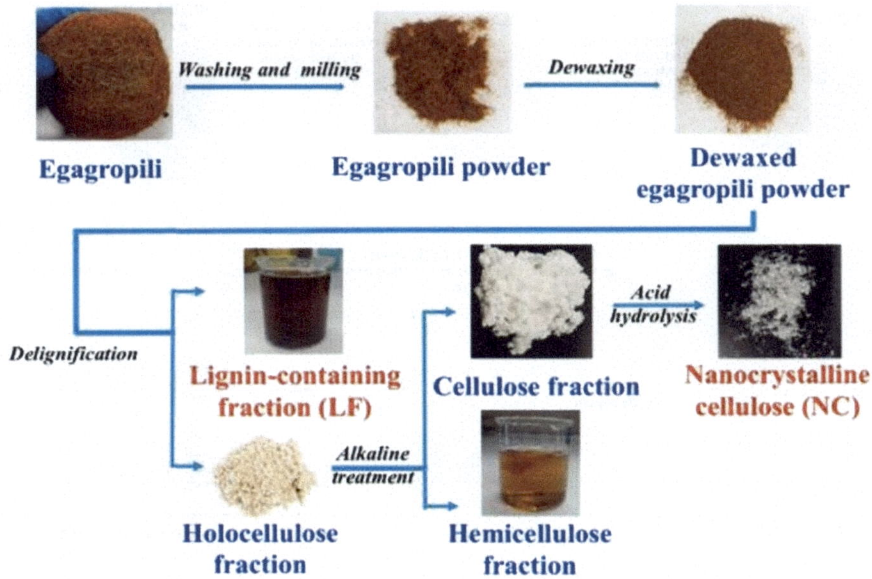

Fig. 8 Extraction procedure for nanocrystalline cellulose and lignin from egagropili [102]

8 Other Applications

P. oceanica and green tea extracts were dipped in peach slices once they had shown the highest total phenolic content, antifungal activity, and antioxidant activity [88]. The production of composites based on polyhydroxyalkanoates and *P. oceanica* fibers is technically feasible, and it represents a valuable solution to the current management problem of seagrass waste accumulation on coastal beaches, providing an environmentally sustainable alternative to landfill disposal [106]. According to HPLC–MS study, *P. oceanica* ethanolic leaf extract (PEE) contains a significant quantity of chicoric acid. PEE was subjected into in vitro cell-free and cell-based studies to see if it could be used in skin care and disease. PEE was found to stimulate fibroblast proliferation and collagen formation, as well as anti-melanogenic and lipolytic effects in adipocytes, suggesting that it could be used in anti-aging, anti-cellulite, and depigmenting products [118].

9 Conclusions

Posidonia fibers are environmentally friendly because they are a renewable resource. Marine plants such as *P. oceanica* can continuously regenerate, making them a sustainable material. In addition, their biodegradability means that they will not harm the environment if disposed of improperly. The chemical composition of *P. oceanica* fibers was analyzed and compared with the chemical composition of other cellulosic fiber sources. The results showed that the fiber content of *P. oceanica* is high in cellulose (62%) and lignin (30%). The fiber content of other cellulosic fiber sources ranged from 43 to 85% cellulose and from 5 to 30% lignin. *P. oceanica* fibers have a high cellulose content and can be used as a source of cellulose for various applications.

Acknowledgements The authors are thankful to the Deanship of Graduate Studies and Scientific Research at University of Bisha for supporting this work through the Fast-Track Research Support Program.

References

1. Abd El-Sayed ES, El-Sakhawy M, El-Sakhawy MAM (2020) Non-wood fibers as raw material for pulp and paper industry. Nord Pulp Pap Res J 35(2):215–230. https://doi.org/10.1515/npprj-2019-0064
2. Abd El-Sayed ES, El-Sakhawy M, El-Sakhawy MA-M (2020) Non-wood fibers as raw material for pulp and paper industry. Nord Pulp Pap Res J 35(2):215–230
3. Abdelhak, M. (2017). Study of Some North African Grasses (Ampelodesma mauritanica and Esparto Grass). In: Grasses: benefits, diversities and functional roles, p 43
4. Aguir C, M'Henni MF (2006) Experimental study on carboxymethylation of cellulose extracted from Posidonia oceanica. J Appl Polym Sci 99(4):1808–1816
5. Agustino S, Mataya B, Senelwa K et al (2011) Non-wood forest products and services for socio-economic development. .... The African Forest ....
6. Alamsjah MA, Sulmartiwi L, Pursetyo KT, Amin MNG, Wardani KAK, Arifianto MD (2017) Modifying bioproduct technology of medium density fiberboard from the seaweed waste Kappaphycus alvarezii and Gracilaria verrucosa. J Indian Acad Wood Sci 14(1):32–45
7. Allègue L, Zidi M, Sghaier S (2015) Mechanical properties of Posidonia oceanica fibers reinforced cement. J Compos Mater 49(5):509–517
8. Allouche FN, Mameri N, Guibal E (2011) Pb(II) biosorption on Posidonia oceanica biomass. Chem Eng J 168(3):1174–1184. https://doi.org/10.1016/j.cej.2011.02.005
9. Alzate-Gaviria L, Domínguez-Maldonado J, Chablé-Villacís R, Olguin-Maciel E, Leal-Bautista RM, Canché-Escamilla G, Caballero-Vázquez A, Hernández-Zepeda C, Barredo-Pool FA, Tapia-Tussell R (2020) Presence of polyphenols complex aromatic "Lignin" in Sargassum spp. from Mexican Caribbean. J Mar Sci Eng 9(1):6
10. Assumpcao RMV (1992) Non-wood fiber utilization in pulping and papermaking—UNIDO's activities. TAPPI non-wood plant fiber pulping progress report, vol 20, pp 191–201
11. Azeez MA (2018) Pulping of non-woody biomass. In: Pulp and paper processing, pp 55–86
12. Bajpai P (2018) Nonwood fiber use in pulp and paper. In: Biermann's handbook of pulp and paper. Elsevier, pp 261–278. https://doi.org/10.1016/B978-0-12-814240-0.00010-0
13. Benito-González I, Göksen G, Pérez-Bassart Z, López-Rubio A, Sánchez R, Alonso JM, Gavara R, Gallur M, Martínez-Sanz M (2021) Pilot plant scale-up of the production of optimized starch-based biocomposites loaded with cellulosic nanocrystals from Posidonia oceanica waste biomass. Food Packag Shelf Life 30:100730
14. Beroual M, Boumaza L, Mehelli O, Trache D, Tarchoun AF, Khimeche K (2021) Physico-chemical properties and thermal stability of microcrystalline cellulose isolated from esparto grass using different delignification approaches. J Polym Environ 29(1):130–142

15. Bettaieb F, Khiari R, Dufresne A, Mhenni MF, Putaux JL, Boufi S (2015) Nanofibrillar cellulose from Posidonia oceanica: properties and morphological features. Ind Crops Prod 72:97–106. https://doi.org/10.1016/j.indcrop.2014.12.060
16. Bettaieb F, Khiari R, Hassan ML, Belgacem MN, Bras J, Dufresne A, Mhenni MF (2015) Preparation and characterization of new cellulose nanocrystals from marine biomass Posidonia oceanica. Ind Crops Prod 72:175–182
17. Bhardwaj NK, Kaur D, Chaudhry S, Sharma M, Arya S (2019) Approaches for converting sugarcane trash, a promising agro residue, into pulp and paper using soda pulping and elemental chlorine-free bleaching. J Clean Prod 217:225–233. https://doi.org/10.1016/j.jclepro.2019.01.223
18. Boadu KB, Ansong M, Afrifah KA, Nsiah-Asante E (2022) Pulp and paper making characteristics of fibers from plantation-grown Oxythenantera abyssinica and Beema bamboo (a tissue cultured clone from Bambusa balcooa). J Nat Fibers 19(11):4198–4209
19. Bonanno G, Di Martino V (2017) Trace element compartmentation in the seagrass Posidonia oceanica and biomonitoring applications. Mar Pollut Bull 116(1–2):196–203
20. Bonhomme D, Boudouresque CF, Astruch P, Bonhomme J, Bonhomme P, Goujard A, Thibaut T (2015) Typology of the reef formations of the Mediterranean seagrass Posidonia oceanica, and the discovery of extensive reefs in the Gulf of Hyeres (Provence, Mediterranean). Travaux Scientifiques Du Parc National de Port-Cros 29:41–73
21. Boudouresque CF, Bernard G, Bonhomme P, Charbonnel E, Diviacco G, Meinesz A, Pergent G, Pergent-Martini C, Ruitton S, Tunesi L (2012) Protection and conservation of Posidonia oceanica meadows. RAMOGE and RAC/SPA, p 202
22. Boudouresque CF, Pergent G, Pergent-Martini C, Ruitton S, Thibaut T, Verlaque M (2016) The necromass of the Posidonia oceanica seagrass meadow: fate, role, ecosystem services and vulnerability. Hydrobiologia 781(1):25–42. https://doi.org/10.1007/s10750-015-2333-y
23. Boulaiche W, Belhamdi B, Hamdi B, Trari M (2019) Kinetic and equilibrium studies of biosorption of M(II) (M = Cu, Pb, Ni, Zn and Cd) onto seaweed Posidonia oceanica fibers. Appl Water Sci 9(8). https://doi.org/10.1007/s13201-019-1062-1
24. Castillo C, Mantecón AR, Sotillo J, Benedito JL, Abuelo A, Gutiérrez C, Hernández J (2014) The use of banquettes of Posidonia oceanica as a source of fiber and minerals in ruminant nutrition. An observational study. Animal 8(10):1663–1666
25. Chadlia A, Farouk MM (2007) Removal of basic blue 41 from aqueous solution by carboxymethylated posidonia oceanica. J Appl Polym Sci 103(2):1215–1225. https://doi.org/10.1002/app.23070
26. Christopoulou N-M, Kalogianni DP, Christopoulos TK (2021) Posidonia oceanica (Mediterranean tapeweed) leaf litter as a source of fluorescent carbon dot preparations. Microchem J 161:105787
27. Cocozza C, Parente A, Zaccone C, Mininni C, Santamaria P, Miano T (2011) Chemical, physical and spectroscopic characterization of Posidonia oceanica (L.) Del. residues and their possible recycle. Biomass Bioenergy 35(2):799–807. https://doi.org/10.1016/j.biombioe.2010.10.033
28. Coletti A, Valerio A, Vismara E (2013) Posidonia oceanica as a renewable lignocellulosic biomass for the synthesis of cellulose acetate and glycidyl methacrylate grafted cellulose. Materials 6(5):2043–2058
29. Çomak B, Bideci A, Bideci ÖS (2018) Effects of hemp fibers on characteristics of cement based mortar. Constr Build Mater 169:794–799
30. da Silva Chagas FD, Lima GC, Dos Santos VIN, Costa LEC, de Sousa WM, Sombra VG, de Araújo DF, Barros FCN, Marinho-Soriano E, de Andrade Feitosa JP (2020) Sulfated polysaccharide from the red algae Gelidiella acerosa: anticoagulant, antiplatelet and antithrombotic effects. Int J Biol Macromol 159:415–421
31. Dural MU, Cavas L, Papageorgiou SK, Katsaros FK (2011) Methylene blue adsorption on activated carbon prepared from Posidonia oceanica (L.) dead leaves: kinetics and equilibrium studies. Chem Eng J 168(1):77–85

32. El Din NGS, El-Sherif ZM (2013) Nutritional value of Cymodocea nodosa and Posidonia oceanica along the western Egyptian Mediterranean coast. Egypt J Aquat Res 39(3):153–165
33. El Khames Saad M, Khiari R, Elaloui E, Moussaoui Y (2014) Adsorption of anthracene using activated carbon and Posidonia oceanica. Arab J Chem 7(1):109–113. https://doi.org/10.1016/j.arabjc.2013.11.002
34. El-Morsy MMS (1980) Studies on the rachises of the Egyptian date palm leaves for hardboard production. Fiber Sci Technol 13(4):317–321
35. El-Morsy MMS, Riad BY, Mohamed MAS (1981) Pulp and paper from the Egyptian date palm leaves. Fiber Sci Technol 14(2):157–161
36. El-Sakhawy M. Lönnberg B, Fahmy Y, Ibrahim AA (1996) Organosolv pulping. 3. Ethanol pulping of wheat straw. Cellulose Chem Technol 30(1–2):161–174
37. Enayati AA, Hamzeh Y, Mirshokraiei SA, Molaii M (2009) Papermaking potential of canola stalks. BioResources 4(1):245–256
38. Fahmy Y, Fahmy TYA, Mobarak F, El-Sakhawy M, Fadl MH (2017) Agricultural residues (wastes) for manufacture of paper, board, and miscellaneous products: background overview and future prospects. Int J ChemTech Res 10(2):424–448. https://doi.org/10.5281/zenodo.546735
39. Fantilli AP, Chiaia B, Gorino A (2016) Fiber volume fraction and ductility index of concrete beams. Cement Concr Compos 65:139–149
40. Ferdous T (2021) Morphological and chemical characteristics of different non-wood species and their effect on pulping. University of Dhaka
41. Ferrero B, Boronat T, Moriana R, Fenollar O, Balart R (2013) Green composites based on wheat gluten matrix and Posidonia oceanica waste fibers as reinforcements. Polym Compos 34(10):1663–1669
42. Fortunati E, Luzi F, Puglia D, Petrucci R, Kenny JM, Torre L (2015) Processing of PLA nanocomposites with cellulose nanocrystals extracted from Posidonia oceanica waste: innovative reuse of coastal plant. Ind Crops Prod 67:439–447. https://doi.org/10.1016/j.indcrop.2015.01.075
43. Gansberger M, Montgomery LFR, Liebhard P (2015) Botanical characteristics, crop management and potential of Silphium perfoliatum L. as a renewable resource for biogas production: a review. Ind Crops Prod 63:362–372
44. Gharehkhani S, Yarmand H, Goodarzi MS, Shirazi SFS, Amiri A, Zubir MNM, Solangi K, Ibrahim R, Kazi SN, Wongwises S (2017) Experimental investigation on rheological, momentum and heat transfer characteristics of flowing fiber crop suspensions. Int Commun Heat Mass Transfer 80:60–69. https://doi.org/10.1016/j.icheatmasstransfer.2016.11.013
45. González-García S, Teresa Moreira M, Artal G, Maldonado L, Feijoo G (2010) Environmental impact assessment of non-wood based pulp production by soda-anthraquinone pulping process. J Clean Prod 18(2):137–145. https://doi.org/10.1016/j.jclepro.2009.10.008
46. Gopal PM, Sivaram NM, Barik D (2018) Paper industry wastes and energy generation from wastes. In: Energy from toxic organic waste for heat and power generation, pp 83–97. https://doi.org/10.1016/B978-0-08-102528-4.00007-9
47. Goswami T, Kalita D, Rao PG (2008) Greaseproof paper from Banana (Musa paradisica L.) pulp fiber
48. Haddar M, Elloumi A, Bradai C, Koubaa A (2024) Characterization of Posidonia oceanica fibers high-density polyethylene composites: reinforcing potential and effect of coupling agent. J Compos Sci 8(7):236
49. Hamdaoui O, Limam O, Ibos L, Mazioud A (2021) Thermal and mechanical properties of hardened cement paste reinforced with Posidonia-oceanica natural fibers. Constr Build Mater 269:121339
50. Heglmeier A, Zidorn C (2010) Secondary metabolites of Posidonia oceanica (Posidoniaceae). Biochem Syst Ecol 38(5):964–970. https://doi.org/10.1016/j.bse.2010.07.001
51. Ilvessalo-Pfäffli M-S (1995) Descriptions of nonwood plant fibers. In Fiber atlas. Springer, pp 307–359

52. Jiménez L, Angulo V, Ramos E, De La Torre MJ, Ferrer JL (2006) Comparison of various pulping processes for producing pulp from vine shoots. Ind Crops Prod 23(2):122–130. https://doi.org/10.1016/j.indcrop.2005.05.001
53. Kaal J, Serrano O, del Río JC, Rencoret J (2018) Radically different lignin composition in Posidonia species may link to differences in organic carbon sequestration capacity. Org Geochem 124:247–256. https://doi.org/10.1016/j.orggeochem.2018.07.017
54. Kaal J, Serrano O, Nierop KGJ, Schellekens J, Martínez Cortizas A, Mateo MÁ (2016) Molecular composition of plant parts and sediment organic matter in a Mediterranean seagrass (Posidonia oceanica) mat. Aquat Bot 133:50–61. https://doi.org/10.1016/j.aquabot.2016.05.009
55. Kesikidou F, Stefanidou M (2019) Natural fiber-reinforced mortars. J Build Eng 25:100786. https://doi.org/10.1016/j.jobe.2019.100786
56. Khadraoui M, Khiari R, Brosse N, Bergaoui L, Mauret E (2022) Combination of steam explosion and TEMPO-mediated oxidation as pretreatments to produce nanofibril of cellulose from Posidonia oceanica bleached fibers. BioResources 17(2)
57. Khaled A, Hessein A, Abdel-Halim AM, Morsy FM (2014) Distribution of heavy metals in seaweeds collected along Marsa-Matrouh beaches, Egyptian Mediterranean Sea. Egypt J Aquat Res 40(4):363–371
58. Khiari R, Belgacem MN (2017) Potential for using multiscale Posidonia oceanica waste: current status and prospects in material science. In: Lignocellulosic fiber and biomass-based composite materials: processing, properties and applications. Elsevier Ltd. https://doi.org/10.1016/B978-0-08-100959-8.00021-4
59. Khiari R, Marrakchi Z, Belgacem MN, Mauret E, Mhenni F (2011) New lignocellulosic fibers-reinforced composite materials: a stepforward in the valorisation of the Posidonia oceanica balls. Compos Sci Technol 71(16):1867–1872
60. Khiari R, Mhenni MF, Belgacem MN, Mauret E (2010) Chemical composition and pulping of date palm rachis and Posidonia oceanica—a comparison with other wood and non-wood fiber sources. Biores Technol 101(2):775–780. https://doi.org/10.1016/j.biortech.2009.08.079
61. Khristova P, Kordsachia O, Khider T (2005) Alkaline pulping with additives of date palm rachis and leaves from Sudan. Biores Technol 96(1):79–85
62. Khrunyk Y, Lach S, Petrenko I, Ehrlich H (2020) Progress in modern marine biomaterials research. Mar Drugs 18(12):589
63. Kissinger M, Fix J, Rees WE (2007) Wood and non-wood pulp production: comparative ecological footprinting on the Canadian prairies. Ecol Econ 62(3–4):552–558
64. Klímek P, Meinlschmidt P, Wimmer R, Plinke B, Schirp A (2016) Using sunflower (Helianthus annuus L.), topinambour (Helianthus tuberosus L.) and cup-plant (Silphium perfoliatum L.) stalks as alternative raw materials for particleboards. Ind Crops Prod 92:157–164
65. Kumar V, Pathak P, Bhardwaj NK (2020) Facile chemo-refining approach for production of micro-nanofibrillated cellulose from bleached mixed hardwood pulp to improve paper quality. Carbohyd Polym 238:116186
66. Kuqo A, Korpa A, Dhamo N (2019) Posidonia oceanica leaves for processing of PMDI composite boards. J Compos Mater 53(12):1697–1703
67. Lafabrie C, Pergent-Martini C, Pergent G (2008) Metal contamination of Posidonia oceanica meadows along the Corsican coastline (Mediterranean). Environ Pollut 151(1):262–268. https://doi.org/10.1016/j.envpol.2007.01.047
68. Laftah WA, Wan Abdul Rahman WA (2016) Pulping process and the potential of using non-wood pineapple leaves fiber for pulp and paper production: a review. J Nat Fibers 13(1):85–102. https://doi.org/10.1080/15440478.2014.984060
69. Larkum AWD, Orth RJ, Duarte CM (2006) Seagrasses: biology, ecology and conservation. Phycologia 45(5):5
70. Law KN, Kokta BV, Mao CB (2001) Fiber morphology and soda-sulphite pulping of switchgrass. Biores Technol 77(1):1–7. https://doi.org/10.1016/S0960-8524(00)00140-1
71. Lee S-W, Lee E-H, Eom A-H (2008) Effects of organic farming on communities of arbuscular myeorrhizal fungi. Mycobiology 36(1):19–23

72. Liew KC, De Ting PB, Tan YF (2018) Physico-mechanical properties of particleboard made from seaweed adhesive and tapioca starch flour. J Indian Acad Wood Sci 15(2):199–203
73. Liu R, Wang L, Liu Q, Wang Z, Li Z, Fang J, Zhang L, Luo M (2018) Depth-resolved distribution of particle-attached and free-living bacterial communities in the water column of the New Britain Trench. Front Microbiol 9:625
74. Lwako KO, Joseph B, Baptist KJ (2013) A review on pulp manufacture from non wood plant materials. Int J Chem Eng Appl 4(3):144–148
75. Madakadze IC, Masamvu TM, Radiotis T, Li J, Smith DL (2010) Evaluation of pulp and paper making characteristics of elephant grass (Pennisetum purpureum Schum) and switchgrass (Panicum virgatum L.). Afr J Environ Sci Technol 4(7):465–470
76. Marrakchi Z, Khiari R, Oueslati H, Mauret E, Mhenni F (2011) Pulping and papermaking properties of Tunisian Alfa stems (Stipa tenacissima)-effects of refining process. Ind Crops Prod 34(3):1572–1582. https://doi.org/10.1016/j.indcrop.2011.05.022
77. Mayer AK, Kuqo A, Koddenberg T, Mai C (2022) Seagrass-and wood-based cement boards: a comparative study in terms of physico-mechanical and structural properties. Compos A Appl Sci Manuf 156:106864
78. Mirpoor SF, Giosafatto CVL, Di Pierro P, Di Girolamo R, Regalado-González C, Porta R (2020) Valorisation of posidonia oceanica sea balls (Egagropili) as a potential source of reinforcement agents in protein-based biocomposites. Polymers 12(12):1–13. https://doi.org/10.3390/polym12122788
79. Mulyantara LT, Harsono H, Maryana R, Jin G, Das AK, Ohi H (2017) Properties of thermomechanical pulps derived from sugarcane bagasse and oil palm empty fruit bunches. Ind Crops Prod 98:139–145
80. Nabais JMV, Laginhas C, Carrott MMLR, Carrott PJM, Amorós JEC, Gisbert AVN (2013) Surface and porous characterisation of activated carbons made from a novel biomass precursor, the esparto grass. Appl Surf Sci 265:919–924
81. Nayak A, Bhushan B (2019) An overview of the recent trends on the waste valorization techniques for food wastes. J Environ Manage 233:352–370. https://doi.org/10.1016/j.jenvman.2018.12.041
82. Nazerian M, Beyki Z, Gargarii RM, Kool F (2016) The effect of some technological production variables on mechanical and physical properties of particleboard manufactured from cotton (Gossypium hirsutum) stalks. Maderas Ciencia Tecnología 18(1):167–178
83. Ncibi MC, Mahjoub B, Seffen M (2007) Kinetic and equilibrium studies of methylene blue biosorption by Posidonia oceanica (L.) fibers. J Hazardous Mater 139(2):280–285. https://doi.org/10.1016/j.jhazmat.2006.06.029
84. Neis FA, de Costa F, de Araújo AT, Fett JP, Fett-Neto AG (2019) Multiple industrial uses of non-wood pine products. Ind Crops Prod 130:248–258. https://doi.org/10.1016/j.indcrop.2018.12.088
85. Oleszek M, Król A, Tys J, Matyka M, Kulik M (2014) Comparison of biogas production from wild and cultivated varieties of reed canary grass. Biores Technol 156:303–306
86. Pasqualini V, Pergent-Martini C, Clabaut P, Pergent G (1998) Mapping of Posidonia oceanica using aerial photographs and side scan sonar: Application off the Island of Corsica (France). Estuar Coast Shelf Sci 47(3):359–367. https://doi.org/10.1006/ecss.1998.0361
87. Pennesi C, Totti C, Beolchini F (2013) Removal of Vanadium(III) and Molybdenum(V) from wastewater using Posidonia oceanica (Tracheophyta) biomass. PLoS ONE 8(10). https://doi.org/10.1371/journal.pone.0076870
88. Piva G, Fracassetti D, Tirelli A, Mascheroni E, Musatti A, Inglese P, Piergiovanni L, Rollini M (2017) Evaluation of the antioxidant/antimicrobial performance of Posidonia oceanica in comparison with three commercial natural extracts and as a treatment on fresh-cut peaches (Prunus persica Batsch). Postharvest Biol Technol 124:54–61. https://doi.org/10.1016/j.postharvbio.2016.10.001
89. Puglia D, Petrucci R, Fortunati E, Luzi F, Kenny JM, Torre L (2014) Revalorisation of Posidonia oceanica as reinforcement in polyethylene/maleic anhydride grafted polyethylene composites. J Renew Mater 2(1):66–76

90. Qin Z, Kwon H, Dunn JB, Mueller S, Wander MM, Wang M (2018) Carbon calculator for land use change from biofuels production (CCLUB). Argonne National Lab. (ANL), Argonne, IL (United States)
91. Roma LC Jr, Martello LS, Savastano H Jr (2008) Evaluation of mechanical, physical and thermal performance of cement-based tiles reinforced with vegetable fibers. Constr Build Mater 22(4):668–674
92. Rosado MJ, Bausch F, Rencoret J, Marques G, Gutiérrez A, Rosenau T, Potthast A, José C (2021) Differences in the content, composition and structure of the lignins from rind and pith of papyrus (Cyperus papyrus L.) culms. Ind Crops Prod 174:114226
93. Rossignolo JA, Duran AJFP, Bueno C, Martinelli Filho JE, Junior HS, Tonin FG (2022) Algae application in civil construction: a review with focus on the potential uses of the pelagic Sargassum spp. biomass. J Environ Manage 303:114258
94. Rossignolo JA, Felicio Peres Duran AJ, Bueno C, Martinelli Filho JE, Savastano Junior H, Tonin FG (2022) Algae application in civil construction: a review with focus on the potential uses of the pelagic Sargassum spp. biomass. J Environ Manage 303. https://doi.org/10.1016/j.jenvman.2021.114258
95. Rousu P, Rousu P, Anttila J (2002) Sustainable pulp production from agricultural waste. Resour Conserv Recycl 35(1–2):85–103
96. Rowell J (1997) Chapter 5. Paper and composites from agro-based resources. In: Chemical composition of fibers, pp 83–134
97. Saeed HAM, Liu Y, Chen H (2018) Exploring Sudanese agricultural residues as alternative fibers for pulp and paper manufacturing. IOP Conf Ser Mater Sci Eng 368(1):12030
98. Saeed HAM, Liu Y, Lucia LA, Chen H (2017) Sudanese agro-residue as a novel furnish for pulp and paper manufacturing. BioResources 12(2):4166–4176. https://doi.org/10.15376/biores.12.2.4166-4176
99. Saleh SMK, Amer AT, Al-Alawi A (2018) Potential ecological risk of heavy metals in surface sediments from the Aden coast, Southern Yemen. J Environ Sci Toxicol Food Technol 12(10):42–55
100. Salehi FA, Rezapour K, Abdollahshirazi A, Effatnejad R (2022) Modelling of techno-economic and environmental analysis for the energy generation technologies from bagasse (case study). Int J Glob Warm 26(2):179–197
101. Sánchez C (2009) Lignocellulosic residues: biodegradation and bioconversion by fungi. Biotechnol Adv 27(2):185–194
102. Sanyang ML, Sapuan SM, Jawaid M, Ishak MR, Sahari J (2016) Effect of sugar palm-derived cellulose reinforcement on the mechanical and water barrier properties of sugar palm starch biocomposite films. BioResources 11(2):4134–4145. https://doi.org/10.15376/biores.11.2.4134-4145
103. Saval JM, Lapuente R, Navarro V, Tenza-Abril AJ (2014) Fire-resistance, physical, and mechanical characterization of particleboard containing oceanic Posidonia waste. Mater Constr 64(314):e019–e019
104. Sawsen C, Fouzia K, Mohamed B, Moussa G (2015) Effect of flax fibers treatments on the rheological and the mechanical behavior of a cement composite. Constr Build Mater 79:229–235
105. Scurlock JMO, Dayton DC, Hames B (2000) Bamboo: an overlooked biomass resource? Biomass Bioenerg 19(4):229–244
106. Seggiani M, Cinelli P, Mallegni N, Balestri E, Puccini M, Vitolo S, Lardicci C, Lazzeri A (2017) New bio-composites based on polyhydroxyalkanoates and Posidonia oceanica fibers for applications in a marine environment. Materials 10(4). https://doi.org/10.3390/ma10040326
107. Seixas M, Cardoso D, Moreira LE, Paciornik S (2024) Influence of elastic properties and nodes on the flexural behaviour of bamboo culms. Adv Bamboo Sci 100100
108. Shabaka SH, Khalil MK, El-Sikaily A, Youssef NA-E (2021) Posidonia oceanica litter along the Mediterranean Coast of Egypt: status and a preliminary assessment of nutrients and trace elements contents. Estuar Coast Shelf Sci 255:107342

109. Shanker AK, Cervantes C, Loza-Tavera H, Avudainayagam S (2005) Chromium toxicity in plants. Environ Int 31(5):739–753
110. Simeone S, De Falco G (2013) Posidonia oceanica banquette removal: sedimentological, geomorphological and ecological implications. J Coast Res 65(special issue):1045–1050
111. Smook GA (1989) Handbook for pulp and paper technologists. Tappi Press
112. Swift DG, Smith RBL (1978) Sisal fiber reinforcement of cement paste and concrete. In: Proceedings of international conference on materials of construction for developing countries. Asian Institute of Technology, Bangkok, pp 193–198
113. Tarchoun AF, Trache D, Klapötke TM (2019) Microcrystalline cellulose from Posidonia oceanica brown algae: extraction and characterization. Int J Biol Macromol 138:837–845. https://doi.org/10.1016/j.ijbiomac.2019.07.176
114. Tolêdo Filho RD, Scrivener K, England GL, Ghavami K (2000) Durability of alkali-sensitive sisal and coconut fibers in cement mortar composites. Cement Concr Compos 22(2):127–143
115. Torres E, Fombuena V, Vallés-Lluch A, Ellingham T (2017) Improvement of mechanical and biological properties of Polycaprolactone loaded with Hydroxyapatite and Halloysite nanotubes. Mater Sci Eng, C 75:418–424
116. Torres FG, Rodriguez S, Saavedra AC (2019) Green composite materials from biopolymers reinforced with agroforestry waste. J Polym Environ 27(12):2651–2673
117. Vacchi M, De Falco G, Simeone S, Montefalcone M, Morri C, Ferrari M, Bianchi CN (2017) Biogeomorphology of the Mediterranean Posidonia oceanica seagrass meadows. Earth Surf Proc Land 42(1):42–54. https://doi.org/10.1002/esp.3932
118. Vasarri M, De Biasi AM, Barletta E, Pretti C, Degl'Innocenti D (2021) An overview of new insights into the benefits of the seagrass Posidonia oceanica for human health. Mar Drugs 19(9):476
119. Wahab MA, Ben Hassine R, Jellali S (2011) Posidonia oceanica (L.) fibers as a potential low-cost adsorbent for the removal and recovery of orthophosphate. J Hazardous Mater 191(1–3):333–341. https://doi.org/10.1016/j.jhazmat.2011.04.085
120. Wen J-L, Sun S-L, Xue B-L, Sun R-C (2013) Quantitative structural characterization of the lignins from the stem and pith of bamboo (Phyllostachys pubescens). Holzforschung 67(6):613–627
121. Wise LE (1946) Chlorite holocellulose, its fractionation and bearing on summative wood analysis and on studies on the hemicelluloses. Paper Trade 122:35–43
122. Wongsa A, Kunthawatwong R, Naenudon S, Sata V, Chindaprasirt P (2020) Natural fiber reinforced high calcium fly ash geopolymer mortar. Constr Build Mater 241:118143
123. Yeniocak M, Goktas O, Erdil YZ, Ozen E, Alma MH (2014) Investigating the use of vine pruning stalks (Vitis Vinifera L. CV. Sultani) as raw material for particleboard manufacturing. Wood Res 59(1):167–176
124. Zhang Z, Malik MZ, Khan A, Ali N, Malik S, Bilal M (2022) Environmental impacts of hazardous waste, and management strategies to reconcile circular economy and eco-sustainability. Sci Total Environ 807:150856

Chemical Composition and Pulping of Fiber from *Posidonia Oceanica*

Fedia Bettaieb, Noureddine Baaka, and Mohamed Taha

Abstract One of the most common species that are native to the Mediterranean Sea is the seagrass *Posidonia oceanica* (L.). The leaves of the *P. oceanica* plant are likely composed of seven bundles, each measuring 8–11 mm in width and 20–80 cm in length. Rhizomes range in thickness from 5 to 10 mm and carry the roots that connect the plant to the substrate and allow nutrients to be absorbed from the sediment. *P. oceanica* leaves have been tried for a variety of traditional purposes. For example, they were utilized as a material to make mattresses. Farmers along the Mediterranean coasts have long utilized the dead leaves of *P. oceanica* as compost. Over the last fifty years, frequent studies have been conducted in Italy, Tunisia, and Greece to make Posidonia-based compost. Nevertheless, multiple studies have been conducted on the dignification of *P. oceanica* in order to create cellulose and convert it into other derivatives. This chapter describes the chemical composition and pulping of fiber derived from *P. oceanica*.

Keywords *Posidonia oceanica* · Valorization · Chemical composition · Pulping · Fibre

1 Introduction

Posidonia (*Posidonia oceanica*) is a coastal marine plant with long narrow leaves that form dense beds [39]. It is a major Mediterranean species that consists of rhizomes on which grow roots and flat leaves about 1 cm wide and up to 1.5 m high. These leaves

F. Bettaieb (✉)
University of Grenoble Alpes, CNRS, Grenoble INP, LGP2, Grenoble, France
e-mail: bettaieb_fedia@yahoo.fr

N. Baaka
Higher Institute of Fashion of Monastir, Monastir, Tunisia

M. Taha
Mechanical Engineering Department, College of Engineering and Technology, Arab Academy for Science, Technology and Maritime Transport, Aswan, Egypt

R. Khiari and M. Jawaid (eds.), *Current Status and Opportunity in Fibre and Composites*, Sustainable Materials and Technology,
https://doi.org/10.1007/978-981-96-4546-6_2

are generally grouped in clusters of 4–8 leaves. *P. oceanica* is mainly located around the Mediterranean basin, along the North African coasts (Egypt, Libya, Tunisia, Algeria, and Morocco) and the European coasts (France, Corsica, Sardinia, Italy, Yugoslavia, Greece, Turkey, and Spain). The total area covered by meadows in the Mediterranean basin has been estimated at between 30,000 and 40,000 km^2. *P. oceanica* meadows form a quasi-continuous belt around the Mediterranean, interrupted locally at the level of estuaries, ports, and other areas of high anthropization [39]. On the scale of the Tunisian Mediterranean region, *P. oceanica* meadows are distributed along the entire Tunisian coast, i.e., about 1300 km.

Additionally, the seas of Tunisia are home to a large number of Posidonia (*P. oceanica*) marine plants. Every season, the sea rejects the debris of this plant in the form of pellets, which accumulate in massive quantities along the coast. Given the importance of the tourist activity in the Tunisian economy, the beaches must be cleared of this debris. The recovery of this waste can therefore present an interesting economic activity allowing the production of products and biosourced materials that is to say from plant biomass. Throughout history, humans have found numerous uses for the dead leaves of *P. oceanica.* The Egyptians used Posidonia balls to make felts for shoes [18, 19]. The Venetians also used Posidonia leaves to package and transport their glassware [17]. Posidonia leaves have also been used for a long time as compost by farmers of the Mediterranean coasts [69].

Today, there are new ways to use *P. oceanica.* In fact, research has been done on Posidonia's delignification in order to create cellulose or cellulosic derivatives [1–3, 48–50]. Khiari et al. tested the papermaking potential and refinability of fibers obtained from seagrass.

2 Posidonia (*P. Oceanica*)

2.1 Description

There are nine species of seagrass, eight of which are found on the coasts of Australia and the ninth on the coasts of the Mediterranean [39]. Table 1 presents the nine species of marine phanerogams.

One important species in the Mediterranean is P. oceanica. It lives exclusively in the sea. It belongs to the family of marine phanerogams and more precisely of the family of Posidoniaceaes. *P. oceanica* is widespread, and its geographical distribution has been studied by many researchers [63, 65]. Aspects related to its ecology and physiology have also been reported [5–7, 20, 42, 43 54–56, 58, 59]. *P. oceanica* was first accurately described in 1623 by G. Bauhin [40] under the name Alga marina. The current name of *P. oceanica* (Linnaeus) Delile was given by Delile in his work entitled "La flore d'Egypte" published in 1824, during the nineteenth century. The structure of Posidonia shoots has been widely described in many works [12, 17, 21, 23, , 37, 38, 40]. They have the same cell structure [24] and morphology as other

Table 1 List of marine phanerogam species [39]

Genus	Species	Reference
Posidonia	*angustifolia*	Kuo [53]
Posidonia	*australis*	Hooker [41]
Posidonia	*coriacea*	Kuo and Cambridge [52]
Posidonia	*denhartogii*	Kuo and Cambridge [52]
Posidonia	*kirkmanii*	Kuo and Cambridge [52]
Posidonia	*ostendfeldii*	Hartog [40]
Posidonia	*sinuosa*	Kuo and Cambridge [52]
Posidonia	*robertsoniae*	Kuo and Cambridge [52]
Posidonia	*oceanica*	Delile [28]

marine phanerogams. The marine phanerogams, *P. oceanica*, all have the same type of morphology. *Posidonia* is formed by creeping or erect stems of woody consistency and generally buried in the sediments. These stems, called rhizomes, have growth nodes and end in groups of leaves called bundles. The leaves are flattened and measure about 1 cm wide and reach 1.5 m in height. These leaves are grouped in bundles of 4 to 8 units. They have at their base a short sheath corresponding to the petiole of terrestrial plants. This is separated from the rest of the leaf by a small woody tongue called a ligule [21].

2.2 *Life Cycle of P. Oceanica*

Molinier and Zevaco [60] described the life cycle of *P. oceanica*, which can be summarized in three phases:

- a latent phase, from October to January, characterized by a very limited growth of leaves;
- a phase of great activity, from February to May, during which the leaves show an accelerated growth. As they develop, small organisms encrust them; they appear white and end up lying down under the weight of these encrustations.
- a phase of slowed growth, from June to September, ending with a massive fall of the leaves which are torn off during the autumn and winter storms.

Contrary to the algae, the posidonies are plants which flower and produce fruits.

The flowering of the Posidonia is not especially spectacular: the flowers are green, most often hidden in the heart of the ribboned leaves network of the plant. This phenomenon is not frequent, flowering occurring irregularly once every two years. To reproduce, the Posidonia has other means, in particular by natural cuttings using fragments of Posidonia broken and transported during storms and which end up rooting again.

Fig. 1 Pelota and leaves of *P. oceanica*

The fruits of the Posidonia reach maturity in 6–9 months. It is in summer that they are detached from the plant and begin to drift with the currents. These fruits have the shape and the dimension of an olive of dark brown to black color. They are commonly called "olive of sea". After a fortnight, the envelope of the fruit is torn, the seed falls then toward the bottom. If the environment gathers the good conditions, then the germination starts (Fig. 1).

2.3 Geographical Distribution

P. oceanica is distributed along the coasts of North Africa (Egypt, Libya, Tunisia, Algeria, and Morocco) and the European coasts (France, Corsica, Sardinia, Italy, Yugoslavia, Greece, Turkey, and Spain). This plant is distributed mainly around the Mediterranean basin except for the large estuaries of the Rhone, Nile, Po, and Ebro rivers [40, 64].

3 Traditional Methods of Exploitation of *P. Oceanica*

Several ways of exploiting the dead leaves of *P. oceanica* have been tested by man since antiquity (and even since prehistory) for different applications.

Thus, the men of the Lazaret cave (Alpes-Maritimes, France) slept on a bedding made of *P. oceanica* leaves [29]. Posidonia leaves were also used as reinforcing material inside mattresses or as bedding for animals. This type of use ceased when it was discovered that these leaves contained phenolic acids. The Egyptians also used seaweed balls as felt for making shoes [18, 19, 34, 72, 73]. They exploited the waste products of the sea grass for sore throats and skin diseases, due to their curative properties. An old botanical manual Cazzuola [22] mentions this plant among the products of the popular pharmacopoeia. It can also mention the fact that the leaves of *P. oceanica* have long been used by the Venetians to pack and transport their delicate and famous glassware. These leaves were known as "Venice straw" [17]. Finally, the coastal populations of North Africa (Egypt, Libya, and Tunisia) and Southern Europe (France, Corsica, Sardinia, Italy, Yugoslavia, Greece, Turkey, and Spain) still used, at the beginning of the twentieth century, the dry leaves of *P. oceanica* for the construction of roofs [33]. Indeed, the sheets of posidonies allow a good thermal insulation [17].

In addition to these uses, there are other traditional applications that continue to be studied. Farmers on the Mediterranean coast have traditionally utilized the decomposed leaves of *P. oceanica* as compost. During the last fifty years, trials have been regularly carried out in Italy, Tunisia, and Greece to produce a compost based on *P. oceanica* leaves, with rather interesting and promising results [67, 69, 71]. Moreover, fresh leaves of *P. oceanica* have a good nutritional value. Indeed, in Italy, leaf powder added to chicken feed has been shown to improve egg laying and egg weight [9, 10, 35]. In Tunisia in the 1920s, attempts to feed leaves to livestock, mixed with fodder, met with mixed success: donkeys and sheep refused to accept it, while horses accepted it [17, 66]. Finally, the fruits of *P. oceanica* washed up on beaches have been consumed by cattle (in Tunisia), pigs (in Corsica), and even by humans, in periods of deficiency [17, 27].

4 Study of the Chemical Composition of the Posidonia

Table 2 provides a summary of the findings regarding the chemical composition of Posidonia. These results are compared to those of other studies conducted in Morocco, Egypt, and Sudan on the composition of date palm leaves and rachis. According to this table, seagrass has a relatively high rate of extractable matter, especially in hot water or in an ethanol-toluene mixture. Moreover, there is a big difference between the ash content of seagrass compared to date palm [14–17]: seagrass has a high content (12%) compared to rachis (5%). Concerning cellulose, the contents are on the other hand very close (40% for the posidonie and 45% for the rachis of date palm).

Following a comparison with other plants (annuals, perennials) and coniferous and deciduous forests (see Table 3), the chemical composition of seagrass was then examined. The following conclusions can be drawn.

Seagrass has extractable matter concentrations that are higher in cold and warm water than in wood (hardwood or softwood), but they are nonetheless similar to those of annual and perennial plants. On the other hand, the contents of extractables in a 1% soda solution are comparable to those found for wood, which generally do not exceed 20%, and lower than those corresponding to non-woody plants, which generally have a level of around 30%. Lastly, seagrass has a level of 10.7% for extractables in the ethanol-toluene solvent system, which is higher than wood but on par with values corresponding to annual and perennial plants. The Klason lignin

Table 2 Chemical analysis of Posidonia according to the literature

	Posidonia	Date palm rachis				Date palm leaves		
	Khiari et al. [50]	Khiari et al. [50]	Khiristova et al. [51]	Bendahou et al. [13–17]	El Morsy et al. [61, 31]	Khristova et al. [51]	Bendahou et al. [13]	Ezzat [31]
Cold water solubility (%)	7.3	5.0	n.d.	n.d.	n.d.	n.d.	n.d.	n.d.
Hot water solubility (%)	12.2	8.1	8.7	n.d.	n.d.	10.8	n.d.	n.d.
Solubility in 1% NaOH (%)	16.5	20.8	25.6	n.d.	n.d.	29.9	n.d.	n.d.
Solubility in ethanol/ toluene (%)	10.7	6.3	12.8$^{(1)}$	4	n.d.	11.7$^{(1)}$	3	5.9$^{(2)}$
Ash (%)	12	5	5.6	2.5	3.4	9.6	6.5	3.9
Klason Lignin	29.8	27.2	23.8$^{(3)}$	14	25.8	31.2$^{(3)}$	27	n.d.
Holocellulose	61.8	74.8	n.d.	72	n.d.	n.d.	59.5	n.d.
Cellulose (%)	40	45	43.1	44	n.d	30.3	33.5	n.d.

Reprinted from Khiari et al. [50] with permission from Elsevier
$^{(1)}$ in ethanol-cyclohexane; $^{(2)}$ in ethanol-benzene; $^{(3)}$ lignin content (Klason lignin + soluble part of lignin); n.d.: not determined

content of seagrass (29%) is relatively high and close to the levels usually found in softwoods. Comparable to hardwoods and softwoods, the holocellulose and cellulose concentrations are also rather high. In conclusion, seagrass has a very high ash content (12%), yet it is nonetheless similar to several "pseudotrunks" of plants including rice, amaranth, and bananas [26]. This ash content may be a characteristic of this plant of marine origin but, despite intensive washing, it may also be related to pollution by sand trapped in the peels. It constitutes a major obstacle for paper applications, especially if it is silica. Indeed, a high silica content causes the appearance of deposits and limits the regeneration of chemicals used during the pulping process.

4.1 Elemental Analysis of P. Oceanica*: Comparison with Rachis of the Date Palm*

Scanning electron microscopy analyses coupled with an X-ray imaging technique were performed to study the elemental analysis of *P. oceanica*.

For date palm rachis, the ash consists mainly of calcium (Ca), chlorine (Cl), potassium (K) and sodium (Na). The amount of silicon found in the rachis ash is very low (0.14% by mass relative to the raw material) and comparable to that found

Table 3 Chemical composition of woody and non-woody plants

	References	C.S	H.S	O.S	1% NaOH	Ash (%)	Holocellulose	Klason lignin (%)	Hemicellulose	Cellulose
Wood from trees or shrubs										
Pinus brutia	Copur et al. [25]	2.2	2.8	1.94	16.1	0.4	75.5	26.1	28.5	47
Pinus pinaster **	Jimenez et al. [44, 45]	n.d.	2	1–2.6	7.9–10.3	0.3–0.5	69–67	26–28	13, 7	56
Eucalyptus globulus**	Jimenez et al. [44, 45]	n.d.	2.8	1.15	12.42	0.6	80.5	19.9	27.7	53
Olivier*	Jimenez et al. [45–46]	15.5	17	10.4	30.0	1.4	65.83	15.64	24.33	41.5
Green Oak	Eugenio et al. [30]	n.d.	n.d.	n.d.	n.d.	n.d.	71.2	16.3	28.3	43
Prosopis alba	Jimenez et al. [45]	n.d.	4.7	4.65	20.8	n.d.	63.6	19.3	22.0	42
Chamaecytisus	Jimenez et al. [45]	n.d.	3	3.43	16.1	n.d.	75.3	14.8	31.7	44
Phragmites	Jimenez et al. [45]	n.d.	5,4	6.36	34.7	n.d.	64.1	2.6	24.4	40
Retama monosperma	Jimenez et al. [45]	n.d.	3.8	5.03	16.9	n.d.	71.7	21.5	29.0	43
Paulownia fortuna	Jimenez et al. [47, 45]	n.d.	9.6	5.5	31.5	n.d.	70.7	22.4	33.3	37
Herbaceous or non-woody plants										
Arundo donax	Shatalov et al. [70]	n.d.	6.7	9.2	n.d.	4.8	61.2	20.9	32.1	29.2
Banana pseudostems	Cordeiro et al. [26]	n.d.	5.4	2.7	n.d.	14	65.2	12.7	25.2	40
Wheat straw	Schott [68]	5.8–11	14	4.6–9.2	41–42.8	4–9	n.d.	11–21	21–28.5	33–45.5
Cynara Cardunculus L*	Antunes et al. [8]	n.d.	10	6	n.d.	8	64	20	26	38

(continued)

Table 3 (continued)

	References	C.S	H.S	O.S	1% NaOH	Ash (%)	Holocellulose	Klason lignin (%)	Hemicellulose	Cellulose
Hibiscus cannabinus)	Manfred [57]	n.d.	n.d.	n.d.	n.d.	1.7–5	n.d.	14.5–18.7	n.d.	31–39
Miscanthus sinensis	Barba et al. [11]	n.d.	9.1	3.1	n.d.	0.7	72.5	19.9	30.3	42
Rice	Alcaide et al. [4]	10.6	13	4.6–5.7	49.1	13–20	n.d.	11–13.5	13–26.2	42–49.8
Barley	Alcaide et al. [4]	16	16	4.7	47	4.9–7	n.d.	7–18	24.5	34–48
Rye straw	Alcaide et al. [4]	8.4	9.4	3.2–5.2	37.4	4- 4.3	n.d.	18.5–19	23–30.5	55
Sorghum	Jimenez et al. [47]	n.d.	21.7	7.99	41.6	4.85	71.7	13.4	29.3	42

Reprinted from Khiari et al. [50] with permission from Elsevier
O.S. stands for solubility in organic solvents; C.S. stands for cold water solubility (%); H.S. for hot water solubility (%); * Average of two or three types; ** Average of eleven varieties. 1%NaOH: Solubility in 1% sodium hydroxide

by Khristova et al. [51]. Therefore, the relatively high ash content is not considered problematic for the pulping process, as it does not significantly affect the yield. However, this is not the case for seagrass, which has a very high silicon content (2.1% by mass in relation to the material).

4.2 Morphological Characterization of the Fibers of the P. Oceanica*: Comparison with Rachis of the Date Palm*

A morphological study was carried out to observe the structure of Posidonia in raw samples. The analysis was conducted using scanning electron microscopy (SEM). A comparison was made between the obtained results and the morphology of date palm rachis. The SEM observation [50] clearly shows that the morphology of the two materials is totally different. The structure of the date palm rachis is very porous compared to that of the Posidonia. It can therefore conclude that its accessibility to solvents or reagents is greater. This can be related to the rate of extractables in a 1% NaOH solution which is higher for the rachis than for the posidonie.

5 Delignification and Physicochemical Characterization

5.1 Effect of Temperature on the Delignification Process

Multiple soda-anthraquinone cookings were conducted on seagrass at temperatures ranging from 150 to 170 °C. Table 4 provides a summary of the experimental conditions and the various characteristics of the obtained pulps. There is a noticeable reduction in delignification yield, extractables and lignin contents, Kappa number, and viscosity as the delignification temperature is raised from 150 °C to 170 °C. At the same time, there is a relative increase in the percentage of holocellulose. On the other hand, the ash content remains constant. This parameter must be taken into consideration since silicon is the major component of Posidonia ashes. Indeed, the presence of this element can cause major problems during the pulping and paper making process (contamination and clogging of filters, abrasion, etc.).

In comparison with the date palm rachis fibers, which have a degree of polymerization (DP) close to 1200, the poidonia fibers have a DP close to 500. This value suggests that their mechanical properties will also be low.

Table 4 Elemental analysis of *P. oceanica* ash

	P. oceanica	Date palm rachis		Date palm leaves	Amarante	"Pseudotrunc" of banana tree
	Khiari et al. [50]	Khiari et al. [50]	Khristova et al. [51]	Khristova et al. [51]	Fiserova et al. [32]	Cordeiro et al. [26]
Chemical composition of ash (mass fraction in %)						
Si	17.7	2.8	1.1	5.8	0.25	2.7
Ca	9.12	21.5	n.d.	n.d.	4.17	7.5
Mg	3.89	3.53	n.d.	n.d.	0.035	4.3
Fe	3,78	240 ppm	n.d.	n.d.	n.d.	n.d.
Cu	< 100 ppm	360 ppm	n.d.	n.d.	0.01	n.d.
K	2.04	10.2	n.d.	n.d.	36.67	33.4
P	0.12	0.7	n.d.	n.d.	n.d.	2.2
S	1.92	1.69	n.d.	n.d.	n.d.	n.d.
C	< 0.3	1.5	n.d.	n.d.	n.d.	n.d.
Cl	0.72	18.6	n.d.	n.d.	n.d.	n.d.
Na	2.49	6.79	n.d.	n.d.	n.d.	n.d.
Quantity of silicon in relation to the raw material (mass fraction in %)						
Si	2.13	0.14	0.06	0.56	0.03	0.38

Comparison with other plants (n.d.: not determined). Reprinted from Khiari et al. [50] with permission from Elsevier

5.2 Density Measurement

Fiber density is a property that depends on the origin of the plant. After cooking, the density of *P. oceanica* fibers is determined. Table 5 summarizes the values obtained for Posidonia as well as data from the literature.

The density of fibers extracted from Posidonia is 1.43. With the exception of the alfa fiber, it should be noted that the density of this fiber is the lowest among the cellulosic fibers presented in Table 5.

Table 5 Measurement of fiber densities from various lignocellulosic materials

Material	Density measurement	Reference
P. oceanica	1.42	Khiari et al. [50]
Date palm rachis	1.43	Khiari et al. [50]
Cotton	1.51	Ghali et al. [36]
Alfa	1.34	Ghali et al. [36]

Table 6 Morphological properties of *P. oceanica* pulp fibers

	P. oceanica	Date palm rachis
Length of fibers* (mm)	0.69[a]; 0,89[b]	0.46[a]; 0.55[b]
Fiber diameter (μm)	22.3	21.3
Fine elements (% in length)	30.8	7.5

Reprinted from Khiari et al. [50] with permission from Elsevier
[a] arithmetic mean; [b] length weighted average

5.3 Analysis by Scanning Electron Microscope and Morfi

In order to have an idea on the individualization and the size of the fibers obtained after cooking, scanning electron microscopy observations have been made.

Generally speaking, the fibers are well individualized. Posidonia fibers have a different morphological structure, spiral. They also have a ribbon-like shape and a reduced length compared to palm rachis fibers. These observations were confirmed by fiber length measurements made using the Morfi. The results, in terms of arithmetic mean length, diameter, and percentage of fine elements, are summarized in Table 6.

The diameters of *P. oceanica* fibers are comparable to those obtained from date palm rachis. However, the average lengths differ: date palm rachis fibers are approximately 60% longer than Posidonia fibers. Consequently, their length to diameter ratio (form factor) is different: it is equal to 40 for date palm rachis. It is equal to 40 for date palm rachis while it is lower for seagrass (about 25). Hardwood and softwood fibers typically have arithmetic mean lengths of 2–3 mm and 1–2 mm, respectively. The corresponding length to diameter ratios is close to 100 and 60. Posidonia fibers have a significantly lower form factor.

6 Papermaking and Others Applications

In this part, we are focused to illustrate the different skind of application using *P. oceanica*. *P. oceanica*'s dead leaves have been tested by humans for diverse purposes since antiquity. In the following, the main traditional applications traditional applications are presented:

- **Manufacture of Metals**: The *P. oceanica* was used as material for the manufacture of metals. This was observed, for example, among the men of the Lazaret cave (Alpes-Maritimes, France) who slept on bedding made from *P. oceanica* leaves [29]. This type of use stopped when it was discovered that these leaves contained phenolic acids (Pergent et al. 1995) which are harmful to the skin.
- **Industrial activity**: *Posidonia* balls were used by the Egyptians to make felts for the manufacture of shoes [18, 19, 34, 72, 73]. The Venetians also used Posidonia leaves to pack and transport their delicate and famous glassware. These Posidonia leaves were known under the name of "straw of Venice" [17]. Until the beginning

of the twentieth century, the coastal populations of North Africa (Egypt, Libya, and Tunisia) and Southern Europe (France, Corsica, Sardinia, Italy, Yugoslavia, Greece, Turkey, and Spain) used the dry leaves of *Posidonia* for the construction of roofs [33]. Indeed, they bring a good thermal insulation [17].

- **Pharmacology**: Thanks to their curative properties, the waste of posidonie was also used against sore throat and skin diseases. In fact, an old botanical manual botany [22] cited the *Posidonia* among the products of the popular pharmacopoeia.
- **Agriculture Activity**: Apart from these uses, the farmers of the Mediterranean coasts have also used the dead leaves of Posidonia for a long time as compost. During the last fifty years, trials have been regularly carried out in Italy, Tunisia, and Greece, in order to produce a compost based on Posidonia leaves [67, 69, 71]. In the same way, fresh leaves are nutritionally beneficial. This supports, in Italy, the use of leaf powder as a supplement to poultry feed, which has boosted egg laying and egg weight [10, 11, 35]. In Tunisia, around the 1920s, attempts to feed leaves to livestock, mixed with fodder, met with mixed success. Donkeys and sheep resisted it, while horses accepted it [17, 66].Moreover, the fruits of Posidonia rejected on the beaches were consumed by the cattle (in Tunisia), the pigs (in Corsica), and even by the man, during periods of deficiency [17, 27].

In addition, Khiari et al. [50] presented in 2010 the complete characterization of raw seagrass materials and delignified fibers, including chemical composition. The effect of delignification conditions (treatment in the presence of soda and anthraquinone) on the properties of the fibrous suspensions produced was presented. In-depth discussions were held regarding the evolution of Kappa number, degree of polymerization (DP), and chemical composition (% residual lignin, holocellulose, etc.). The outcomes were then contrasted with those of other pulps, both conventional and non-conventional. Finally, Khiari et al. [50, 48] tested the paper potential and refinability of fibers obtained from seagrass. The physical properties of the papers produced were also compared to those of other types of papers [50]. The use of Posidonia fibers in papermaking was studied by Khiari et al. [50]. Fiber suspensions obtained from Posidonia were tested to produce the paper and/ handsheets. Due to the low strength at wet forms. This can be explained by the low degree of polymerization of the fibers and their morphological properties, in particular the length/diameter ratio. It is therefore difficult to imagine the use of Posidonia even for paper production. However, these fibers result in shapes with very poor mechanical properties but excellent softness. The use of these fibers in the mixture is therefore conceivable, provided the proportions used are optimized.

References

1. Aguir C, Khalfaoui M, Laribi N, Mhenni MF (2009) Preparation and characterization of new succinic anhydride grafted Posidonia for the removal of organic and inorganic pollutants. J Hazard Mater 172:1579–1590

2. Aguir C, Mhenni MF (2006) Experimental study on carboxymethylation of cellulose extracted from *Posidonia oceanica*. J Appl Polym Sci 98:1808–1816
3. Aguir C, Mhenni MF (2007) Removal of basic blue 41 from aqueous solution by carboxymethylated *Posidonia oceanica*. J Appl Polym Sci 103:1215–1225
4. Alcaide LJ, Parra IS, Baldovin F (1990) Characterization of Spanish agricultural residues with a view to obtaining cellulose pulp. Tappi J 73(5):173–176
5. Alcoverro T, Cerbian E, Ballesteros E (2001) The photosynthetic capacity of the seagrass *P. oceanica*: influence of nitrogen and light. J Exp Mar Biol Ecol 261:107–120
6. Alcoverro T, Manzanera M, Romero J (2001) Annual metabolic carbon balance of the seagrass *P. oceanica*: the importance of carbohydrate reserves. Mar Ecol Prog Ser 211:105–116
7. Alcoverro T, Romero J, Duarte CM, López NI (1997) Spatial and temporal variations in nutrient limitation of seagrass *P. oceanica* growth in the NW Mediterranean. Mar Ecol Prog Ser 146:155–161
8. Antunes A, Amaral E, Belgacem MN (2000) *Cynara cardunculus L.*: chemical composition and soda-anthraquinone cooking. Ind Crops Prod 12:85–91
9. Baldissera-Nordio C, Gallarati Scotti G, Rigoni M, Saibene G, Toschi A (1968) Utilizzazione dei vegetali marini nell'alimentazione dei polli. Terra Pugliese 17(12):1–8
10. Baldissera-Nordio C, Gallarati-Scotti GC, Rigoni M (1967) Valore nulritivo e possibilita di utilizzazione zootecnica di *Posidonia oceanica*. Atti Convegno nazionale sulla Attivita subacquee italie 1:21–28
11. Barba C, De la Rosa A, Vidal T, Colom JF, Farriol X, Montane D (2002) TCF bleached pulps from *Miscanthus sinensis* by the impregnation rapid steam pulping (IRSP) process. J Wood Chem Technol 22(4):249–266
12. Bay D (1984) A field study of the growth dynamics and productivity of *P. oceanica* (L.) Delile in
13. Bendahou A, Dufresne A, Kaddami H, Habibi Y (2007) Isolation and structural characterization of hemicelluloses from palm of *Phoenix dactylifera L.* Carbohyd Polym 68:601–608
14. Bendahou A, Habibi Y, Kaddami H, Dufresne A (2009) Physico-chemical characterization of palm from *Phoenix dactylifera*–L, preparation of cellulose whiskers and natural rubber-based nanocomposites. J Biobased Mater Bioenergy 3:81–90
15. Bendahou A, Kaddami H, Dufresne A, Kheli B, Taourirte M, Raihane M, Issartel N, Sautereau H, Gérard JF, Sami N (2006) Short palm tree fibers–Thermoset matrices composites. Compo Part A Appl Sci Manuf 37:1413–1422
16. Bendahou A, Kaddami H, Sautereau H, Raihane M, Erchiqui F, Dufresne A (2008) Short palm tree fibers polyolefin composites. Effect of filler content and coupling agent on physical properties. Macromol Mater Eng 239:140–148
17. Boudouresque CF, Meinesz A (1982) Découverte de l'herbier de Posidonie. Cahiers du Parc National de Port-Cros 4:1–79
18. Cannon JFM (1979) An experimental investigation of Posidonia balls. Aquat Bot 6(4):407–410
19. Cannon JFM (1985) Seaballs and lakeballs–an old Mediterranean theme with a new Irish variation. Watsonia 15:177–181
20. Capiomont A, Sandmeier M, Caye G, Meinesz A (1996) Enzyme polymorphism in *P. oceanica*, a seagrass endemic to the Mediterranean. Aquat Bot 54(4):265–277
21. Caye G (1980) Sur la morphogenèse et le cycle végétatif de *P. oceanica* (L. Delile), Thèse, Université d'Aix-Marseille, France
22. Cazzuola F (1980) Le piante utili e nocive agli uomini e agli animali che crescono spontanee e coltivate in Italia, mastère (SBL0063657) Turin, Italie
23. Cinelli F, Fresi E, Lorenzi C, Mucedola A (1995) La *Posidonia oceanica*. Un contributo per la salvaguardia del principale ecosistema marino del Mediterraneo. Rivista Maritima 1–271
24. Colombo PM, Rascio N, Cinelli F (1983) *P. oceanica* (L.) Delile: a structural study of the photosynthetic apparatus. PSZNI Mar Ecol 4:133–145
25. Copur Y, Tozluoglu A (2008) A comparison of kraft, PS, kraft-AQ and kraft-NaBH4 pulps of Brutia pine. Biores Technol 99(5):909–913

26. Cordeiro N, Belgacem MN, Torres IC, Mourad JCVP (2004) Chemical composition and pulping of banana pseudo-stems. Ind Crops Prod 19:147–154
27. Cuenod A (1954) Flore analytique et synoptique de la Tunisie. S.E.F.A.N Publ, Tunis, pp 1–267
28. Delile (1813) http://www.tela-botanica.org/eflore/BDNFF/4.02/nn/52120
29. Delumley H, Pillard B, Pillard F (1969) L'habitat et les activités de l'homme du Lazaret. Une cabanne acheuléenne de la grotte. Paleo 335–349
30. Eugenio ME, Alaejos J, Diaz MJ, Lopez F, Vidal T (2006) Evaluation of Holm oak (Quercus Ilex) wood as alternative source for cellulose pulp. Cellul Chem Technol 40(1–2):53–61
31. Ezzat S (1974) Leaves of date palm tree (*Phoenix dactylifera*) as a technical feasable source of raw material for paper production. Cellul Chem Technol 8:627–634
32. Fiserova M, Gigac J, Majtnerova A, Szeiffova G (2006) Evaluation of annualplants (*Amaranthus caudatus* L., *Atriplex hortensis* L., *Helianthus tuberosus* L.) for pulp production. Cellul Chem Technol 40(6):405–412
33. Le Floch E (1983) Contribution à une étude ethnobotanique de la flore tunisienne. Ministère de l'Education Nationale et de la Recherche Scientifique publ, Tunis
34. Font Quer P (1990) Plantas Medicinales. In: El Dioscorides Renovado, 12th edn. Editorial Labor, Barcelona, p 1033
35. Gallarati-Scotti G (1968) Primi risultati di prove di alimentazione con farina di *Posidonia oceanica*. Avicoltura 3:85–93
36. Ghali L (2004) Caractérisations physique et mécanique des fibres d'Alfa. ENIM université de Monastir, Mastère
37. Gobert S, Lejeune P, Lepoint G, Bouquegneau JM (2005) C, N, P concentrations and requirements of flowering *P. oceanica* shoots. Hydrobiologia 533:253–259
38. Gobert S (2002) Variations spatiales et temporelles de l'herbier à *P. oceanica* (L.) Delile. Thèse de Doctorat, Faculté des Sciences, Université de Liège, Belgique
39. Green EP, Short FT (2003) World atlas of seagrasses: present status and future conservation. Thesis, The University of California Press, 0-520-24047-2
40. Hartog CD (1970) The sea-grasses of the world. North Holland publishing Co, Amsterdam, Pays-Bas, pp 1–275
41. Hooker JD (1858) The botany of the Antarctic voyage of the H.M. discovery ships Erebus and Terror. III. Flora Tasmaniae 2(6):43
42. Invers O, Pérez M, Romero J (2002) Seasonal nitrogen speciation in temperate seagrass *P. oceanica* (L.) Delile. J Exp Mar Biol Ecol 273:219–240
43. Invers O, Kraemer GP, Pérez M, Romero J (2004) Effects of nitrogen addition on nitrogen metabolism and carbon reserves in the temperate seagrass *P. oceanica*. J Exp Mar Biol Ecol 303(1):97–114
44. Jimenez L, Lopez F (1990) Characterization of Spanish agricultural residues with a view to obtaining cellulose pulp. Tappi J 73(8):173–176
45. Jimenez L, Rodriguez A, Perez A, Moral A, Serrano L (2008) Alternative raw materials and pulping process using clean technologies. Ind Crops Prod 28(1):11–16
46. Jimenez L, Sanchez I, Lopez F (1992) Olive wood as a raw material for paper manufacture. Tappi J 11:89–91
47. Jiménez L, Lopez F, Martınez C (1993) Paper from sorghum stalks. Holzforschung 47:529–533
48. Khiari R, Marrakchi Z, Belgacem MN, Mauret E, Mhenni MF (2011) New lignocellulosic fibres-reinforced composite materials: a step forward in the valorisation of the *Posidonia oceanica* balls. Compos Sci Technol 71:187–1867
49. Khiari R, Mauret E, Belgacem MN, Mhenni F (2011) Tunisian date palm rachis used as an alternative source of fibres for papermaking applications. BioResources 6:265–281
50. Khiari R, Mhenni MF, Belgacem MN, Mauret E (2010) Chemical composition and pulping of date palm rachis and *Posidonia oceanica*—a comparison with other wood and non-wood fibre sources. Biores Technol 101:775–780
51. Khristova P, Kordsachia O, Khider T (2005) Alkaline pulping with additives of date palm rachis and leaves form Sudan. Biores Technol 96:79–85

52. Kuo J, Cambridge ML (1984) A taxonomic study of the *Posidonia ostenfeldii* complex (Posidonaeae) with description of four new Australian seagrasses. Aquat Bot 20:267–295
53. Kuo J (1978) Morphology, anatomy and histochemistry of Australian seagrasses of genus *Posidonia* Konig (Posidoniaceae).1. Leaf blade and leaf sheath of *Posidonia australis Hook.* Aquat Bot 5:171–190
54. Lepoint G, Defawe O, Gobert S, Dauby P, Bouquegneau JM (2002) Experimental evidence for N recycling in the leaves of the seagrass *P. oceanica*. J Sea Res 48:173–179
55. Lepoint G, Millet G, Dauby P, Gobert S, Bouquegneau JM (2002) An annual nitrogen budget of the seagrass *P. oceanica* as determined by in situ uptake experiments. Mar Ecol Prog Ser 237:87–96
56. Lepoint G, Vangeluwe D, Eisinger M, Paster M, VanTreeck P, Bouquegneau JM, Gobert S (2004) Nitrogen dynamics in *P. oceanica* cuttings: implications for transplantation experiments. Mar Pollut Bull 48:465–470
57. Manfred J (1993) Non-wood plant fibres, will there be a come-back in paper-making?*. Ind Crops Prod 2:51–57
58. Marbà N, Hemminga MA, Mateo MA, Duarte CM, Mass YEM, Terrados J (2002) Carbon and nitrogen translocation between seagrass ramets. Mar Ecol Prog Ser 226:287–300
59. Micheli C, Paganin P, Peirano A, Caye G, Meinesz A, Bianchi CN (2005) Genetic variability of *P. oceanica* (L.) Delile in relation to local factors and biogeographic patterns. Aquat Bot 82(3):210–221
60. Molinier R, Zevaco C (1962) Etudes écologiques et biocénotiques dans la baie du Brusc (Var). Bulletin de l'Institut Océanographique de Monaco, 1234
61. El Morsy MMS (1980) Studies on the rachis of Egyptian date palm leaves for hardboard production. Fibre Sci Technol 13(4):317–321
62. El Morsy MMS, Riad BY, Mohamed MAS (1981) Pulp and paper from the Egyptian date palm leaves. Fibre Sci Technol 14(2):157–161
63. Pasqualini V, Pergent-Martini C, Clabaut P, Pergent G (1998) Mapping of *P. oceanica* using aerial photographs and side-scan sonar: application of the island of Corsica (France). Estuar Coast Shelf Sci 47:359–367
64. Phillips RC, Meñez EG (1988) Seagrasses. Smithsonian contributions to the marine sciences, vol 34. Smithsonian, Institution Press, Washington, p 105
65. Piazzi L, Acunto S, Cinelli F (2000) Mapping of *P. oceanica* beds around Elba Island (western Mediterranean) with integration of direct and indirect methods. Oceanologica Acta 23(3):339–346
66. Pottier J (1929) Etude sur les possibilités d'utilisation des plantes marines tunisiennes pour la nourriture du bétail. Annales de l'Institut Oceanographique 6(3):321–362
67. Saidane A, Dewaele N, Van Develde R (1979) Contribution a l'étude du compostage de plantes marines en vue de la préparation d'un amendement organique et d'un substrat horticole. Bulletin de l'Institut National Scientifique et Technique d'Océanographie et de Pêche- Salammbô-, 6(1–4):133–150
68. Schott S (2000) Valorisation de la pâte de paille de blé, obtenue par le procédé d'explosion en phase vapeur, pour une application en papeterie, thèse, Institut Polytechnique de Grenoble, France
69. Seri F, De Gara L, Caretto S, Leo L, Santamaria P (2004) Influence of an increased NaCl concentration on yield and quality of cherry tomato grown in Posidonia *Posidonia oceanica (L. Delie)*. J Sci Food Agric 84:1885–1890
70. Shatalov AA, Quilho T, Pereira H (2001) Arundo donax L. reed: new perspectives for pulping and bleaching. 1. Raw material characterization. Tappi J 84(1):96
71. Sordina JB (1951) Marine plants of Greece and their use in agriculture. Prak Hells Hydrobiologie Inst 5:73–124
72. Tackholm V, Drar M (1954) Flora of Egypt. Vol. III. Angiospermae, part Monocotyledones: Liliaceae-Musaceae. Bull Facul Sci Egypt 3:30
73. Weddel (1877) Note sur les aegagropiles de mer. Actes Congrès international botanistes-Amsterdam, pp 58–62

Preparation and Characterization of Cellulose from *Posidonia oceanica* Deposits and Its Applications

Madelyn N. Moawad, Abeer A. M. El-Sayed, and Soha H. Shabaka

Abstract *Posidonia oceanica*, an endemic seagrass species in the Mediterranean Sea, generates substantial amounts of stranded deposits on beaches. These deposits, though a nuisance, represent a valuable source of renewable material. Seagrassess are characterized by high lignocellulosic content, similar to land plants. In terms of sustainability, many researchers focused on the use of *P. oceanica* waste, which has many industrial, agricultural, and environmental applications. The current work discusses the methods of preparation, characterization, and different applications of lignocellulosic material extracted from *P. oceanica* stranded deposits, addressing their benefits as a readily available and renewable resource.

M. N. Moawad · A. A. M. El-Sayed · S. H. Shabaka (✉)
National Institute of Oceanography and Fisheries (NIOF), Cairo, Egypt
e-mail: sh.shabaka@niof.sci.eg

R. Khiari and M. Jawaid (eds.), *Current Status and Opportunity in Fibre and Composites*, Sustainable Materials and Technology,
https://doi.org/10.1007/978-981-96-4546-6_3

Graphical Abstract

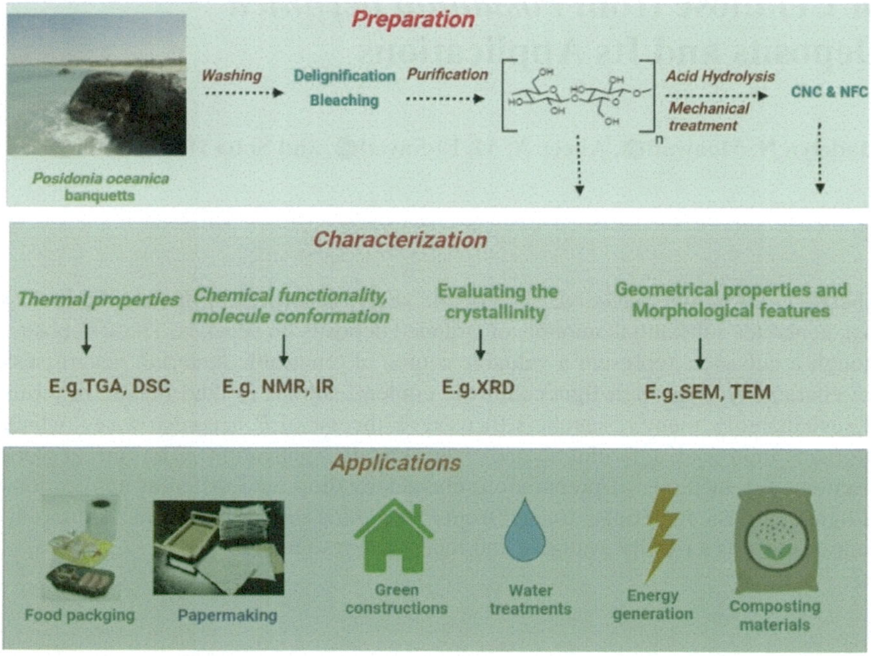

Keywords Posidonia oceanica · Biomass valorization · Sustainable applications

1 Introduction

Posidonia oceanica (L.) Delile, a seagrass species unique to the Mediterranean Sea, creates vast underwater meadows that stretch from the coastal shallows to depths of around 40 m. These meadows are ecologically crucial due to their significant environmental contributions. The economic value of the services and benefits provided by *P. oceanica* is substantial, estimated at between 25.3 and 45.9 million euros per year, or 283–513 euros per hectare annually [8].

The shoots of *P. oceanica* consist of roots, leaves with a length of 15–100 cm, and erect or creeping rhizomes covered with thick sheaths and a system of roots. Over time, *P. oceanica* develops a complex structure of rhizomes, roots, and sheaths that raises the seabed, creating what's known as "dead matte." During autumn, when leaves and rhizomes die and detach from the meadows, they wash up on beaches, where they accumulate into formations called "banquettes." [7]. Despite their natural occurrence, these beach deposits are seen as an inconvenience by Mediterranean beachgoers and are often removed and discarded as waste [2, 38]. These stranded deposits have attracted the interest of scientists for the exploitation of the energy content of their lignocellulosic materials [34, 37, 41].

Holocellulose is the water-insoluble carbohydrate fraction of the cell wall of seagrasses. It consists of cellulose and hemicellulose, which are major constituents of seagrasses lignocellulosic material. Cellulose gives strength and stability to the cell wall, it constitutes 20–77%, of the dried seagrass material and mostly concentrated in the leaves [33]. Hemicellulos is the second most abundant polymer in the cell wall of cellulosic contents, it constituted 21% of dried *P. oceanica* balls (*egagropili*) [23], and the highest value of 38% were reported for *Zostera marina* [11]. Hemicellulose connects to other plant components in two ways: it binds to cellulose through physical intermixing and hydrogen bonds, while forming covalent bonds with lignin [1]. Lignin consists of phenylpropane monomers that are cross-linked to form condensed macromolecules, and its primary function is to provide structural rigidity to plants [33]. Among all seagrass families, the Posidoniaceae species have the highest lignin content [33], generally seagrasses have a lower lignin content than land plants, which makes the process of delignification for cellulose isolation easier more economic. Holocellulose has been calculated to be 61.8% of *P. oceanica* dry mass and cellulose contributes 40% [23], a value that makes the fibers of *P. oceanica* a renewable cellulose source, which in turn provides economic and environmental benefits [22].

Cellulose extraction involves several chemical procedures, including alkaline extraction, bleaching, acid hydrolysis, and chlorination [25]. The crystalline structure of cellulose is so important that multiple analytical techniques must be used to monitor it throughout chemical treatment. These include X-ray diffraction (XRD), solid-state nuclear magnetic resonance spectroscopy (ss NMR), infrared spectroscopy (IR), Raman spectroscopy, sum-frequency-generation spectroscopy (SFG), differential scanning calorimetry (DSC), and various physicochemical tests. Because cellulose has such a complex structure, these characterization techniques are used to study cellulose microfibrils after other polysaccharides have been systematically removed [35].

Due to its outstanding mechanical strength and barrier properties, cellulose is widely used in packaging applications—both as a primary material and as an additive to enhance the performance of other biopolymers [4].

The production of cellulose from seagrasses and particularly from *P. oceanica* stranded deposits could be a solution to the accumulated deposits, which is a burden to solid waste management systems. This study reviews the diverse methods used to prepare and characterize cellulose extracted from beached *P. oceanica* deposits, as well as its various applications.

2 Preparation of Holocellulose and Cellulose from *P. oceanica* Waste

2.1 Preparation of Holocellulose

Delignifying agents are various chemicals used to break down lignin and isolate holocellulose from biomass. The chlorination method is based on cooking the dried *P. oceanica* with sodium chlorite ($NaClO_2$) and acetic acid or acetone [3, 6]. Following modern green chemistry principles, hydrogen peroxide has emerged as an environmentally friendly alternative to chlorine-based chemicals for bleaching purposes [10]. Cooking the dried seagrass material with a combination of NaOH and Na_2SO_3 at 140 °C for 5 h and 30: 70 mixture compositions yielded the best result for the extraction of holocellulose [1]. According to Aslam et al. [1], NaOH works to break down lignin, while sulfite ions speed up this delignification process by breaking lignin into smaller pieces that form stable sodium salts. This process has an added benefit: it reduces NaOH's reaction with carbohydrates, which leads to higher yields [1].

2.2 Preparation of Cellulose

Before isolating cellulose from *P. oceanica* leaves, a dewaxing step is typically performed. This involves treating the leaves in a toluene/ethanol mixture (2:1 volume ratio) for 6 h, then filtering and washing them with ethanol for 30 min [12, 27]. The solution will turn to green color due to the removal of the chloroplasts [27]. Microwave-assisted alkali treatment has been applied to the red seaweed *Gelidiella acerosa* which was found to reduce the heating time and eliminate the dewaxing process [39]. A pretreatment step is often applied to eliminate calcium carbonate adhering to the surface of seagrasses (tube worms, calcareous algae, etc.), and it involves washing with 3% v/v acetic acid and stirring for two hours at a constant temperature of 80 °C in NaOH (5% w/v) [36].

The isolation of cellulose is generally conducted in two steps; delignification and bleaching, using alkali and bleaching agent, and removal of hemicellulose for purification. The alkaline treatment increases the roughness of the fiber surface, while peroxide treatment reduces the fibers' hydrophilicity. Together, these changes improve both the fibers' wettability and their interaction with the matrix material [36]. In the alkali-anthraquinone process method, the isolation involves adding NaOH and anthraquinone ($C_{14}H_8O_2$) and cooking at a constant temperature, and bleached using 100 ml of sodium hypochlorite solution [6, 29]. For the elimination of hemicellulose from the isolated fibers, the bleached fibers are mixed with NaOH or KOH solution in a hot water bath at 80 °C under mechanical stirring [6, 12].

The cellulose nanocrystals (CNC) are usually prepared from the isolated cellulose or holocellulose of *P. oceanica* [6, 27], which is obtained by the hydrolysis with sulfuric acid (9 mL: 1 g) with vigorous stirring, dilution with deionized water, and centrifugation of the suspension to concentrate the cellulose crystals, and to remove the excess acid. The precipitate undergoes a series of purification steps: it's first rinsed, then centrifuged, and dialyzed in deionized water until reaching a neutral pH. The final step involves sonicating the resulting suspension. On the other hand, various mechanical methods, such as high-pressure homogenizers, microfluidizers, ultrafine grinders, and the ultrasonic technique, were used to recover nanofibrillar cellulose (NFC) from various lignocellulosic materials [5].

3 Structural Characterization of Cellulose Molecule Extracted from *P. oceanica*

In nature, cellulose exists as cellulose I, which has two different forms: cellulose Iα and cellulose Iβ. While cellulose Iα is mainly found in simpler organisms such as bacteria and algae, cellulose Iβ is the dominant form in more complex plants. At the molecular level, cellulose consists of glucose molecules joined together in linear chains through β-1,4-glycosidic bonds [35]. The β,D-glucopyranose unit in cellulose has a distinctive six-membered arm-chair structure, with all methane (C–H) groups positioned axially and both hydroxyl (OH) and hydroxymethyl (CH_2OH) groups arranged equatorially. Although cellulose is made up of glucose units, its physical and chemical properties are much more complex and can vary significantly based on three factors: the plant source, the methods used for extraction and purification, and the state of aggregation. One of cellulose's most distinctive characteristics is its inability to dissolve in water. This insolubility is caused by the crystallization of cellulose chains during their synthesis in the biological systems. When viewed in cross section, natural cellulose chains appear as flat ribbons due to their crystalline packing arrangement, which enables them to pack tightly together. These ribbons have a specific structure: C–H groups are exposed along their flat sides, while hydroxyl groups are positioned along their thin edges. The ribbons stack together in a precise way—their flat surfaces connect through van der Waals forces, while the hydroxyl groups form hydrogen bonds with neighboring chains, creating extended sheets. This network of interactions results in the formation of thread-like crystalline structures known as cellulose microfibrils. As a result, cellulose cannot be synthesized in laboratory [24]. Crystallinity measures the degree of structural order in a material, calculated as the ratio of crystalline regions to the total volume (which includes both crystalline and amorphous regions) [35]. The degree of crystallinity plays a crucial role in determining the mechanical properties of cellulose and its derivatives, particularly their strength and stiffness. Additionally, the balance between crystalline and amorphous regions in cellulose affects how accessible and reactive the material is to chemical modifications [35]. As a consequence, crystalline cellulose is used in a variety of engineering applications. However, the cell walls of producing organisms contain crystalline cellulose as well as integrated matrix polymers like hemicellulose and lignin (amorphous components), together referred as "lignocellulosic biomass" [24], that affects cellulose crystallinity. Therefore, curial purification processes involving chemical pulping on a large scale are required [25].

In this section, we will go through some of the emerging imaging, diffraction/scattering, and spectroscopic techniques as well as different physicochemical assays that could be used to characterize *P. oceanica* cellulose.

3.1 Microscopic Imaging Techniques

Various microscopic imaging methods—including transmission electron microscopy (TEM), scanning electron microscopy (SEM), and atomic force microscopy (AFM)—have made it possible to directly observe cell wall components in lignocellulosic biomass, revealing both their network-like structure and how different components interact with each other [35] were the main techniques employed to examine the size and shape of the dispersed structures.

3.1.1 Transmission Electron Microscopy (TEM)

TEM was used to explore the geometrical properties of CNC and NFC resulting from *P. oceanica* leaves and balls [5, 6]. Digital image analysis (Image J) of TEM micrographs was used to determine the size of the associated nanoparticles. According to Bettaieb et al. [6], the CNC from *P. oceanica* exhibited well-dispersed elongated rod-like structure with an average diameter and length ranging from 6.8 to 8.1 nm and 276 to 520 nm, respectively.

The high aspect ratio of nanofibrillated cellulose (NFC) extracted from *P. oceanica* makes it well-suited for incorporation into polymer matrices, offering a valuable way to utilize this abundant agricultural waste material. TEM data indicated wide heterogeneous distribution of nanosized fibrils both in width (2–21 nm) and in length [5].

While electron microscopy has revealed important insights about cellulose microfibrils, its sample preparation process, which typically requires drying, can create artifacts by altering the natural structure of cellulose—causing microfibrils to collapse and clump together. This limitation makes TEM less reliable for studying microfibril morphology and diameter in cell walls [35].

3.1.2 Scanning Electron Microscopy (SEM)

SEM sample preparation is easier than TEM since electron transparent materials and heavy metal stains are not required [35]. SEM technique was adopted by several authors to investigate the morphological features of unbleached and bleached *P.*

oceanica [6, 14, 27, 36]. SEM imaging reveals that unbleached leaves contain three main cellular components: vascular tissue, epidermis, and parenchyma. Their cell walls vary in thickness from 1 to 5 μm, with the thickness depending on the amount of lignin, ash, and hemicelluloses present. The lumen (the inner cavity of these cells) measures between 8 and 20 μm in width [5, 27]. Calcium carbonate was found on the surface of an untreated leaf [27]. On the other hand, the unbleached balls of *P. oceanica* were mostly made up of sclereid cells (or fibers) with a diameter of around 20 μm and a wall thickness of about 10 μm [5].

Upon chemical treatment, lignin, hemicelluloses, and ashes are eliminated resulting in a reduction of cell element dimensions and lumen width [14, 27]. The cell elements are no longer recognizable [27], and the bleached fibers have a smooth and clean surface that explicates the diameter reduction [14].

Luzi et al. [27] conducted morphological investigation on CNC using SEM obtained after acid hydrolyzing procedure at 45 °C for 30 min. Based on the SEM analysis results, the treatment process was extended to include a second step with a longer duration (60 min at 45 °C). This modification was made to ensure the reaction would complete fully and produce cellulose materials at a precise nanoscale level. The SEM images underlined reduction in the presence of unreacted material and the effectiveness of the new hydrolysis procedure with a consequent high yield. According to morphological investigation, individual cellulose nanocrystals with an acicular structure of 5–10 nm in diameter and 200–450 nm in length, and an aspect/ratio of 40–45, are found in the aqueous suspension [27]. The diameters of microfibrils measured by SEM agree with those estimated by AFM [42]. Nevertheless, SEM usually involves dehydration or critical-point drying and deposition of a conductive coating to avoid charging, which might result in distortions. Consequently, the approach is frequently employed in conjunction with other microscopic and spectroscopic methods [35].

3.1.3 Atomic Force Microscopy (AFM)

Scanning force microscopy (SFM), also called atomic force microscopy (AFM), offers an alternative to electron microscopy (SEM and TEM) for observing cellulose microfibrils. This technique can visualize structures from the micrometer down to sub-nanometer scale, with the advantage of being able to do so under biologically relevant conditions. AFM allows highly improved spatial resolution (0.1–100 nm) with minimal sample preparation for direct examining of sample surfaces [35].

AFM (atomic force microscopy) was used to examine the morphological characteristics of cellulose nanocrystals (CNC) [14]. The AFM images confirm the success of the acid hydrolysis treatment, showing that the aqueous suspension contains individual needle-shaped cellulose nanocrystals. These nanocrystals measure 180 nm in length and 4.9 nm in diameter, resulting in an aspect ratio of 36.7 [14]. The aspect ratios are more accurately assessed by TEM and AFM morphological observations than those observed by nano-Zetasizer. Since the nano-Zetasizer approach takes into account spherical particles whose diameter corresponds to the length of the nano-rods, the detected size of nano-rod is higher in length than those determined by TEM and AFM [6].

3.2 Diffraction/Scattering Techniques

Beyond direct imaging techniques, researchers use X-ray diffraction (XRD) or wide-angle X-ray scattering (WAXS) to measure microfibril dimensions. XRD has become the preferred method for measuring cellulose crystallinity because it is both reliable and precise, while requiring minimal sample preparation. This technique determines crystallinity by measuring the proportion of crystalline cellulose mass relative to the total sample mass [35].

X-ray diffraction measurements were conducted in both untreated and treated *P. oceanica* [5, 6]. The XRD analysis of both untreated and treated fibers shows three main peaks at approximately 16°, 22°, and 34° (2θ), which are characteristic of the (110), (002), and (004) planes found in cellulose type I. The persistence of these peaks after treatment indicates that the crystalline structure of cellulose type I remains unchanged by the chemical processing.

In the raw material's XRD pattern, the amorphous regions are indicated by low diffracted intensity around 18° (2θ). The analysis also detected peaks indicating common sea biomass contaminants: silica, calcium carbonate, and calcium oxalate. After treatment, the crystallinity showed improvement, likely due to the removal of amorphous components (lignin and hemicelluloses). Additionally, the intensity of peaks associated with the contaminants (silica, calcium carbonate, and oxalate) decreased [5, 6]. Similar findings were obtained from WAXS analysis conducted in both untreated and treated fibers of *P. oceanica* [36]. While microscopy and diffraction/scattering methods can provide structural information, they cannot capture the dynamic aspects of cellulose structure needed to explain certain characteristics. To address this limitation, researchers use various spectroscopic techniques—including NMR, IR, Raman, and more recently, neutron spectroscopy—which have proven effective at providing detailed insights into cellulose's structure [35].

3.3 Spectroscopy Analysis

While NMR and IR spectroscopy are valuable tools for analyzing purified cellulose structure, they have limited usefulness when studying native cellulose or lignocellulosic biomass because other cell wall components interfere with their spectral readings [35].

3.3.1 Infrared Spectroscopy (IR)

Infrared (IR) spectroscopy is a vibrational spectroscopic technique that reveals information about chemical functionality, molecular conformation, and hydrogen bonding. In crystalline cellulose, it can be used to study both intra- and intermolecular hydrogen bonds. The spectrum shows two particularly significant bands:

- The "crystallinity band" (1420–1430 cm^{-1}), which corresponds to symmetric CH_2 bending vibration
- The "amorphous band" (893–898 cm^{-1}), which represents C–O–C stretching at β-(1 → 4)-glycosidic linkages [35].

The IR spectrum shows several characteristic bands.

The O–H stretch band appears between 3200 and 3600 cm^{-1} as a broad peak, reflecting both the complex hydrogen bond networks in the crystal and water molecules in amorphous regions. Around 2900 cm^{-1}, strong bands indicate C–H stretch vibrations. The hydroxymethyl group of cellulose shows two shoulder peaks: one at 2840–2880 cm^{-1} (symmetric CH_2 stretch) and another at 2920–2970 cm^{-1} (asymmetric CH_2 stretch).

Between 1450 and 1480 cm^{-1}, there's a deformation vibration (scissoring mode) of the CH_2 group on C6. The region between 1200 and 1450 cm^{-1} contains bands created by multiple movements: C–C–H bending, CH_2 rocking and wagging, and C–O–H bending modes.

Different regions of the IR spectrum reveal specific molecular movements in cellulose.

The region between 950 and 1150 cm^{-1} shows peaks from skeletal ring stretches and stretching of both glycosidic and alcoholic C–O bonds. Below 950 cm^{-1}, the spectrum reflects bending movements of skeletal bonds (C–C–C, C–O–C, O–C–C, and O–C–O). In the lower region of 400–800 cm^{-1}, the peaks likely correspond to torsion or bending modes of the six-membered ring in relation to the glycosidic bonds. Identifying which functional groups correspond to specific peaks is relatively straightforward when dealing with small, isolated compounds. However, in the case of cellulose, it is a little more complicated. This is because all bands are highly coupled over multiple covalent bonds, especially at wave numbers below 1450 cm^{-1} [24]. Unfortunately, due to interference from other components, IR spectroscopy has difficulty quantifying the crystallinity present in lignocellulosic biomass or matrix polymer [35].

For instance, the absorption band appearing between 1700 and 1750 cm^{-1} can be attributed to several sources: acetyl and carbonyl groups from carboxylic ester linkages found in hemicelluloses and lignin, as well as wax and natural fats. This band was eliminated in treated fibers, indicating that the chemical treatment has removed the majority of noncellulosic components [27, 36]. The band appearing in the 1200–1250 cm^{-1} region, which corresponds to the C–O stretching vibration of acetyl groups in lignin and hemicelluloses, becomes less intense after treatment [27, 36]. The disappearance of the signal at 1511 cm^{-1}, which is characteristic of lignin's aromatic ring, confirms that the chemical treatment successfully removed the lignin component [14, 27]. The spectrum of raw *P. oceanica* displayed two notable features: a distinct peak at 877 cm^{-1} and a strong band at 1420 cm^{-1} [36]. These signals indicated the presence of calcium carbonate impurities, which was consistent with the WAXS pattern findings. Following treatment, the spectral changes showed the removal of these impurities: the band at 877 cm^{-1} vanished and the band in the 1290–1570 cm^{-1} region significantly decreased in intensity [36].

The cellulose I structure of the produced CNC (cellulose nanocrystals) was confirmed by the enhanced visibility of several characteristic signals: those at 1430 and 1420 cm^{-1}, 1334, 1316, 1158, 1061, and 893 cm^{-1} [14].

3.3.2 Solid-State Cross-Polarization Magic Angle Spinning Nuclear Magnetic Resonance Spectroscopy (^{13}C CP MAS NMR)

NMR spectroscopy offers qualitative and quantitative information about the chemical surrounding of an atom in a sample. Chemically equivalent carbons placed at magnetically nonequivalent places can be distinguished using this technique [35]. However, they can vary in crystalline solids due to the packing configuration. Advances in solid-state ^{13}C NMR techniques have expanded the applications of NMR spectroscopy. Specifically, when cross-polarization magic angle spinning (CP/MAS) ^{13}C NMR is used to study native cellulose structure, it can distinguish between cellulose Iα and Iβ forms by examining the multiplicity of the C4 resonance peak in the 88–90 ppm range [35]. A key advantage of NMR spectroscopy is that resonance peak intensities directly correlate with the carbon atom density in the sample, allowing straightforward conversion of peak intensities into relative concentrations of different carbon types. In a typical cellulose NMR spectrum, the carbon signals appear in specific regions:

C1 appears at 102–108 ppm.
C4 appears at 81–93 ppm.
C2, C3, and C5 cluster between 70 and 81 ppm.
C6 appears between 60 and 70 ppm.

The position of these peaks from downfield is determined by how many electronegative atoms are in the nearest and next-nearest neighbor positions to each carbon. For example:

C1 appears furthest downfield because it's most influenced by oxygen atoms (O1 and O5 as nearest neighbors, O2 as next-nearest).
C6 appears furthest upfield because it's least influenced by oxygen atoms (O6 as nearest neighbor, O5 as next-nearest). The other carbons (C2, C3, C4, and C5) appear between these two extremes [24].

The crystalline and paracrystalline components are underlined by CP MAS NMR [10]. The broad, undefined peak in the upfield region (81–86 ppm) primarily indicates the presence of amorphous cellulose. Part of this component, called paracrystalline cellulose, likely comes from chains on crystal surfaces that are less ordered than those in the crystal core. By comparing the relative intensities of this broad peak (81–86 ppm) with the sharp peaks (86–91 ppm), researchers can estimate both the crystallinity of cellulose and the surface-to-volume ratio of cellulose microfibrils [24].

Coletti et al. [10] study the crystalline components of *P. oceanica* before and after extraction procedure using ^{13}C CP MAS NMR. The two spectra show minor change in composition with lignin signals (110–140 ppm) only seen in the spectrum of untreated material. The signals appearing between 60 and 90 ppm, which correspond to cellulose components, show an increase in the crystalline content after chemical treatment of the biomass.

3.4 Thermal Properties

3.4.1 Thermogravimetric Analysis (TGA)

The natural fibers display a multistep thermal degradation in thermogravimetric analysis (TGA). Thermal degradation of untreated lignocellulosic materials occurs at temperatures ranging from 200 to 600 °C, corresponding to the thermal disintegration of their primary components (hemicelluloses, lignin, and cellulose). Hemicelluloses began to decompose at temperatures ranging from 200 to 310 °C. Cellulose breaks down at high temperatures between 315 and 400 °C, with the fastest rate of decomposition occurring around 350 °C. In contrast, lignin degrades gradually across the entire temperature range of fiber decomposition [36].

According to studies by Fortunati et al. [14] and Luzi et al. [27], thermogravimetric analysis was used to study the thermal properties of *P. oceanica* materials at different processing stages: unbleached leaves, bleached leaves, and hydrolyzed cellulose nanocrystals (CNC).

The unbleached materials undergo multiple stages of thermal degradation, indicating the presence of different components. The first stage occurs around 100–150 °C, involving the removal of moisture. The second stage appears around 255–294 °C, representing hemicellulose degradation. This peak is absent in both bleached and CNC materials. The lower thermal stability of hemicellulose can generally be attributed to its acetyl groups [14, 27].

Luzi et al. [27] recorded higher thermal stability of cellulose with degradation peak at 344 °C for CNC than that recorded for unbleached fibers (333 °C) and bleached cellulose materials at 320 °C. According to Fortunati et al. (2015), acid-treated crystals (CNC) showed decreased thermal stability, with cellulose degradation occurring at 291 °C. This lower thermal stability likely resulted from two factors: the presence of residual sulfate groups from acid hydrolysis and the increased proportion of short cellulose chains. Furthermore, the cellulose degradation peak for bleached fibers was seen to shift to a higher temperature of 356 °C compared to 335 °C for unbleached fibers.

Generally, the onset temperature greater than 250 °C indicates that these natural fibers are suitable for processing with thermoplastic polymer matrices without incurring thermal degradation [37].

3.4.2 Differential Scanning Calorimetry (DSC)

Different authors, on the other hand, adopted differential scanning calorimetry (DSC) to investigate the thermal stability of various polymer/*Posidonia* fiber composites [14, 23, 36].

3.5 Physicochemical Analysis

Various methods were used to measure Klason lignin, Kappa number, and ash content, as described below:

Klason Lignin

The acid-insoluble lignin was determined by acid hydrolysis of holocellulose and/or cellulose fibers [6]. The insoluble residue (lignin) was rinsed, dried, and weighed (C). Pyrolysis in a furnace was used to obtain the inorganic material, and the remaining ashes were weighed (C_0) [13].

Kappa Number

The Kappa number, which measures the degree of delignification and ranges from 1 to 100, was determined by measuring how much potassium permanganate solution was consumed by moisture-free fiber at 25 °C [6].

Ash Yield

The measurement was performed by calculating the solid residue that remained after burning the test material at 525 $\pm$ 25°C for a minimum of 4 h [6].

In addition to spectroscopic techniques, crystallinity can be measured using various physicochemical methods: the Updegraff method [40], iodine adsorption [19], enthalpy of wetting, and water sorption methods [20]. However, when compared to the crystallinity determined by XRD, physicochemical approaches usually yield a greater crystallinity results. The compositional and structural variability of cell walls is one possible cause of this mismatch that makes access to noncrystalline components more difficult. Therefore, any assumption of a direct relationship between crystallinity and the studied physicochemical parameters would be proven incorrect [35].

4 Applications of Cellulose from *P. oceanica*

4.1 Environmental Applications

The biocomposites based on polyhydroxyalkanoates (PHAs) and fibers of *P. oceanica* up to 30 wt % as a thermoplastic matrix are technically feasible and represent an eco-sustainable alternative to the disposal of *P. oceanica* wastes [37]. These renewable resource-based composites could replace traditional nonbiodegradable plastics in coastal habitat restoration and preservation, particularly for transplanting tools and structures.

4.2 Industrial Applications

4.2.1 Food Packaging

While thermoforming is the most widely used processing technique in food packaging, poly(3-hydroxybutyrate) (PHB) presents challenges for this method due to two factors: it has low melting elasticity when in sheet form, and it requires very specific processing conditions within a narrow temperature range [16]. When a rigid fiber is added to a highly crystalline polymer, it reduces the polymer's ability to be thermoformed. According to Sánchez-Safon et al. [36], they created composites by melt blending PHB with *P. oceanica* fibers at a concentration of 20 phr. These fibers proved effective as fillers in PHB composites, strengthening the material while maintaining PHB's original crystallinity and breakdown rate. The trade-off was that adding fibers led to lower thermal stability and decreased water barrier properties.

Fortunati et al. [14] created two types of poly(lactic acid) (PLA) nanocomposite films using cellulose nanocrystals from *P. oceanica*: one with regular CNC and another with surfactant-modified nanocrystals (s-CNC). These materials exhibited unique properties in terms of their mechanical strength, thermal behavior, optical characteristics, and migration patterns. The surfactant improved how well the CNCs dispersed throughout the polymer matrix, which enhanced both crystal formation and plasticity in the material. The successful development of these biodegradable nanocomposites from waste biomass suggests they could be valuable for food packaging applications.

4.2.2 Food Preservation

According to Benito-Gonzalez et al. [4], cellulosic aerogels (HCAG) made from *P. oceanica* waste demonstrated exceptional absorption capabilities for both water and oil, reaching 1500–1900%. When loaded with hydrophilic and hydrophobic extracts, these aerogels helped reduce both lipid oxidation and oxymyoglobin degradation in red meat over a 10-day storage period. These findings suggest that HCAG derived from *P. oceanica* waste could be used to create antioxidant bioactive absorption pads for food packaging preservation [4].

4.2.3 Papermaking

Various research studies have examined how suitable different seagrass species are for paper production by analyzing their fiber structure and chemical makeup (Syed et al. 2016). According to Khiari et al. [23], holocellulose, lignin, and cellulose extracted from *P. oceanica* had similar chemical structures to those encountered in softwood and hardwood. However, the high ash content and low degree of polymerization (DP) (about 370) may be considered severe drawbacks in the pulping and papermaking context [23].

4.2.4 Green Construction

Adding a combination of seagrass and refined wood fibers to strengthen gypsum mortar offers a green and sustainable alternative for construction materials, making it especially beneficial in regions that experience frequent earthquakes [26]. According to Kuqo and Mai [26], using the lignocellulosic material from *P. oceanica* as filler can have numerous advantages in the local building sector. When natural cellulosic fibers extracted from *P. oceanica* were added to gypsum, they enhanced its durability and structural properties. Tests revealed that adding these fibers improved impact resistance by 57% compared to plain plaster without reinforcement. In addition, Olacia et al. [32] examined the suitability of using the cellulosic fiber from *P. oceanica* as reinforcement in adobe bricks in comparison with the traditional straw fibers. The results showed that bricks with long seagrass fibers presented outstanding behavior in terms of flexural and compressive strength, suggesting it as a sustainable and value-added construction product [32].

Hamdaoui et al. [18] explored whether *P. oceanica* fibers could serve as effective loose-fill insulation materials for buildings in Mediterranean regions. The analysis revealed that *P. oceanica* fibers possess thermal properties comparable to traditional insulation materials, with similar thermal conductivity and diffusivity values. Additionally, these fibers demonstrated a notably high mass heat capacity of 2533 J per kilogram-Kelvin (Hamdaoui et al. 2018). Moreover, Garcia-Garcia et al. [15] made composite fiberboards from *P. oceanica* as environmental-friendly product by thermo-compression using a bio-based epoxy binder resin. These innovative fiberboards showed promise as a potential substitute for traditional particle boards and fiberboards made from conventional natural fibers.

4.2.5 Reinforcement Agents

Khiari et al. [21] prepared new lignocellulosic particles obtained from *P. oceanica*. When these lignocellulosic bio-based components were added as reinforcement to commercial biodegradable thermoplastic matrix, they enhanced its thermal and mechanical properties. Consequently, researchers used these materials in combination with BIOPLAST GF 106 matrix to develop various composite films. The use of *P. oceanica* fibers and deinking paper sludge (DPS) as reinforcement agents to produce composites material for industrial applications was studied by Haddar et al. [17]. Different ratios of these fillers were combined with a high-density polyethylene (HDPE) thermoplastic matrix to create various binary and hybrid composites. This hybridization of *P. oceanica* natural fibers with DPS waste resulted in composites with optimized thermal and mechanical properties. Nanofibrillar cellulose (NFC) from *P. oceanica* was isolated by Bettaieb et al. [5], which was utilized to prepare NFC-reinforced nanocomposites based on a rubber matrix. Using dynamic mechanical analysis of the reinforcing potential of the resulting NFC have shown promising mechanical properties.

4.3 Water Treatment

Research by Ncibi et al. [30] examined how effectively *P. oceanica* fibers could remove phenol from water. The study found optimal biosorption occurred at pH 5.2, achieving 2 mg/g capacity. When phenol concentration was increased from 10 to 150 mg/L, the fibers' biosorption capacity improved significantly, rising from 0.71 to 3.08 mg/g.

A study by Ncibi et al. [31] evaluated how *P. oceanica* fibers could remove Alpacide yellow (a metal complex textile anionic dye) from water solutions. The researchers discovered peak biosorption (4.68 mg/g) occurred at pH 2, and efficiency improved when they increased the biomass amount to 10 g/L with an initial dye concentration of 50 mg/L. Their data analysis showed the process followed pseudo-second-order and Elovich equations, indicating a chemisorption mechanism. The equilibrium results fit both Redlich–Peterson and Langmuir isotherm models, suggesting the dye molecules formed a single layer on the *P. oceanica* surface, with adsorption sites showing similar attraction levels to the textile pollutant.

Coletti et al. [10] developed two cellulose derivatives from *P. oceanica*'s α-cellulose: cellulose acetate (CA) and glycidyl methacrylate grafted cellulose (GMA-C). Their tests showed GMA-C could remove β-naphthol from water with 47% efficiency. When they modified GMA-C by hydrolyzing its glycidyl group into a glycerol group, its performance improved dramatically, achieving 92% efficiency in removing p-nitrophenol from water.

4.4 Energy Generation

The fuel properties of lignin and cellulose of *P. oceanica* were studied by Voca et al. [41]. The study demonstrated that *P. oceanica*'s lignin and cellulose components are suitable for energy generation through combustion, due to several favorable characteristics: high ash content, high heating value, and low levels of sulfur, nitrogen, and carbon.

4.5 Agricultural Applications

P. oceanica wastes have long been used as a composting material together with sludge for agriculture purposes [9]. Mariles et al. [28] found that *When P. oceanica fibers are bound together using the biopolymer chitosan, they create effective agricultural substrates. This chitosan treatment improved performance, resulting in higher germination rates and reducing average germination time by approximately one day compared to untreated fibers.*

5 Conclusion

Exploiting the cellulosic content available in the *P. oceanica* litter represents the best solution for the disposal of biological waste in an economical and environmentally friendly manner. The recycling of these important wastes and their exploitation in industrial, environmental, and agricultural applications as alternative materials achieve several goals of sustainable development. The current work provided a review of most of the methods used to extract and characterize cellulose extracted from *P. oceanica*, as well as the associated applications, in the hope that it will benefit other Mediterranean countries that do not exploit these important raw materials and also provide an opportunity for scientists to optimize the different methods of extraction and applications.

References

1. Aslam U, Javed SH, Kazmi M, Feroze N (2014) Extraction of holocellulose from bagasse using delignifying agents: a comparative study. J Chem Soc Pak 36:73–77
2. Aydin M, Cavas L, Merdivan M (2012) An alternative evaluation method for accumulated dead leaves of Posidonia oceanica (L.) Delile on the beaches: removal of uranium from aqueous solutions. J Radioanal Nucl Chem 293(2):489–496. https://doi.org/10.1007/S10967-012-1782-2
3. Basu P (2018) Analytical techniques. In: Biomass gasification, pyrolysis torrefaction, pp 479–495. https://doi.org/10.1016/B978-0-12-812992-0.00023-6

4. Benito-González I, López-Rubio A, Galarza-Jiménez P, Martínez-Sanz M (2021) Multifunctional cellulosic aerogels from Posidonia oceanica waste biomass with antioxidant properties for meat preservation. Int J Biol Macromol 185:654–663. https://doi.org/10.1016/j.ijbiomac.2021.06.192
5. Bettaieb F, Khiari R, Dufresne A, Mhenni MF, Putaux JL, Boufi S (2015) Nanofibrillar cellulose from Posidonia oceanica: properties and morphological features. Ind Crops Prod 72:97–106. https://doi.org/10.1016/j.indcrop.2014.12.060
6. Bettaieb F, Khiari R, Hassan ML, Belgacem M, Bras J, Dufresne A, Mhenni MF (2015) Preparation and characterization of new cellulose nanocrystals from marine biomass Posidonia oceanica. Ind Crops Prod 72:175–182. https://doi.org/10.1016/j.indcrop.2014.12.038
7. Boudouresque CF, Pergent G, Pergent-Martini C, Ruitton S, Thibaut T, Verlaque M (2016) The necromass of the Posidonia oceanica seagrass meadow: fate, role, ecosystem services and vulnerability. Hydrobiologia 781:25–42. https://doi.org/10.1007/S10750-015-2333-Y
8. Campagne CS, Salles JM, Boissery P, Deter J (2015) The seagrass Posidonia oceanica: ecosystem services identification and economic evaluation of goods and benefits. Mar Pollut Bull 97:391–400. https://doi.org/10.1016/J.MARPOLBUL.2015.05.061
9. Castaldi P, Melis P (2002) Composting of Posidonia oceanica and Its use in agriculture. In: Insam H, Riddech N, Klammer S (eds) Microbiology of Composting. Springer, Berlin, Heidelberg. https://doi.org/10.1007/978-3-662-08724-4_35
10. Coletti A, Valerio A, Vismara E (2013) *Posidonia oceanica* as a renewable lignocellulosic biomass for the synthesis of cellulose acetate and glycidyl methacrylate grafted cellulose. Materials (Basel). 6:2043–2058. https://doi.org/10.3390/ma6052043
11. Davies P, Morvan C, Sire O, Baley C (2007) Structure and properties of fibres from sea-grass (Zostera marina). J Mater Sci 42(13):4850–4857. https://doi.org/10.1007/S10853-006-0546-1
12. Fabra MJ, Seba-Piera I, Talens-Perales D, López-Rubio A, Polaina J, Marín-Navarro J (2020) Revalorization of cellulosic wastes from Posidonia oceanica and Arundo donax as catalytic materials based on affinity immobilization of an engineered β-galactosidase. Food Hydrocoll 103. https://doi.org/10.1016/j.foodhyd.2019.105633
13. Ferrero B, Boronat T, Moriana R, Fenollar O, Balart R (2013) Green composites based on wheat gluten matrix and posidonia oceanica waste fibers as reinforcements. Polym Compos 34:1663–1669. https://doi.org/10.1002/pc.22567
14. Fortunati E, Luzi F, Puglia D, Petrucci R, Kenny JM, Torre L (2015) Processing of PLA nanocomposites with cellulose nanocrystals extracted from Posidonia oceanica waste: innovative reuse of coastal plant. Ind Crops Prod 67:439–447. https://doi.org/10.1016/j.indcrop.2015.01.075
15. Garcia-Garcia D, Quiles-Carrillo L, Montanes N, Fombuena V, Balart R (2018) Manufacturing and characterization of composite fiberboards with *Posidonia oceanica* wastes with an environmentally-friendly binder from epoxy resin. Materials 11:35–50. https://doi.org/10.3390/ma11010035
16. González-Ausejo J, Sanchez-Safont E, Lagaron JM, Olsson RT, Gamez-Perez J, Cabedo L (2017) Assessing the thermoformability of poly (3-hydroxybutyrate-co-3-hydroxyvalerate)/poly (acid lactic) blends compatibilized with diisocyanates. Polym Test 62:235–245. SN Appl Sci 1:1414–1429. https://doi.org/10.1007/s42452-019-1420-506.026
17. Haddar M, Elloumi A, Koubaa A, Bradai Ch, Migneault S, Elhalouani F (2018) Synergetic effect of Posidonia oceanica fibres and deinking paper sludge on the thermo-mechanical properties of high density polyethylene composites. Ind Crops Prod 121:26–35. https://doi.org/10.1016/j.indcrp.2018.04.075
18. Hamdaoui O, Ibos L, Mazioud A, Safi M, Limam O (2018) Thermophysical characterization of Posidonia oceanica marine fibers intended to be used as an insulation material in Mediterranean buildings. Constr Build Mater 180:68–76 https://doi.org/10.1016/j.conbuildmat.2018.05.195
19. Hessler LE, Power RE (1954) The use of iodine adsorption as a measure of cellulose fiber crystallinity. Text Res J 24:822–827. https://doi.org/10.1177/004051755402400906
20. Ioelovich M (2016) Physicochemical methods for determination of cellulose crystallinity. ChemXpress 9:245–251

21. Khiari R, Marrakchi Z, Belgacem MN, Mauret E, Mhenni F (2011) New lignocellulosic fibres-reinforced composite materials: A stepforward in the valorisation of the Posidonia oceanica balls. Compos Sci Technol 71:1867–1872. https://doi.org/10.1016/j.compscitech.2011.08.022
22. Khiari R, Mhenni MF, Belgacem MN, Mauret E (2010) Chemical composition and pulping of date palm rachis and Posidonia oceanica—A comparison with other wood and non-wood fibre sources. Bioresour Technol 101:775–780. https://doi.org/10.1016/J.BIORTECH.2009.08.079
23. Khiari R, Ikhlass F, Farouk MM (2010) Study of liquids absorption and retention capacities of new cellulosic materials and sodium cellulose carboxylmethylate prepared from Posidonia. Fibers Polym 114(11):593–597. https://doi.org/10.1007/S12221-010-0593-X
24. Kim SH, Lee CM, Kafle K (2013) Characterization of crystalline cellulose in biomass: basic principles, applications, and limitations of XRD, NMR, IR, Raman, and SFG. Korean J Chem Eng 30:2127–2141. https://doi.org/10.1007/s11814-013-0162-0
25. Kumar Gupta P, Sai Raghunath S, Venkatesh Prasanna D, Venkat P, Shree V, Chithananthan C, Choudhary S, Surender K, Geetha K (2019) An update on overview of cellulose, its structure and applications. In: Cellulose. IntechOpen. https://doi.org/10.5772/intechopen.84727
26. Kuqo A, Mai C (2021) Mechanical properties of lightweight gypsum composites comprised of seagrass *Posidonia oceanica* and pine (*Pinus sylvestris*) wood fibers. Constr Build Mater 282:122714–122723. https://doi.org/10.1016/j.conbuildmat.2021.122714
27. Luzi F, Fortunati E, Puglia D, Petrucci R, Kenny JM, Torre L (2016) Modulation of acid hydrolysis reaction time for the extraction of cellulose nanocrystals from Posidonia oceanica leaves. J Renew Mater 4:190–198. https://doi.org/10.7569/JRM.2015.634134
28. Mariles B-A, Jaime G-P, Eva B-B, Ignacio M, Pablo D-G (2019) Fibers of the seagrass *Posidonia oceanica* as substrate for germination of lentil seeds. SN Appl Sci 1:1414–1420. https://doi.org/10.1007/s42452-019-1420-5
29. Moral A, Aguado R, Roldán R, Tijero A, Ballesteros M, (2019) Soda-anthraquinone pulping and cationization of Posidonia oceanica. BioResources 14:9228–9243. https://doi.org/10.15376/biores.14.4.9228-9243
30. Ncibi MC, Borhane M, Seffen M (2006) Biosorption of phenol onto Posidonia oceanica (L.) seagrass in batch system: equilibrium and kinetic modelling. Can J Chem Eng 84:495–500. https://doi.org/10.1002/cjce.5450840411
31. Ncibi MC, Mahjoub B, Ben Hamissa AM, Ben Mansourc R, Seffen M (2009) Biosorption of textile metal-complexed dye from aqueous medium using *Posidonia oceanica* (L.) leaf sheaths: mathematical modelling. Desalination 243:109–121. https://doi.org/10.1016/j.desal.2008.04.018
32. Olacia E, Pisello AL, Chiodo V, Maisano S, Frazzica A, Cabeza LF (2019) Use of seagrass fibres in adobe bricks. In: IOP conference series: Earth and Environmental Science. https://doi.org/10.1088/1755-1315/225/1/012051
33. Pfeifer L, Classen B (2020) The cell wall of seagrasses: fascinating, peculiar and a blank canvas for future research. Front Plant Sci 11. https://doi.org/10.3389/fpls.2020.588754
34. Rammou E, Mitani A, Ntalos G, Koutsianitis D, Taghiyari HR, Papadopoulos AN (2021) The potential use of seaweed (posidonia oceanica) as an alternative lignocellulosic raw material for wood composites manufacture. Coatings 11:1–9. https://doi.org/10.3390/coatings11010069
35. Rongpipi S, Ye D, Gomez ED, Gomez EW (2019) Progress and opportunities in the characterization of cellulose—an important regulator of cell wall growth and mechanics. Front Plant Sci 9. https://doi.org/10.3389/fpls.2018.01894
36. Sánchez-Safont EL, Aldureid A, Lagarón JM, Gámez-Pérez J, Cabedo L (2018) Biocomposites of different lignocellulosic wastes for sustainable food packaging applications. Compos Part B Eng 145:215–225. https://doi.org/10.1016/j.compositesb.2018.03.037
37. Seggiani M, Cinelli P, Mallegni N, Balestri E, Puccini M, Vitolo S, Lardicci C, Lazzeri A (2017) New bio-composites based on polyhydroxyalkanoates and *Posidonia oceanica* fibres for applications in a marine environment. Materials (Basel). 10:326. https://doi.org/10.3390/ma10040326
38. Shabaka SH, Khalil MK, El-Sikaily A, Youssef NAE (2021) Posidonia oceanica litter along the Mediterranean Coast of Egypt: status and a preliminary assessment of nutrients and trace

elements contents. Estuar Coast Shelf Sci 255:107342. https://doi.org/10.1016/J.ECSS.2021.107342

39. Singh S, Gaikwad KK, Park SI, Lee YS (2017) Microwave-assisted step reduced extraction of seaweed (Gelidiella aceroso) cellulose nanocrystals. Int J Biol Macromol 99:506–510. https://doi.org/10.1016/j.ijbiomac.2017.03.004
40. Updegraff DM (1969) Semimicro determination of cellulose in biological materials. Anal Biochem 32:420–424. https://doi.org/10.1016/S0003-2697(69)80009-6
41. Voca N, Grubor M, Peter A, Kricka T (2019) Evaluation of Posidonia oceanica waste as a biomass source for energy generation. Bioenergy Res 12:1104–1112. https://doi.org/10.1007/s12155-019-10032-7
42. Zheng Y, Wang X, Chen Y, Wagner E, Cosgrove DJ (2018) Xyloglucan in the primary cell wall: assessment by FESEM, selective enzyme digestions and nanogold affinity tags. Plant J 93:211–226. https://doi.org/10.1111/tpj.13778

Posidonia Oceanica: Chemical Composition, Fractionation Methods, and Applications

Rim Mnafki, Ramzi Khiari, Younes Moussaoui, and Jalel Labidi

Abstract However, *P. oceanica* is found in two shapes: leaves decomposed and transported by clouds to form giant benches at the seaside that protect beaches from erosion. Balls that are the result of roots and rhizome fragments that are transported and formed by the waves along the shoreline are thought to be the most significant and well-studied seagrass species in the Mediterranean Sea; much work has been put into combining all of the spatial data that is currently available to provide a complete distribution of meadows in the basin. Several studies have focused on fractionating this seagrass to highlight its cellulose and lignin richness and achieve its value. The purpose of this review was to provide a small-scale evaluation of the current and known historical distribution of *P. oceanica;* the fractionation methods are used to separate *P. oceanica* components, its characteristics, the chemical composition, and the main routes and applications for valorizing *P. oceanica* for medical, industrial, and agricultural uses as documented in recent scientific studies published in recent years.

R. Mnafki
Laboratory for the Application of Materials to the Environment, Water, and Energy (LR21ES15), Faculty of Sciences of Gafsa, University of Gafsa, Gafsa, Tunisia

R. Mnafki · Y. Moussaoui
Facultyof Sciences of Gafsa, University of Gafsa, Gafsa, Tunisia

R. Khiari (✉)
Department of Textile, Higher Institute of Technological Studies of Ksar Hellal, 5070 Ksar Hellal, Tunisia
e-mail: khiari_ramzi2000@yahoo.fr

CNRS, Grenoble INP, LGP2, Université Grenoble Alpes, F-3800 Grenoble, France

Y. Moussaoui
Organic Chemistry Laboratory (LR17ES08), Faculty of Sciences of Sfax, University of Sfax, Sfax, Tunisia

J. Labidi
Department of Chemical and Environmental Engineering, Biorefinery Processes Research Group, University of Basque Country (UPV/EHU), Donostia-San Sebastian, Spain

R. Khiari and M. Jawaid (eds.), *Current Status and Opportunity in Fibre and Composites*, Sustainable Materials and Technology,
https://doi.org/10.1007/978-981-96-4546-6_4

Keywords *Posidonia oceanica* · Chemical composition · Fractionation methods · Applications

1 Introduction

Lignocellulosic fibers are the most often utilized biodegradable reinforcing components for composite materials. Growing interest has been shown in the use of lignocellulosic materials to create new composites in recent years [1, 2]. Their affordability and environmental friendliness contribute to their popularity. Although the utilization of these fibers is desirable from bothtechnical and environmental perspectives, it's also critical to take fibrous wastes into account as prospective candidates for green composites [3, 4]. Almond husks, olive pits, sawdust, and other fibrous materials can be found as wastes from various industrial processes [5, 6]. Also, large quantities of biomass, such as algae waste, are continuously produced, and some of their final uses include composting, the production of fuel, and the conversion to thermal energy [7, 8]. Among these algae wastes, we are interested in one of the most productive ecosystems on the planet "*Posidonia oceanica*" (*P. oceanica*). This marine magnoliophyte, also known as a phanerogam, is unique to the Mediterranean and may be found practically anywhere. *P. oceanica* has been shown to be an ideal source for the extraction of lignocellulosic fractions with the potential for use in packaging materials or as fillers to improve the properties of other biopolymers [9, 10]. Because *P. oceanica* is considerably different from other lignocellulosic materials, it is highly advised to try to make it into a resource that offers both economic advantages and favorable environmental feedback [11]. In the creation of one of the Mediterranean region's most productive ecosystems, where a complex trophic network arises, *P. oceanica* plays a significant role. In particular, the grasslands of *Posidonia oceanica* create a favorable substrate for several epiphytic organisms, help consolidate the seabed, and prevent erosion of the shoreline. They also significantly contribute in providing shelter for juvenile forms and larval stages as well as mooring for eggs [12].

P. oceanica is a long-lived seagrass (4–30 years) that serves a purpose similar to that of wetlands and tropical forests in that it is crucial to the sea's natural balance. Numerous fish and other aquatic animal species find their homes and food in the *P. oceanica* fields (meadows) [13]. *P. oceanica* ranges in depth from the surface to 40 m, depending on the clarity of the water [14]. It appears as beams that are typically between 20 and 80 cm long and between 4 and 8 broadsheets that are around 1 cm wide and organized in roots, rhizomes, and leaves [15, 16] covering about 1.5% of the total Mediterranean Sea surface [17, 18]. Multiple measurement methods (oxygen, 14C, leaf marking, and lepidochronology) reveal a *P. oceanica* leaf production between 300 and 2000 g m^{-2} per year [19, 20], with a seasonal leaf withering and rhizome detachment. The majority of this process takes place in the fall, when the biggest loss of seagrass material also happens to coincide with an increase in wave activity and transport.

P. oceanica leaves (POL), rhizomes, and roots often accumulate along the Mediterranean coast, forming imposing masses on the beaches called "banquettes" that can range in thickness from a few centimeters to several meters and slow down the erosion of the beach [21, 22]. *P. oceanica* usually reproduces sexually, while stem extension may occasionally result in asexual reproduction. The sexual annual reproductive cycle begins in the autumn with pollination, and after a period of six to nine months, its full fruits release seeds that can sink to the seafloor and grow roots to create a new plant. The meadows of *P. oceanica* create a significant quantity of waste each year in the form of leaves, rhizomes, and roots that might act as biological sediment on the seafloor [23]. When conditions are right, the floating dead leaves of *P. oceanica (POL)* that reach the coast produce dense deposits along with other debris and deposits. When conditions are right, they can consolidate and produce a very compact and strong structure known as "banquettes," which are crucial for preventing erosion of the coast [24–26].

P. oceanica is comparable to softwoods like pine and spruce in that it has a very high lignin concentration, making it one of the richest lignocellulosic biomasses [19]. In reality, lignocellulosic biomass contains three main biopolymers: cellulose, hemicellulose, and lignin, which links cellulose and hemicellulose to form a hard structure, decreasing even enzymatic accessibility to the biomass's carbohydrate fraction and ultimately its energy valorization and significant ecological value because they generate large amounts of oxygen, provide food, shelter, and protection for a variety of marine species, and act as a natural barrier to reduce wave energy and stabilize the coast profile [27, 28], as well as smaller amounts of other organic substances, including pectin and proteins [29]. In addition to its immeasurable ecological significance for the Mediterranean marine environment, it has traditionally been used for a variety of purposes in the architectural, industrial, and commercial sectors, as well as in the medical field. The awareness of the advantages of the phytocomplex for human health has recently been increased by a growing body of evidence in addition to the knowledge passed down via tradition [30, 31]. As a result, researchers are becoming more interested in *P. oceanica*'s potential as a source of unexplored phytotherapeutic compounds.

The world's energy consumption has been rising steadily over the past few decades as a result of the increasing human population and industrialization. Additionally, the International Energy Outlook 2017 [32] forecasts a rise in global energy consumption of 28%, or from 575 to 736 quadrillion British thermal units, between 2015 and 2040. Similar estimates have been made in a report by the International Energy Agency, which projects a 35% rise in energy consumption between 2010 and 2035 [33]. The European Commission has set two challenging goals to be achieved by 2030 in order to properly safeguard the environment. The first is to lower GHG emissions by at least 40% from their levels in 1990, and the second is to increase the amount of energy produced from renewable sources by at least 27%. This explains the recent focus on renewable energy sources [34] that has been so intense. The fibrous beach-cast *P. oceanica* wastes may also be ideal for burning with energy recovery since their heating capacities seem to be comparable to those of peat, wood, and straw. Beach-cast *P. oceanica* waste can also be used in the composting process with other biomasses that have characteristics that work against each other. For example, ash, carbon (C) concentrations, and the C/N ratio are all good things. Bleached *P. oceanica* fibers in particular should be utilized as bulking agents and may be easily combined, for instance, with grass clippings and organic waste from fruit and vegetable markets [35].

2 Geographical Répartition

The Mediterranean region nearly entirely contains *Posidonia oceanica.* About 1.5% of the Mediterranean Sea's surface is covered with *Posidonia oceanica* meadows, which also exist in 16 other Mediterranean nations [36] but do not extend into the Sea of Marmara or the Black Sea. Finally, it is rare or absent in the extreme north of the Adriatic [37–39] and along the coasts of Languedoc, between the Camargue and Port-la-Nouvelle [40]. Spanish, French/Monaco, Italian, Albanian, Tunisian, Egyptian, and Turkish historical data on seagrass beds were available. The percentage of regression (total historical area compared to total current area) of *P. oceanica* meadows and the time range of data are presented for each nation in Table 1, which includes a comprehensive list of compiled research and thorough information.

P. oceanica meadows were discovered to be extensively distributed over Spain's islands and continental shoreline, with a total estimated size of 172,699 hectares. *P. oceanica* covered a total of 94,030 acres along the French coast and was found there, more or less consistently, over the continental shores and islands (including Corsica). With the exception of the main river mouths, the *P. oceanica* meadows in Italy encompassed 337,611 ha and were distinguished by a rather constant distribution throughout the continental and insular Tyrrhenian, Ionian, and South-Western Adriatic Sea shores. Except for a spotty distribution in the northern sector, meadows were absent from the Adriatic beaches in the north and central-western regions. Additionally, several maps and point data were used to illustrate the distribution of beds along the Tunisian beaches. The Gulfs of Gabes and Tunis, the Galite Archipelago, Zembra Island, and other coastal locations, totaling 518,685 hectares, have all been reported to have *P. oceanica* [41]. In total, *P. oceanica* is a Mediterranean endemic, i.e., strictly confined to this sea. It cannot endure temperatures below 10 °C and above 29 °C [42]. Posidonia is absent from coastal lagoons, in front of the mouths of coastal rivers, in the middle of certain bays, and at the level of groundwater outflows because it is afraid of desalination (33) and oversalination (> 41) [43].

Concerning the etymology, *Oceanica* (= oceanic) derives from a mix-up with the eelgrass Zostera marina, which is found in the Atlantic Ocean in Portugal, along the shores of Spain's Basque Country, and on the Canary Islands. Posidonia is named after "Poseidon," the Greek deity of the sea. There are several vernacular names for *Posidonia* in the Mediterranean.

The nine species declared in Table 2 that make up the genus *Posidonia* have linear leaves and a persistent leaf sheath that remains on the plant after the leaf blade has

Table 1 Lengths of coastline with the known current and historical presence of *Posidonia oceanica*, the percentage of regression, and the time range of the data

Country	*PO* total current area (ha)	*PO* total historical area (ha)	*PO* regression (%)	Time range of data
Tunisia	518.685	531.844	2%	1972–2010
Italy	337.611	395.298	25%3	1990–2005
Turkey	287	–	–	2009
Spain	172.669	222.254	29%1	1993–2011
France/Monaco	94.030	96.783	9%2	1980–2011
Greece	44.939	–	–	2011
Croatia	31.437	–	–	2010
Cyprus	9.040	–	–	2008
Slovenia	9	–	–	2004
Malta	5.860	–	–	2002
Montenegro	–	–	–	2004
Albania	4.803	5.710	16%	2007–2008
Algeria	4.072	–	–	2010
Libya	1.235	–	–	2011
Syria, Lebanon	Absent	–	–	2003
Egypt	–	–	–	2006
Morocco	–	–	–	2006

Data referred to Andalucia, Murcia, Valenciana, the Balearic Islands, the continental coasts of France/Monaco, Liguria, Tuscany, Latium, Apulia, and Sardinia regions, polygons and point data: surface area refers to polygons only, point data only: no surface area is available; confirmed *PO* absence

fallen off. They have hermaphrodite inflorescences and are monoecious. There is no perianth [49].

Table 2 List of marine phanerogam species

Type species	Références
Posidonia Angustifolia	[44]
Posidonia Australis	[45]
Posidonia Coriácea	[46]
Posidonia Denhartogii	[46]
Posidonia Kirkmanii	[46]
Posidonia Ostendfeldii	[47]
Posidonia Sinuosa	[46]
Posidonia Robertsoniae	[46]
Posidonia Oceanica	[48]

3 Morphology

The overall appearance of raw fibers reveals a typical lignocellulosic fiber structure. A collection of fibrils connected by hemicellulose makes up each fiber, which has an average diameter of 100 μm. These hollow fibrils come in a variety of sizes and forms. Because lignin, pectin, wax, and vegetable oils are present, these fibers have a rough surface, as may be seen by looking at their surfaces [50]. Lefebvre et al. [48] have gone into great detail about the structure of *P. oceanica* and have concluded that *P. oceanica* balls (POB) exist in two forms: a spherical form (ball-shaped) and an ellipsoidal shape (egg-shaped). Three unique forms of POB were identified: heterogeneous, homogeneous, or with or without a specific nucleus. An intermediate type was also identified. The first kind was heterogeneous, with a virtually whole portion of *P. oceanica*'s rhizome serving as the nucleus. The second form, in contrast, was homogenous and had a thick fibrous center made primarily of bundles of thin fibers called rhizome fibers. We postulate that homogeneous aegagropiles are the end product of the heterogeneous type's disintegration. In fact, both forms of aegagropiles appeared to have their nuclei in the same plant organ, the rhizome, but to varying degrees [48]. Additionally, each of them displayed a series of concentric layers with varied textures and variations in the thickness and content of the mineral sand grains and plant fibers. The center nucleus of the diverse aegagropiles was white in color and frequently had an identifiable portion of rhizome, which occasionally erupted at one or both ends of the long axis. Multiple clusters of black pieces or fibers took the place of the rhizomic component in the intermediate POB.

The core of the homogenous agropiles was muddled. Both varieties of *P. oceanica* balls (POB) have the same concentric fibrous layers: the deep layer was defined by a larger proportion of sand particles together with fibers, giving it a white color and a more compact look, while the median layer seemed comparatively rich in fibers and the surface layer consisted of sparsely scattered fibers [48]. Ncibi et al. [51] have detailed the textural characteristics of *P.oceanica* balls via SEM technique (Fig. 1). The most important finding is that the raw biomass possesses the usual compact lignocellulosic fibrous structure. Furthermore, the transverse slice demonstrates that *P. oceanica* is not porous at all when it is in its natural state. Indeed, the Posidonia is made up of a number of holocellulosic microfibers that are connected by lignin, much like all lignocellulosic-based fibers [51].

Bettaieb et al. [52] used SEM spectroscopy to study the shape of *P. oceanica* balls (POB) and leaves (POL) in detail. They concluded that, unlike *P. oceanica* balls (Fig. 2a and b), the structure of *PO* leaves (Fig. 2c and d) is very porous. Sclereid cells (or fibers) with a diameter of around 20 m and a wall thickness of about 10 m made up the majority of the *PO* balls. However, in the case of *P. oceanica* leaves, multiple distinct cell components, including parenchyma, vascular tissue, epidermis, and fiber cells, may be observed. Depending on where in the leaf they were located, the fibers had thick cell walls (between 2 and 5 m) and a wide lumen (between 10 and 20 m), with the outer cells being thicker than the inner ones [52]. It is obvious that the overall appearance of raw fibers displays a typical lignocellulosic fiber structure.

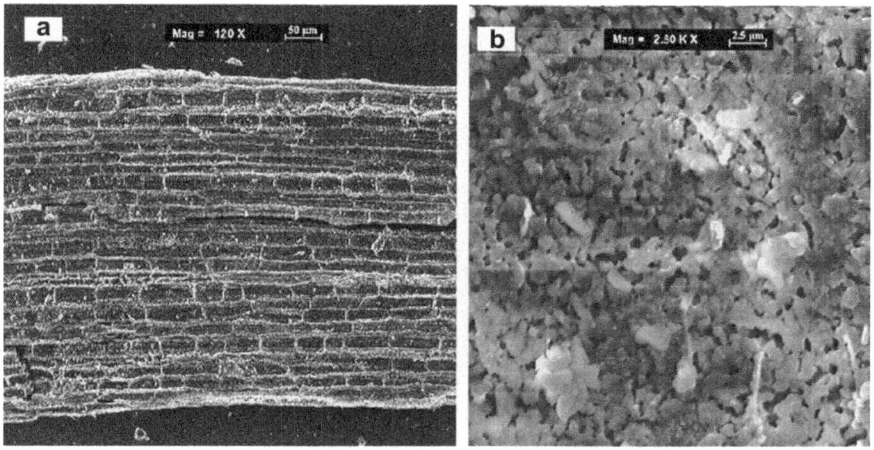

Fig. 1 SEM images of raw *P. oceanica* fibers: external (**a**) and transversal section (**b**) micrographs [51]

Each fiber has a 100-μm average diameter and is made up of a collection of fibrils connected by hemicellulose. Figures demonstrate the hollow nature of these fibrils, which come in a variety of sizes and forms. Because lignin, pectin, wax, and vegetable oils are present, the surface of these fibers is rough, as can be seen by looking at them up closely [53].

4 Chemical Composition

Although land-based plants make up the majority of the initial raw materials used, marine plants, such as *P. oceanica*, could be a viable alternative because their fibers share characteristics with other lignocellulosic materials. It has been shown that *P. oceanica* is the best source for extracting lignocellulosic fractions with promising characteristics [54]. *P. oceanica* exists in two different forms: balls and leaves, both of whose chemical compositions have been extensively studied. Table 3 presents a comparative analysis of the average values and standard deviations of a few selected parameters for the chemical characteristics of the *PO*B and POL. Posidonia residues were distinguished by high moisture (with a percentage of water content greater than 80%), and organic content, according to the data shown in this table.

POL has chemical components that reveal cellulose to be the main constituent at 45% and lignin amount at 19%. But POB composition indicated an important percentage of lignin in terms of 32%.

Due to its very high lignin concentration, *P. oceanica* is one of the lignocellulosic biomasses with the most lignin, which is the same as softwoods [62]. In actuality,

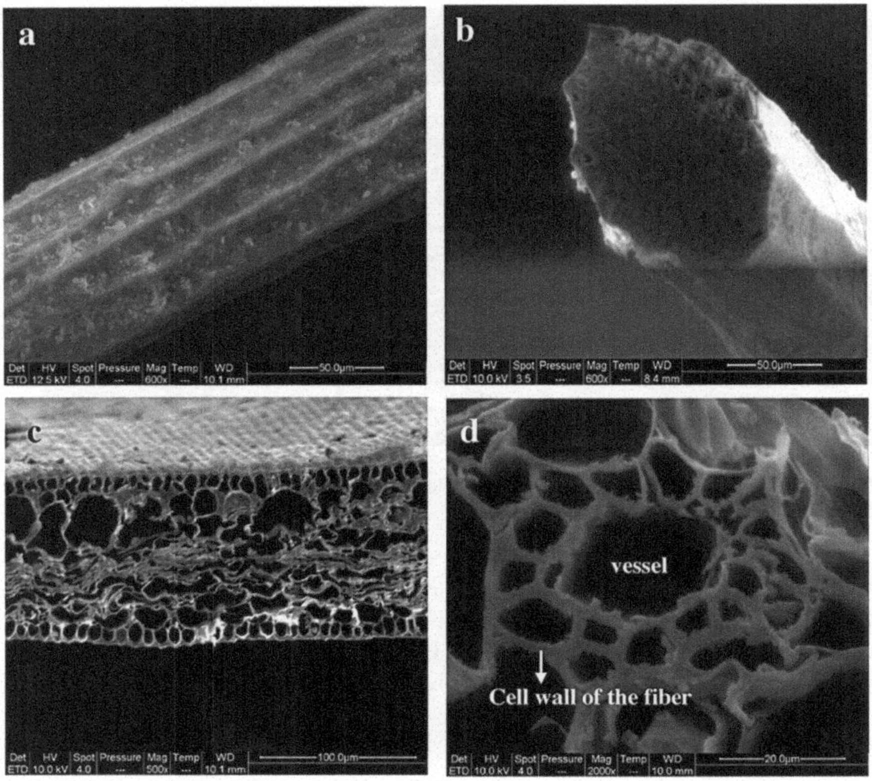

Fig. 2 Macroscopic observation of the *P. oceanica* balls (**a**, **b**) leaves (**c**, **d**)

lignocellulosic biomass contains three main biopolymers, namely cellulose, hemicellulose, and lignin, together with a comparatively minor amount of additional organic substances, including pectin and proteins [29]. However, due to its high lignin concentration (more than 30% of the organic content), this biomass is among the richest in lignin, comparable to softwoods [62]. The plant's lignin offers structural stability and defends it from microbial assault. The samples of POB had the lowest ash concentrations, whereas the samples of leaves had the highest ash values. Through vacuolar ion sequestering and cytosolic osmolyte buildup, it appears that the leaf tissues tend to ingest larger quantities of salts than the fibers [63]. As a result, ash compounds are preferentially removed throughout the degradation processes of the fibers. This fact, along with the fact that the leaves had a lot more ash than the fibrous ones, suggested that the elements might be accumulating differently in the *P. oceanica* tissues. However, this biomass is one of the richest in lignins, comparable to softwoods like pine and spruce, due to the high proportion of lignin (i.e., more than 30% of the organic content) [62]. The structural stability of the plant is provided by lignin, which also shields it from microbial assault.

Table 3 Average values and standard deviations of a few selected parameters for the chemical characteristics of the *PO*B and POL

P. oceanica Origin	*P. oceanica* part	Chemical Composition	References
Valencia, Spain	Leaves	• % Ash: 13 • % Extractives: 8 • % Lignin: 18% • % Hemicelluloses: 35 • % Cellulose: 24	[55]
Izmir, Turkey	Balls	• % Lignin: 23.8 • % Hemicellulose: 13.6 • % Cellulose: 42.9	[56]
Chott-Meriem (Sousse)	Leaves	• % Ash: 12.2 • % Extractives: 1.8 • % Holocellulose: 59	[57]
Calpe, Alicante (Spain)	Leaves	• % Ash: 13 • % Lignin: 18 • % Holocellulose: 59	[58]
Monastir (Marina),	Leaves	Monosaccharide (mol %): • Galactose 25.38 • Glucose 24.37 • Arabinose 21.64 • Rhamnose 11.83 • Xylose 10.88 • Fucose 5.9	[59]
Annaba, Algeria	Balls	• % Ash: 12 • % Extractives: 10.7 • % Lignin: 29.8 • % Hemicellulose: 21.8 • % Cellulose: 40	[60]
Monastir (Tunisia)	Balls	• Moisture content: 5.52 • % Ethanol-Toluene extractives: 7.02 • % Ash:12.03 • % Klason Lignin: 29.3 • % Cellulose: 44.2 • % Hemicelluloses: 18.3	[61]

In order to tackle the issue of their buildup on widely used beaches and provide an alternative solution to the landfill, the chemical analysis of beach-cast *P. oceanica* residues appears to offer numerous potentials which uses of *P. oceanica* trash according to their individual properties. The physical characteristics of the beached *P. oceanica* fibers, particularly the more recent ones, including the quality of the organic matter and relatively low levels of the main heavy metals and trace elements, clearly indicate their potential use in agriculture as parts of the growth medium in soilless systems. Following the chemical composition obtained using several techniques, FTIR analysis allowed a good analysis of the structure of the biomass in its two forms. A FTIR study of POB was also carried out. The asymmetric and symmetric C–H stretching vibrations of aliphatic compounds like CH, CH_2, and CH_3 groups were responsible for the two peaks at 2918 and 2852 cm^{-1} [64]. The peaks at 1605 and 1432 cm^{-1} were attributed to the carboxylate (COO–)'s asymmetric and symmetric C–O stretching, respectively. Stretching and O–H deformation of carboxylic groups (COOH) might possibly contribute to the peak at 1432 cm^{-1} and the peak at 1605 cm^{-1}, respectively. As a result, contributions from the vibrations of aromatic and aliphatic carboxylates as well as aromatic C=C structures may be present in these spectral regions suggestive of carboxylates [65]. The narrow peak at 1063 cm^{-1} was attributed to the C–O vibrations linked to polysaccharide structures [66], whereas the minor peaks at 1273 and 1170 cm^{-1} and the wide peak at 1063 cm^{-1} were likely caused by C–O stretching vibrations. The peak at 898 cm^{-1}, together with the peaks at 1170 and 1063 cm^{-1}, corresponds to cellulose [35, 67, 68]. This peak may be the result of out-of-phase asymmetric ring stretching. According to the FTIR analysis, the unprocessed Posidonia biomass leaves presented two peaks at 2918 and 2850 cm^{-1}, which correlate to the CH_2 asymmetrical and symmetrical stretching that is often seen in fatty acids [69]. It is also important to note that the band at 1729 cm^{-1}, which corresponds to the esters and acetyl groups from hemicelluloses and lignin [70], overlapped with the band corresponding to bound water (located at approximately 1630 cm^{-1}), as well as the distinctive bands of cellulose, such as those at 1430, 1375, and 1318 cm^{-1}, which correspond to the CH2 scissoring, CH bending, and CH2 rocking, in line with the CC, CO, CH stretching and COH bending modes [71, 72].

5 Fractionation and Applications

5.1 Fraction Technologies and Methods

Because of its abundant supply and renewable status, *P. oceanica* has gained interest for usage as a feedstock in the production of chemicals and fuels. The utilization of thermochemical and biochemical conversion technologies has led to the acceptance of *P. oceanica* as a potential alternative energy source [73]. According to relevant literature, several studies employing various techniques have been carried out to determine the components of fiber.

Duarte et al. [74] studied the extraction of the bioactive compounds using the Supercritical Fluid Extraction (SFE) method, which is based on the idea of extracting fluids with characteristics of both liquids and gases when they are in supercritical conditions with temperature and pressure elevated above their critical points [74]. Different methods were used to estimate the *P. oceanica*'s chemical composition during fractionation: The assessment of the extractives is done using the following solutions: hot water (TAPPI T207), 1% sodium hydroxide solution (TAPPI T212), and ethanol-toluene TAPPI(TAPPI T204), cold and hot water (T207 cm-08), 1% sodium hydroxide solution (T212 om-07), and ethanol–toluene (T204 cm-07. Ash content was identified by using (T211 om-07). Additionally, the quantities of lignin, holocellulose, and cellulose were determined using the corresponding standard techniques T222 om-06, method of Wise et al. (1946) [75], T203 cm-99, and T236 om-06. Due to the difficulties of separating lignin from the other components of this biomass, the NREL standard approach may also be used to fractionate the components. This method may be the most effective for fractionation.

5.2 *Applications*

5.2.1 Human Health

P. oceanica is described as a source of chemicals that might be used for a variety of purposes in human health due to its high secondary metabolite content [76]. Many studies have demonstrated the effective anticancer activities of *P. oceanica* [77]. *P. oceanica* uses this capacity to boost antioxidant enzyme activity as a defensive strategy against a variety of environmental stresses [78].

Ayad et al. revealed that *P. oceanica* leaf extracts exhibited antibacterial effects on both gram-positive and -negative bacteria, with a focus on Pseudomonas aeruginosa and Staphylococcus aureus microbacteries with variable activity against other used bacterial strains [79].

Farid et al. in 2018 [80], have also valued the antiviral properties of PO, which guarantee the contribution of the *P. oceanica* ball in removing 45% of the H5N1 virus infection.

Due to its antioxidant properties, *P. oceanica* seagrass may be used to prevent cellular damage and illnesses linked to oxidative stress. Chicoric acid-rich *P. oceanica* leaf extract has been demonstrated to affect human dermal fibroblast proliferation and collagen production, as well as to have anti-melanogenic effects on the human melanoma cell line MeWo and lipolytic effects on human subcutaneous preadipocytes. These effects were attributed to the extract's antioxidant power [81]. Considering that oxidative stress is a major factor in the development of wrinkles, skin aging, undesirable hyperpigmentation, and cellulite, *P. oceanica* extract's antioxidant activity may be linked to protection against these conditions [82].

5.2.2 Food Packaging

Piva et al. and Gonzales et al. were interested in the potential uses of *P. oceanica* seagrass in food science since it is an antibacterial and antioxidant agent. A number of researches have examined the possible uses of *P. oceanica* in the field of food preservation by lowering and removing infections and spoiling bacteria in food [83, 84]. *P. oceanica* has been shown to be the best source for cellulose and lignocellulosic fraction extraction, with promising features for the creation of cellulose-based packaging materials [85]. Although cellulose alone possesses intriguing features for food packaging applications, treatment with acid hydrolysis digests the amorphous domains to produce extremely fine nanocellulose [86].

According to Dufresne [87], crystal nanocellulose as well as nanofibres have an appealing mix of qualities, including biocompatibility, high elastic modulus, a high thermal stability, and great optical clarity, to allow them to be used to produce high-barrier films [81]. Films produced from POB cellulose showed a clear improvement in terms of properties and tensile strength when compared to benchmark biopolymers like thermoplastic corn starch [88] and polylactic acid (PLA) [89]. This makes the films a promising replacement for food packaging. Their properties allow them to be on par with those of petroleum-based polymers, such as oriented polystyrene and polyethylene terephthalate, which are frequently used in food packaging applications. In particular, the POB films outperformed these materials in terms of stiffness [53, 61].

5.2.3 Eco-Friendly Building Materials

P. oceanica balls are regarded as garbage, but they may be a significant source of fiber. Their use in the building industry to create bio-insulation materials may simultaneously minimize organic waste and energy use [100]. The use of these fibers in cement matrix has been investigated by Allegue et al. [101]. They discovered that the mechanical characteristics of the composite (ductility and flexural strength) were improved. To the best of our knowledge, the findings demonstrated that the *P. oceanica* fiber-based cement had better mechanical strength and toughness in comparison to some other wood-based cement due to its higher compatibility. The *P. oceanica* cements also displayed a more uniform structure and superior water resistance and flame retardance qualities [102].

P. oceanica fibers exhibit thermal characteristics that are equivalent to those of commercial insulating materials, with a focus on high bulk heat capacity. Because they are readily available and renewable, they may also be a promising environmentally friendly loose-fill insulating material in the building industry. Additionally, their preparation and extraction did not require a great deal of sophistication. Additionally, it is important to note that these measurements were done in a controlled laboratory environment and that the variance in temperature and humidity in actual building usage settings was not considered [103].

5.2.4 Biocomposite Preparation

Plastic pollution has emerged as a major worldwide hazard, mostly to marine ecosystems. Therefore, switching to biodegradable materials from oil-derived plastics is necessary for pursuing a more sustainable way of living. In this regard, polymer matrices made from bio-renewable polysaccharides and proteins are of considerable interest.

6 Conclusion

The most significant species in the Mediterranean Sea is *P. oceanica*, which belongs to the Posidoniaceae family. This chapter was established with the goal of highlighting how, up until recently, scientists have described this Mediterranean angiosperm and gained a thorough understanding of its structure in order to use it effectively in a variety of fields, including industry and agriculture, and only indirectly to benefit human health as recipients of favorable environmental conditions. This summary of *P. oceanica*'s characteristics enables the scientific community to regard this plant as a prospective resource for a variety of uses, notwithstanding its ecological importance.

They have proven that they might be used in a range of applications and industries due to their structural and physical characteristics. In contrast, their cellulose constituents have previously been applied in biotechnological methods to produce additional useful compounds, such as biofuels. These fibers might readily serve as the foundation for bioplastics and biocomposites, as well as insulating materials for structures, due to their mechanical and physical qualities. Additionally, their separated lignin fractions might be employed as biomaterial reinforcement with ease.

References

1. Mujtaba M, Fernandes Fraceto L, Fazeli M, Mukherjee S, Savassa SM, Araujo de Medeiros G, do Espírito Santo Pereira A, Mancini SD, Lipponen J, Vilaplana F (2023) Lignocellulosic biomass from agricultural waste to the circular economy: a review with focus on biofuels, biocomposites and bioplastics. In: Journal of cleaner production, vol 402. Elsevier Ltd. https://doi.org/10.1016/j.jclepro.2023.136815
2. Lendvai L (2023) Lignocellulosic agro-residue/polylactic acid (PLA) biocomposites: rapeseed straw as a sustainable filler. Clean Mater 9. https://doi.org/10.1016/j.clema.2023.100196
3. Mukherjee T, Kao N (2011) PLA Based biopolymer reinforced with natural fibre: a review. J Polym Environ 19(3):714–725. https://doi.org/10.1007/s10924-011-0320-6
4. Kalia S, Avérous L, Njuguna J, Dufresne A, Cherian BM (2011) Natural fibers, bio- and nanocomposites. In: International journal of polymer science, vol 2011. Hindawi Limited. https://doi.org/10.1155/2011/735932
5. Kumagai S, Matsuo Y (2013) Composite produced from rice husk and chopped carbon fiber without using any binders. Ind Crops Prod 43(1):640–647. https://doi.org/10.1016/j.indcrop.2012.08.009
6. Yamsaemgsung W, Sombatsompop N (2009) Interfacial adhesion and molecular diffusion in melt lamination of wood sawdust/ebonite NR and EPDM. Polym Compos 30(3):248–256. https://doi.org/10.1002/pc.20568
7. Vilar VJP, Botelho CMS, Boaventura RAR (2008) Copper removal by algae Gelidium, agar extraction algal waste and granulated algal waste: kinetics and equilibrium. Biores Technol 99(4):750–762. https://doi.org/10.1016/j.biortech.2007.01.042
8. Chauhan M, Gupta M, Singh B, Singh AK, Gupta VK (2012) Pine needle/isocyanate composites: dimensional stability, biological resistance, flammability, and thermoacoustic characteristics. Polym Compos 33(3):324–335. https://doi.org/10.1002/pc.22151
9. Benito-González I, López-Rubio A, Martínez-Sanz M (n.d.) Potential of lignocellulosic fractions from Posidonia 1 oceanica to improve barrier and mechanical properties of 2 bio-based packaging materials 3 4 5

10. Fortunati E, Luzi F, Puglia D, Petrucci R, Kenny JM, Torre L (2015) Processing of PLA nanocomposites with cellulose nanocrystals extracted from Posidonia oceanica waste: innovative reuse of coastal plant. Ind Crops Prod 67:439–447. https://doi.org/10.1016/j.indcrop.2015.01.075
11. Khiari R, Mhenni MF, Belgacem MN, Mauret E (2011) Valorisation of vegetal wastes as a source of cellulose and cellulose derivatives. J Polym Environ 19(1):80–89. https://doi.org/10.1007/s10924-010-0207-y
12. 2009 IEEE workshop on environmental, energy, and structural monitoring systems: proceedings (2009). IEEE
13. Balata G, Tola A (2017) Cost-opportunity analysis of the use of Posidonia oceanica as a source of bio-energy in tourism-oriented territories. Cas Alghero J Clean Prod 172:4085–4098. https://doi.org/10.1016/j.jclepro.2017.02.072
14. Plis A, Lasek J, Skawińska A, Kopczyński M (2014) Thermo-chemical properties of biomass from Posidonia oceanica. Chem Pap 68(7):879–889. https://doi.org/10.2478/s11696-013-0532-4
15. Gobert S, Cambridge M, Velimirov B, Pergent G, Lepoint G, Bouquegneau JM, Dauby P, Walker DI (2006) Biology of *Posidonia*. In: Larkum AWD, Orth RJ, Duarte CM (eds) Seagrasses: biology, ecology and conservation. Springer, Dordrecht, pp 387–408
16. Vacchi M, de Falco G, Simeone S, Montefalcone M, Morri C, Ferrari M, Bianchi CN (2017) Biogeomorphology of the Mediterranean Posidonia oceanica seagrass meadows. Earth Surf Proc Land 42(1):42–54. https://doi.org/10.1002/esp.3932
17. Cantasano N (2021) Deposition dynamics of Posidonia oceanica "Banquettes" on Calabrian Sandy Beaches (Southern Italy). Coasts 1(1):25–30. https://doi.org/10.3390/coasts1010002
18. Bonhomme D, Boudouresque CF, Astruch P, Bonhomme J, Bonhomme P, Goujard A, Thibaut T (2015) Typology of the reef formations of the Mediterranean seagrass Posidonia oceanica, and the discovery of extensive reefs in the Gulf of Hyères (Provence, Mediterranean). In: 41-Sci. Rep. Port-Cros natl. Park, vol 29
19. Bay D (1984) A feld study of the growth dynamics and productivity of Posidonia oceanica (L.) delile in calvi bay, corsica. In: Aquatic botany, vol 20
20. Pergent G, Pergenl-Martini C (1991) Leaf renewal cycle and primary production of Posidonia oceanica in the bay of Lacco Ameno (Ischia, Italy) using lepidochronological analysis
21. Bockel T, Marre G, Delaruelle G, Agel N, Boissery P, Guilhaumon F, Mouquet N, Mouillot D, Guilbert A, Deter J (2024) Early signals of Posidonia oceanica meadows recovery in a context of wastewater treatment improvements. Mar Pollut Bull 201. https://doi.org/10.1016/j.marpolbul.2024.116193
22. Mutlu E, Duman GS, Karaca D, Özvarol Y, Şahin A (2024) A broad-scale biometrical response of pristine posidonia oceanica meadow to winter and summer conditions in the north-eastern mediterranean coast. Thalassas 40(1):477–498. https://doi.org/10.1007/s41208-024-00666-6
23. Alomar C, Compa M, Fagiano V, Concato M, Deudero S (2024) Posidonia oceanica egagropiles: good indicators for plastic pollution in coastal areas? In: Regional studies in marine science, vol 77. Elsevier B.V. https://doi.org/10.1016/j.rsma.2024.103653
24. Fonseca MS, Koehl MAR, Kopp BS (2007) Biomechanical factors contributing to self-organization in seagrass landscapes. J Exp Mar Biol Ecol 340(2):227–246. https://doi.org/10.1016/j.jembe.2006.09.015
25. Cañada JA (2024) Ambivalence in environmental care: marine care ethics and more-than-human relations in the conservation of seagrass Posidonia oceanica. J Agric Environ Ethics 37(2). https://doi.org/10.1007/s10806-024-09924-2
26. Mateo MÁ, Sánchez-Lizaso JL, Romero J (2003) Posidonia oceanica "banquettes": a preliminary assessment of the relevance for meadow carbon and nutrients budget. Estuar Coast Shelf Sci 56(1):85–90. https://doi.org/10.1016/S0272-7714(02)00123-3
27. Balata G, Tola A (2018) Cost-opportunity analysis of the use of Posidonia oceanica as a source of bio-energy in tourism-oriented territories. the case of Alghero. J Clean Prod 172:4085–4098. https://doi.org/10.1016/j.jclepro.2017.02.072

28. Seggiani M, Cinelli P, Mallegni N, Balestri E, Puccini M, Vitolo S, Lardicci C, Lazzeri A (2017) New bio-composites based on polyhydroxyalkanoates and Posidonia oceanica fibres for applications in a marine environment. Materials 10(4). https://doi.org/10.3390/ma1004 0326
29. Dahadha S, Amin Z, Bazyar Lakeh AA, Elbeshbishy E (2017) Evaluation of different pretreatment processes of lignocellulosic biomass for enhanced biomethane production. Energy Fuels 31(10):10335–10347. https://doi.org/10.1021/acs.energyfuels.7b02045
30. Cornara L, Pastorino G, Borghesi B, Salis A, Clericuzio M, Marchetti C, Damonte G, Burlando B (2018) Posidonia oceanica (L.) delile ethanolic extract modulates cell activities with skin health applications. Marine Drugs 16(1). https://doi.org/10.3390/md16010021
31. Vasarri M, Barletta E, Ramazzotti M, Degl'Innocenti D (2020) In vitro anti-glycation activity of the marine plant Posidonia oceanica (L.) Delile. J Ethnopharmacol 259. https://doi.org/10.1016/j.jep.2020.112960
32. International Energy Outlook 2017 (2017). www.eia.gov
33. Aliane A, Abboudi S, Seladji C, Guendouz B (2016) An illustrated review on solar absorption cooling experimental studies. In: Renewable and sustainable energy reviews, vol 65. Elsevier Ltd., , pp 443–458 https://doi.org/10.1016/j.rser.2016.07.012
34. Ashraf MT, Fang C, Bochenski T, Cybulska I, Alassali A, Sowunmi A, Farzanah R, Brudecki GP, Chaturvedi T, Haris S, Schmidt JE, Thomsen MH (2016) Estimation of bioenergy potential for local biomass in the United Arab Emirates. Emir J Food Agric 28(2):99–106. https://doi.org/10.9755/ejfa.2015-04-060
35. Cocozza C, Parente A, Zaccone C, Mininni C, Santamaria P, Miano T (2011) Chemical, physical and spectroscopic characterization of Posidonia oceanica (L.) Del. residues and their possible recycle. Biomass Bioenergy 35(2):799–807. https://doi.org/10.1016/j.biombioe.2010.10.033
36. Giakoumi S, Sini M, Gerovasileiou V, Mazor T, Beher J, Possingham HP, Abdulla A, Cinar ME, Dendrinos P, Gucu AC, Karamanlidis AA, Rodic P, Panayotidis P, Taskin E, Jaklin A, Voultsiadou E, Webster C, Zenetos A, Katsanevakis S (2013) Ecoregion-based conservation planning in the Mediterranean: dealing with large-scale heterogeneity. PLoS ONE 8(10). https://doi.org/10.1371/journal.pone.0076449
37. Boudouresque CF, Astruch P, André S, Belloni B, Blanfuné A, Charbonnel É, Cheminée A, Cottalorda JM, Dupuy de la Grandrive R, Marengo M, Monnier B, Pergent G, Pergent-Martini C, Perret-Boudouresque M, Ruitton S, Taupier-Letage I, Thibaut T (2024) The heatwave of summer 2022 in the North-Western Mediterranean Sea: some species were winners. Water (Switzerland) 16(2). https://doi.org/10.3390/w16020219
38. Ćosović V, Šanjek R, Hadžić E, Rakarić M, Ištuk Ž, Šušmelj K, Čermelj B, Žvab Rožič P (2023) Impact of submarine karst sulfur springs on benthic foraminiferal assemblage in sediment of northern Adriatic Sea. J Soils Sediments 23(12):4132–4148. https://doi.org/10.1007/s11368-023-03614-8
39. Bettoso N, Faresi L, Pitacco V, Orlando-Bonaca M, Aleffi IF, Lipej L (2023) Species richness of benthic macrofauna on rocky outcrops in the adriatic sea by using species-area relationship (SAR) tools. Water (Switzerland) 15(2). https://doi.org/10.3390/w15020318
40. Cunha AH, Assis JF, Serrão EA (2013) Seagrasses in Portugal: a most endangered marine habitat. Aquat Bot 104:193–203. https://doi.org/10.1016/j.aquabot.2011.08.007
41. Telesca L, Belluscio A, Criscoli A, Ardizzone G, Apostolaki ET, Fraschetti S, Gristina M, Knittweis L, Martin CS, Pergent G, Alagna A, Badalamenti F, Garofalo G, Gerakaris V, Louise Pace M, Pergent-Martini C, Salomidi M (2015) Seagrass meadows (Posidonia oceanica) distribution and trajectories of change. Sci Rep 5. https://doi.org/10.1038/srep12505
42. Mutlu E, Akçalı B (2022) Seasonal macro-floral distribution on soft bottoms in a gulf of a metropolitan providence, Izmir, Aegean Sea. NWSA Acad J 17(3):86–103. https://doi.org/10.12739/nwsa.2022.17.3.5a0168
43. Hachani MA, Ziadi B, Langar H, Sami DA, Turki S, Aleya L (2016) The mapping of the Posidonia oceanica (L.) Delile barrier reef meadow in the southeastern Gulf of Tunis (Tunisia). J Afr Earth Sc 121:358–364. https://doi.org/10.1016/j.jafrearsci.2016.05.030

44. Kuo J (1978) Morphology, Anatomy and histochemistry of Australian seagrasses of genus *Posidonia* Konig (Posidoniaceae).1. Leaf blade and leaf sheath of *Posidonia australis Hook.* Aquat Bot 5:171–190
45. Cantasano N (2022) The phylogenetic history of the seagrass species Posidonia oceanica Int J Environ Agric Sci Short Commentary Volume 2 *Corresponding Author. https://www.researchgate.net/publication/366513208
46. Cantasano N (2023) Back to the sea: the long and winding road of the seagrass species *posidonia oceanica*. Adv Environ Eng Res 04(01):1–8. https://doi.org/10.21926/aeer.2301003
47. Hartog CD (1970) The sea-grasses of the world. North Holland publishing Co, Amsterdam, pp 1–275
48. Lefebvre L, Compère P, Léonard A, Plougonven E, Vandewalle N, Gobert S (2021) Mediterranean aegagropiles from Posidonia oceanica (L.) Delile (1813): a first complete description from macroscopic to microscopic structure. Marine Biol 168(3). https://doi.org/10.1007/s00227-021-03833-y
49. Rende SF, Bosman A, Menna F, Lagudi A, Bruno F, Severino U, Montefalcone M, Irving AD, Raimondi V, Calvo S, Pergent G, Tomasello A (2022) Assessing seagrass restoration actions through a micro-bathymetry survey approach (Italy, Mediterranean Sea). Water (Switzerland) 14(8). https://doi.org/10.3390/w14081285
50. Colinart T, Pajeot M, Vinceslas T, Hellouin de Ménibus A, Lecompte T (2022) How reliable is the thermal conductivity of biobased building insulating materials measured with hot disk device? Bio-Based Build Mater 1:287–292. https://doi.org/10.4028/www.scientific.net/cta.1.287
51. Ncibi MC, Jeanne-Rose V, Mahjoub B, Jean-Marius C, Lambert J, Ehrhardt JJ, Bercion Y, Seffen M, Gaspard S (2009) Preparation and characterisation of raw chars and physically activated carbons derived from marine Posidonia oceanica (L.) fibres. J Hazard Mater 165(1–3):240–249. https://doi.org/10.1016/j.jhazmat.2008.09.126
52. Bettaieb F, Khiari R, Dufresne A, Mhenni MF, Putaux JL, Boufi S (2015) Nanofibrillar cellulose from Posidonia oceanica: properties and morphological features. Ind Crops Prod 72:97–106. https://doi.org/10.1016/j.indcrop.2014.12.060
53. Hamdaoui O, Ibos L, Mazioud A, Safi M, Limam O (2018) Thermophysical characterization of Posidonia Oceanica marine fibers intended to be used as an insulation material in Mediterranean buildings. Constr Build Mater 180:68–76. https://doi.org/10.1016/j.conbuildmat.2018.05.195
54. Mirpoor SF, Giosafatto CVL, di Pierro P, di Girolamo R, Regalado-González C, Porta R (2020) Valorisation of posidonia oceanica sea balls (Egagropili) as a potential source of reinforcement agents in protein-based biocomposites. Polymers 12(12):1–13. https://doi.org/10.3390/polym12122788
55. Benito-González I, López-Rubio A, Gavara R, Martínez-Sanz M (2019) Cellulose nanocrystal-based films produced by more sustainable extraction protocols from Posidonia oceanica waste biomass. Cellulose 26(13–14):8007–8024. https://doi.org/10.1007/s10570-019-02641-4
56. Pilavtepe M, Celiktas MS, Sargin S, Yesil-Celiktas O (2013) Transformation of Posidonia oceanica residues to bioethanol. Ind Crops Prod 51:348–354. https://doi.org/10.1016/j.indcrop.2013.09.020
57. Ncibi MC, Ranguin R, Pintor MJ, Jeanne-Rose V, Sillanpää M, Gaspard S (2014) Preparation and characterization of chemically activated carbons derived from Mediterranean Posidonia oceanica (L.) fibres. J Anal Appl Pyrol 109:205–214. https://doi.org/10.1016/j.jaap.2014.06.010
58. Benito-González I, López-Rubio A, Martínez-Sanz M (2018) Potential of lignocellulosic fractions from Posidonia oceanica to improve barrier and mechanical properties of bio-based packaging materials. Int J Biol Macromol 118:542–551. https://doi.org/10.1016/j.ijbiomac.2018.06.052
59. Salem Y ben, Abdelhamid A, Hammi KM, le Cerf D, Bouraoui A, Majdoub H (2017) Microwave-assisted extraction and pharmacological evaluation of polysaccharides from Posidonia oceanica. Biosci Biotechnol Biochem 81(10):1917–1925. https://doi.org/10.1080/09168451.2017.1361808

60. Tarchoun AF, Trache D, Klapötke TM (2019) Microcrystalline cellulose from Posidonia oceanica brown algae: extraction and characterization. Int J Biol Macromol 138:837–845. https://doi.org/10.1016/j.ijbiomac.2019.07.176
61. Restaino OF, Giosafatto CVL, Mirpoor SF, Cammarota M, Hejazi S, Mariniello L, Schiraldi C, Porta R (2023) Sustainable exploitation of Posidonia oceanica sea balls (Egagropili): a review. In: International journal of molecular sciences, vol 24, issue 8. Multidisciplinary Digital Publishing Institute (MDPI). https://doi.org/10.3390/ijms24087301
62. Das N, Jena PK, Padhi D, Kumar Mohanty M, Sahoo G (2023) A comprehensive review of characterization, pretreatment and its applications on different lignocellulosic biomass for bioethanol production. In: Biomass conversion and biorefinery, vol 13, issue 2. Springer, pp 1503–1527. https://doi.org/10.1007/s13399-021-01294-3
63. Touchette BW (2007) Seagrass-salinity interactions: physiological mechanisms used by submersed marine angiosperms for a life at sea. J Exp Mar Biol Ecol 350(1–2):194–215. https://doi.org/10.1016/j.jembe.2007.05.037
64. Francesca P, Sara M, Luigi T (2008) New biosorbent materials for heavy metal removal: product development guided by active site characterization. Water Res 42(12):2953–2962. https://doi.org/10.1016/j.watres.2008.03.012
65. Artz RR, Chapman SJ, Jean Robertson A, Potts JM, Laggoun-Défarge F, Gogo S, Comont L, Disnar J-R, Francez A-J, Laggoun F, Artz R, Biology S (2008) FTIR spectroscopy can predict organic matter quality in regenerating cutover peatlands. FTIR spectroscopy can predict organic matter quality in 2 regenerating cutover peatlands 3 (Running title: Whole soil FTIR on peat) 4 5 Rebekka. In: Soil biology and biochemistry, vol 40, issue 2. https://insu.hal.science/insu-00173352
66. Grube M, Lin JG, Lee PH, Kokorevicha S (2006) Evaluation of sewage sludge-based compost by FT-IR spectroscopy. Geoderma 130(3–4):324–333. https://doi.org/10.1016/j.geoderma.2005.02.005
67. Izquierdo M, Marzal P, Gabaldón C, Silvetti M, Castaldi P (2012) Study of the Interaction Mechanism in the Biosorption of Copper(II) Ions onto Posidonia oceanica and Peat. Clean: Soil, Air, Water 40(4):428–437. https://doi.org/10.1002/clen.201100303
68. Coletti A, Valerio A, Vismara E (2013) Posidonia oceanica as a renewable lignocellulosic biomass for the synthesis of cellulose acetate and glycidyl methacrylate grafted cellulose. Materials 6(5):2043–2058. https://doi.org/10.3390/ma6052043
69. Abidi N, Cabrales L, Haigler CH (2014) Changes in the cell wall and cellulose content of developing cotton fibers investigated by FTIR spectroscopy. Carbohyd Polym 100:9–16. https://doi.org/10.1016/j.carbpol.2013.01.074
70. Sun XF, Xu F, Sun RC, Fowler P, Baird MS (2005) Characteristics of degraded cellulose obtained from steam-exploded wheat straw. Carbohyd Res 340(1):97–106. https://doi.org/10.1016/j.carres.2004.10.022
71. Oh SY, Yoo D il, Shin Y, Seo G (2005) FTIR analysis of cellulose treated with sodium hydroxide and carbon dioxide. Carbohydrate Res 340(3):417–428. https://doi.org/10.1016/j.carres.2004.11.027
72. Khiari R, Marrakchi Z, Belgacem MN, Mauret E, Mhenni F (2011) New lignocellulosic fibres-reinforced composite materials: a stepforward in the valorisation of the Posidonia oceanica balls. Compos Sci Technol 71(16):1867–1872. https://doi.org/10.1016/j.compscitech.2011.08.022
73. Masmoudi G, Dhaouadi H (2024) A review on adsorption of textile dyes onto an unconventional biosorbent: marine waste of Posidonia Oceanica. In: Chemistry Africa. Springer. https://doi.org/10.1007/s42250-024-01015-z
74. Duarte K, Justino CIL, Gomes AM, Rocha-Santos T, Duarte AC (2014) Green analytical methodologies for preparation of extracts and analysis of bioactive compounds. In: Comprehensive analytical chemistry, vol 65. Elsevier B.V, pp 59–78. https://doi.org/10.1016/B978-0-444-63359-0.00004-5
75. Thomas LH, Martel A, Grillo I, Jarvis MC (2020) Hemicellulose binding and the spacing of cellulose microfibrils in spruce wood. Cellulose 27(8):4249–4254. https://doi.org/10.1007/s10570-020-03091-z

76. Vasarri M, de Biasi AM, Barletta E, Pretti C, Degl'innocenti D (2021) An overview of new insights into the benefits of the seagrass posidonia oceanica for human health. In: Marine drugs, vol 19, issue 9. MDPI. https://doi.org/10.3390/md19090476
77. Liu W, Li Q, Hu J, Wang H, Xu F, Bian Q (2019) Application of natural products derivatization method in the design of targeted anticancer agents from 2000 to 2018. In: Bioorganic and medicinal chemistry, vol 27, issue 23. Elsevier Ltd. https://doi.org/10.1016/j.bmc.2019.115150
78. Sureda A, Box A, Terrados J, Deudero S, Pons A (2008) Antioxidant response of the seagrass when epiphytized by the invasive macroalgae. Marine Environ Res 66(3). https://doi.org/10.1016/j.marenvres.2008.05.009ï
79. Piva G, Fracassetti D, Tirelli A, Mascheroni E, Musatti A, Inglese P, Piergiovanni L, Rollini M (2017) Evaluation of the antioxidant/antimicrobial performance of Posidonia oceanica in comparison with three commercial natural extracts and as a treatment on fresh-cut peaches (Prunus persica Batsch). Postharvest Biol Technol 124:54–61. https://doi.org/10.1016/j.postharvbio.2016.10.001
80. Benito-González I, López-Rubio A, Martínez-Abad A, Ballester AR, Falcó I, González-Candelas L, Sánchez G, Lozano-Sánchez J, Borrás-Linares I, Segura-Carretero A, Martínez-Sanz M (2019) In-depth characterization of bioactive extracts from posidonia oceanica waste biomass. Marine Drugs 17(7). https://doi.org/10.3390/md17070409
81. Benito-González I, López-Rubio A, Gavara R, Martínez-Sanz M (n.d.) Cellulose nanocrystal-based films produced by more sustainable extraction protocols from 2 Posidonia oceanica waste biomass
82. Chen YW, Lee HV, Juan JC, Phang SM (2016) Production of new cellulose nanomaterial from red algae marine biomass Gelidium elegans. Carbohyd Polym 151:1210–1219. https://doi.org/10.1016/j.carbpol.2016.06.083
83. Dufresne A (2006) Comparing the mechanical properties of high performances polymer nanocomposites from biological sources. J Nanosci Nanotechnol 6(2):322–330. https://doi.org/10.1166/jnn.2006.906
84. Martínez-Sanz M, Lopez-Rubio A, Lagaron JM (2013) High-barrier coated bacterial cellulose nanowhiskers films with reduced moisture sensitivity. Carbohyd Polym 98(1):1072–1082. https://doi.org/10.1016/j.carbpol.2013.07.020
85. José Fabra M, López-Rubio A, Ambrosio-Martín J (n.d.) Improving the barrier properties of thermoplastic corn starch-based films containing bacterial cellulose nanowhiskers by means of PHA electrospun 2 coatings of interest in food packaging
86. Mathew AP, Oksman K, Sain M (2005) Mechanical properties of biodegradable composites from poly lactic acid (PLA) and microcrystalline cellulose (MCC). J Appl Polym Sci 97(5):2014–2025. https://doi.org/10.1002/app.21779
87. Auras RA, Singh SP, Singh JJ (n.d.) Evaluation of oriented poly(lactide) polymers with existing PET and oriented PS for fresh food service containers
88. Khiari R, Mhenni MF, Belgacem MN, Mauret E (2010) Chemical composition and pulping of date palm rachis and Posidonia oceanica—a comparison with other wood and non-wood fibre sources. Biores Technol 101(2):775–780. https://doi.org/10.1016/j.biortech.2009.08.079
89. Allègue L, Zidi M, Sghaier S (2015) Mechanical properties of Posidonia oceanica fibers reinforced cement. J Compos Mater 49(5):509–517. https://doi.org/10.1177/0021998314521254
90. Farid MM, Marzouk MM, Hussein SR, Elkhateeb A, Abdel-Hameed ES (2018) Comparative study of Posidonia oceanica L.: LC/ESI/MS analysis, cytotoxic activity and chemosystematic significance. J Mater Environ Sci 9(6):1676–1682. https://doi.org/10.26872/jmes.2018.9.6.187

Carboxymethylcellulose Synthesis and Advanced Applications: A Comparative Analysis of *Posidonia oceanica* and *Ziziphus lotus* Sources

Sara Saad, Manel Elakremi, Ramzi Khiari, Elimame Elaloui, Ridha ben Salem, and Younes Moussaoui

Abstract The quest for sustainable and environmentally friendly alternatives in materials science and biotechnology has driven exploration into the rich diversity of natural resources globally. This chapter focuses on carboxymethylcellulose (CMC), a water-soluble cellulose derivative with significant applications in various industries due to its versatility and biodegradability. The selection of the cellulose source for CMC production plays a pivotal role in influencing its properties and applications. This study delves into the synthesis and diverse applications of CMC across multiple fields, with a specific focus on CMC derived from two distinctive plant species: *Posidonia oceanica* and *Ziziphus lotus*.

Keywords Chemical composition · Carboxymethylcellulose (CMC) · *Posidonia oceanica* · *Ziziphus lotus*

S. Saad · E. Elaloui
Faculty of Sciences of Gafsa, Laboratory for the Application of Materials to the Environment, Water, and Energy (LR21ES15), University of Gafsa, Gafsa, Tunisia

S. Saad · M. Elakremi · E. Elaloui · Y. Moussaoui (✉)
Faculty of Sciences of Gafsa, University of Gafsa, Gafsa, Tunisia
e-mail: y.moussaoui2@gmail.com

M. Elakremi · R. Salem · Y. Moussaoui
Faculty of Sciences of Sfax, Organic Chemistry Laboratory (LR17ES08), University of Sfax, Sfax, Tunisia

R. Khiari
Department of Textile, Higher Institute of Technological Studies of Ksar Hellal, 5070 Ksar Hellal, Tunisia

CNRS, Grenoble INP, LGP2, Université Grenoble Alpes, F-3800 Grenoble, France

R. Khiari and M. Jawaid (eds.), *Current Status and Opportunity in Fibre and Composites*, Sustainable Materials and Technology,
https://doi.org/10.1007/978-981-96-4546-6_5

1 Introduction

The search for sustainable and eco-friendly alternatives in materials science and biotechnology has encouraged researchers to investigate the vast diversity of natural resources available worldwide [61]. One such initiative is the investigation of biopolymers, which are natural, renewable substances with numerous applications in various industries. Carboxymethylcellulose (CMC), a water-soluble cellulose derivative, has received considerable attention because of its versatility and biodegradability. CMC, which is derived from cellulose, a fundamental component of plant cell walls, has applications in the food, medicine, cosmetics, textiles, and other industries [9, 32, 76, 78].

The cellulose source for CMC production is an important factor that can have a considerable impact on its properties and applications (Das et al. 2020). While cellulose may be derived from a wide range of plant materials, two exciting sources have emerged in the area of sustainable materials research: *Posidonia oceanica* [46] and *Ziziphus lotus* [75]. Because these two plant species are from different geographical locations and have varied botanical qualities, the resulting CMC may have distinctive properties.

The purpose of this study was to offer concise information about the synthesis and diverse applications of CMC across multiple fields and to explore the state of CMC generated from *Posidonia oceanica* and *Ziziphus lotus*, revealing their distinct features, chemical compositions, and prospective applications. We will identify the latent potential of these two raw materials as sustainable alternatives to standard cellulose sources by an in-depth investigation of these raw materials. This study not only adds to the growing body of knowledge in the field of biopolymers but also emphasizes the necessity of biodiversity protection and sustainable resource utilization.

2 Background on Carboxymethylcellulose

CMC is an anionic, water-soluble derivative of cellulose, a linear polysaccharide of anhydroglucose. 1,4-Glycosidic linkages are connected the repeating units [80]. The main molecular difference between CMC and cellulose is the anionic carboxymethyl groups ($-CH_2COOH$) in the CMC structure that replace some of the hydroxyl groups present in the original cellulose structure (Fig. 1) [57]. The synthesis of CMC began in 1918. However, the first depiction of the commercial manufacture of these crucial polymeric materials was made in Germany in the early 1920s [54].

The production of CMC was first relied on specific wood-based plants because the archetypal celluloses were mostly taken from wood and other plant-based precursors that naturally had a high percentage of cellulose fibers [84]. However, numerous researchers in the literature have recently presented cellulose-containing polymers as effective alternatives in this context. Several plant-derived precursors have captured

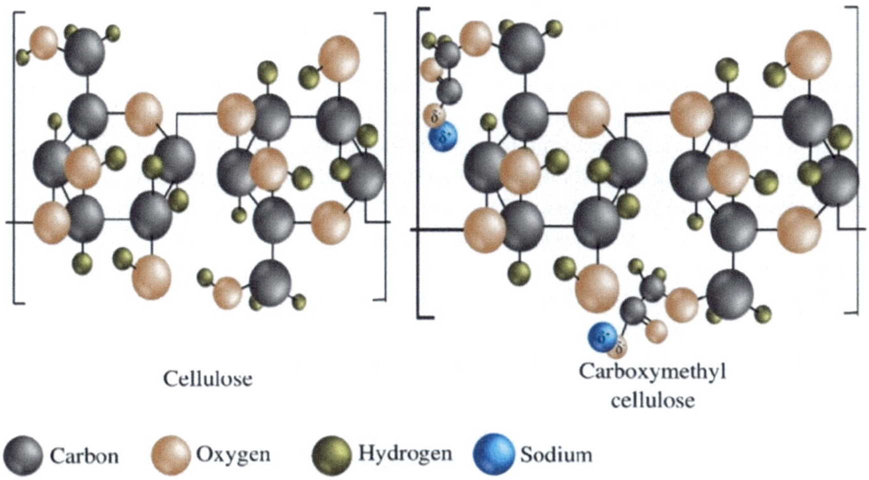

Fig. 1 Major structural difference between cellulose and carboxymethylcellulose

the interest of researchers for potential use in the large-scale or commercial production of CMC, for example, sugarcane bagasse [33], banana pseudo-stem, sago palm [70], corn stalks [81], corn husk [59], cacao pod husks [34], and pineapple peel [18] along with certain waste materials, such as waste cotton linters from textile industries [39], knitted rags [25], wastepaper [44], cotton gin residues [36], and discarded textiles [13].

2.1 Characteristics of CMC

This part focuses on the properties and factors that directly influence the behavior of CMC in its applications and the characteristics of the final CMC products. Specifically, rheology, viscosity, and degree of substitution (DS) play a central role. Rheology, for instance, determines the physical properties and how CMC products flow and respond to pressure variations. However, it is important to note that the rheological properties, including stress–strain flow behavior, pseudoplasticity, viscoelasticity, and thixotropy, are primarily influenced by viscosity. Viscosity, in turn, is closely connected to the DS of CMC. Therefore, the overall attributes of CMC for various applications can be defined by considering these three key parameters: rheology, viscosity, and DS.

2.1.1 Rheological Properties

Within the context of discussing the characterization of substances, rheology plays a crucial role in establishing a connection between the study of how substances flow and deform when subjected to external forces. Furthermore, the examination of rheology in materials provides a comprehensive understanding of the flow properties, including thixotropy, pseudoplasticity, viscoelasticity, and stress–strain flow behavior, within the flow system. It is worth noting that these rheological behaviors or properties are intricately linked to various aspects of polymer systems, such as their structure, particle size, concentration, shape, and surface characteristics. In the case of CMC, the study of its structural properties reveals intriguing and complex flow behaviors when subjected to stress–strain forces. These behaviors have a direct impact on the diverse applications of CMC, such as its use in food packaging, film production, and material coating, among others [1]. The thixotropic, pseudoplastic, and viscoelastic characteristics of CMC are directly relevant to applications like suspension injection, paint formulation, adhesive production, food processing, and cosmetics.

Stress–Strain Flow Behavior

Evaluating the stress–strain relationship of materials helps determine their flow behavior and suitability for specific conditions. As a polymer derivative, CMC often exhibits characteristics resembling those of a non-Newtonian fluid. In low concentrations, CMC may exhibit properties akin to a Newtonian fluid or a viscous flow, in line with the investigation conducted by Ghannam and Esmail in [31].

Pseudoplastic and Viscoelastic Behavior

The pseudoplastic or viscoelastic properties of materials describe their flow behavior and response to stress during various applications. Pseudoplasticity denotes the shear-thinning behavior exhibited by fluid materials. CMC's pseudoplasticity arises primarily from the easy orientation of its long-chain structure under shear stress. The viscosity gradually decreases as CMC undergoes stress, resulting in an increase in its flow behavior index. Factors like the DS value, molecular weight, and temperature can influence CMC's pseudoplastic behavior. Specifically, pseudoplasticity is more pronounced with higher molecular weight and DS values ranging from 0.9 to 1.2 [23]. CMC's pseudoplastic behavior finds applications in various industries, including food processing like condensed milk, mayonnaise and cosmetics, such as lotion and toothpaste, where it contributes to desirable flow properties.

Thixotropy

Thixotropy stands as a crucial property among non-Newtonian fluids and exhibits behavior that contrasts with rheopexy. CMC, characterized by its long-chain polymer structure, displays thixotropy, forming intricate three-dimensional structures through intermolecular interactions. Essentially, thixotropy refers to the transformation observed in shear-thinning systems, where a high-consistency gel transitions into a low-consistency solution. After a prolonged period of rest, the thixotropic solution reverts to a gel-like state. During thixotropy, CMC traces a time-dependent hysteresis loop through two curves, akin to an upward and a downward curve [28]. The upward curve signifies CMC's high consistency or viscosity, reflecting an increase in shear rate as stress approaches the desired value. Once the desired shear rate is reached, the fluid's viscosity or consistency diminishes with shear rate, as indicated by the downward curve in the hysteresis loop. The extent of thixotropy in CMC can sometimes be influenced by factors, such as its DS value or concentration.

2.1.2 Viscosity

When dissolved in water, CMC (carboxymethylcellulose) creates a thick solution due to its polymer structure and strong water solubility. The solution's viscosity—which measures how much the fluid resists flowing—comes from the friction between moving fluid layers. This viscosity is key to understanding how CMC solutions behave and why they work well as thickeners and emulsifiers in various applications, whether in their sodium (Na-CMC) or acid (H-CMC) forms. Viscosity is especially vital because it influences and is influenced by the concentration of CMC. In essence, higher concentrations of CMC maintain high viscosity in the applied medium, while lower concentrations result in reduced viscosity. Multiple factors affect how viscous a CMC solution becomes. These include characteristics of the original cellulose (like its particle size, molecular weight, and degree of substitution) and the conditions used during CMC synthesis (such as the concentration of sodium hydroxide used, the temperature of the reaction, and the pH of the solution) [90].

2.1.3 Degree of Substitution

In the realm of cellulose chemistry, the pivotal parameters for discerning chemical activity revolve around the reactive groups inherent to cellulose. The DS introduces three reactive hydroxyl (–OH) groups into cellulose's anhydroglucose units. Typically, the DS range is expressed from 0 to 3 for each unit. DS, in technical terms, denotes the average count of substitutions of reactive groups by other active molecules within the polymer chain during the synthesis of CMC from cellulose. DS assumes paramount importance in shaping the properties of CMC. Key attributes, such as solubility, emulsification, thickening capacity, acid resistance, viscosity, stability, and salt tolerance, are markedly contingent on the DS value [3]. To illustrate, CMC is entirely insoluble but swells under a DS value of 0.4, while complete solubility is achieved beyond 0.4 [66]. Furthermore, DS values significantly impact properties, such as particle size reduction, viscosity increase, salt or alcohol tolerance, and heightened hygroscopicity [66].

2.2 Sources of Carboxymethylcelluose

2.2.1 Carboxymethylcellose Derived from Various Plant-Based Precursor Materials

Historically, terrestrial cellulose precursors have predominantly served as the primary source for CMC production. Nonetheless, the coexistence of other constituents, such as lignin, pectin, hemicellulose, and minerals alongside cellulose, necessitates excessive energy consumption and costly procedures for their elimination through extensive pre-treatment steps. As a result, the utilization of conventional terrestrial cellulose sources in CMC manufacturing has gradually waned due to their limited availability and the substantial expenses associated with extensive pre-treatment processes. Many researchers have reported a moderate cellulose content ranging from 31 to 60% in various agricultural by-products and waste materials, including leaves, fruit peels, straws and corn cobs [49]. Consequently, there has been a growing interest in recent years in harnessing these resources for commercial CMC production, tailoring the resulting CMC to specific applications based on its properties. It is important to note that CMCs synthesized from different plant-based or agricultural precursors may exhibit varying physicochemical and morphological characteristics, such as degree of substitution (DS), rheological properties, viscosity, and water and oil retention capacity, among others. Consequently, their suitability for specific applications may differ. To enhance their existing properties or introduce new ones, researchers have turned to cotton fibers, which consist of approximately 95% cellulose, for CMC production [7, 50, 72]. Several years ago, Chen and Lou [16] created technical grade CMCs using abundant waste materials that displayed diverse viscosity properties and DS values. This variability in properties made it a highly promising candidate for industrial CMC production for various applications. For example, corncobs are often considered waste material after *Zea mays* seeds are removed, and maize is grown abundantly in various regions across the world. As reported by Singh and Singh [82], India alone produced approximately 16,780,000 tons of maize cereals annually in 2011. Consequently, the researchers recognized the potential of this abundant waste material for CMC synthesis and successfully produced a high-quality CMC product with a moderate DS of 1.18. Another low-cost precursor for CMC, derived from the fibrous, low-protein by-product of the palm oil cultivation industry, was proposed. In 2015, it was reported that roughly 164 million tons of oil palm fronds were produced annually [85]. In this study, the author synthesized a high-purity CMC product with a high production yield of 170.1% and a DS of 1.1 from this cost-effective precursor using the conventional alkylation-etherification process. Water hyacinth, an invasive aquatic plant that proliferates abundantly, can also serve as an effective precursor for CMCs [78]. Additionally, a highly invasive weed known as *Lantana camara L.*, which can have detrimental effects on land productivity and overall biodiversity, was identified as a valuable precursor for producing non-Newtonian pseudoplastic CMC with high viscosity and a DS of 1.22 by Mondal et al. [59].

2.2.2 Carboxymethylcellulose Derived from Alternative Source Materials

In addition to traditional plant-based cellulose precursors, certain waste materials from textile industries and everyday household and office items including office waste paper, paper sludge, and discarded textiles are readily available, often at no cost, and can be effectively employed in CMC synthesis. The production of CMC and its various composites for a wide array of applications using these waste materials not only reduces production expenses but also plays a pivotal role in curbing environmental pollution. For instance, cotton linters, which are a by-product of cotton cleaning factories, have long been recognized as a valuable resource. Cotton linters can be employed to synthesize various grades of CMC products for diverse applications; for example, Jahan et al. [39] reported a method for producing water-soluble CMC from cotton linters, achieving a commendable DS along with high viscosity.

Hivechi et al. [37] showed a way to make CMC from cotton linters that is both cheap and good for the environment. They used ultrasonic and microwave radiation to make the usual alkylation-etherification process easier. In Fakrul Alam and Mondal [25], researchers have found innovative ways to produce CMC from textile waste materials. In one case, knitted rags from textile factories—which are rich in α-cellulose (98%)—were successfully converted into CMC. Building on this eco-friendly approach, Li and colleagues demonstrated in 2020 that even ordinary office waste paper could be transformed into CMC, creating an economical dust suppressant that is better for the environment, Li et al. [52].

2.3 *Synthesis Method of CMC*

The traditional method for making CMC follows a two-step alkylation-etherification process, which has remained fundamentally unchanged since its inception, though researchers have adjusted chemical ratios and reaction conditions based on their specific starting materials.

Starting materials for CMC can come from many sources—including leaves, waste paper, pulp, stems, microorganisms, and textile waste like cotton linters and knitted rags. However, these raw materials are not pure cellulose—they often contain other components like lignin, pectin, ash, and various minerals. This means the amount of usable cellulose varies depending on the source.

The actual carboxymethylation process happens in two main stages:

First, purified α-cellulose is mixed with an alkali (typically sodium hydroxide) and allowed to react. During this mercerization step, the hydroxyl (–OH) groups on each anhydroglucose unit are converted to sodium oxide (–ONa) groups.

Second, these –ONa groups undergo etherification to form the final carboxymethyl groups. This step is particularly sensitive—the concentration of alkali, the ratio between alkali and cellulose, and the reaction temperature must all be precisely controlled. Even small changes in these conditions can significantly affect the degree of substitution (DS) and other properties of the final CMC product. Additionally, an inert solvent, such as ethanol [38], 2-propanol [21], and isopropyl alcohol [88], is introduced during this stage, serving both as a diluent and a swelling agent to facilitate the penetration of the associated reagents into the cellulose structure.

In the subsequent stage, an etherification process is initiated by introducing an etherifying agent into the reaction vessel. Sodium monochloroacetate (NaMCA) is a commonly employed etherifying agent for this purpose [15]. Nevertheless, certain researchers have opted for alternative reagents, such as diazomethane [71], deviating from the traditional etherifying agent. The mercerized cellulose is consistently agitated in the presence of the reagent for a specified duration. It is crucial to maintain an optimal temperature for achieving the best outcomes, necessitating careful temperature control throughout the reaction. This step can is described by the following reaction (II) in Fig. 2.

Following mercerization and etherification, a solid suspension of CMC is generated. Subsequently, after confirming the purity of the product, the suspension undergoes centrifugation and filtration using an appropriate filtration medium [86]. The product is then refined and subjected to additional post-treatment procedures,

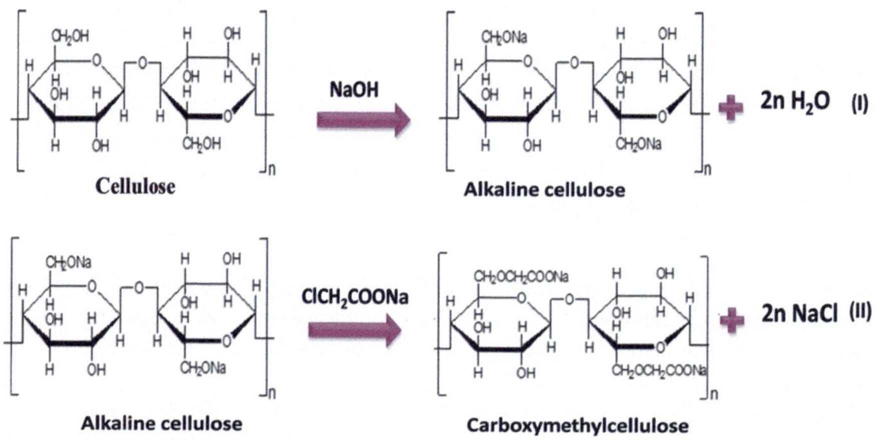

Fig. 2 Synthesis reaction of carboxymethylcellulose

including neutralization with weak acids like acetic acid [8], alcohol washing for dehydration and desalting [67], and acetone rinsing [9]. Finally, the resulting solid product is dried at a specific temperature, with oven drying commonly preferred over sun drying to ensure precise temperature control.

2.4 Importance of Carboxymethylcellulose in Various Industries

2.4.1 Utilization in the Textile Industry

Over the years, a variety of polysaccharide-based thickeners like guar gum, tamarind, corn starch, and tapioca starch have found extensive application in textile printing with various dyes and fabrics. Recently, CMC has emerged as an exceptionally effective thickening agent for paints and textile varnishes, primarily due to its exceptional water-absorbing properties. CMC derived from lignocellulosic waste has been employed as a thickener in vat dyes for textile printing [63]. By adjusting the thickening properties of vat dyes, optimal viscosity is achieved, resulting in high-quality paint. Additionally, it helps retain dye particles in the printing area, alongside other chemicals or printing assistants, and serves as an effective vehicle for dye materials, promoting their binding to textile specimens. CMC has gained significant traction in the textile industry owing to its high viscosity and film-forming characteristics. CMC's ability to dissolve in water makes it ideal for multiple textile applications, from printing pastes and sizing to finishes and lubricants. In particular, crude CMC excels as a sizing agent for various textile materials, including filament threads and yarns. It is especially attractive for textile sizing because it is minimally toxic, biocompatible, water-soluble, and easily washed away when no longer needed.

The versatility of CMC is further demonstrated through its hybrid applications. When combined with starch, it improves how well starch-based sizes stick to cotton and cotton-blend materials. CMC has also been modified into various forms for sizing applications, such as grafted versions with polyvinyl alcohol or polymethyl methacrylate [66]. When working with silk fabrics, CMC plays multiple important roles in the printing process. As a pre-treatment and leveling agent, it helps ensure colors stay fast and uniform during digital printing. During the textile printing process, CMC works alongside other agents to control the viscosity of colors, while also making dyes more water-attracting and helping them penetrate deeper into the fabric. An et al. [5], resulting in sharper prints and better color quality. For example, Jeong and Chun [41] utilized a CMC/sodium alginate/dextrin mixture for digital textile printing. Additionally, Kolmana et al. [48] made an important advance by creating specialized complexes that combine CMC with silica and polyelectrolytes for textile painting. The CMC component helps these complexes stick better to cotton fibers, which results in stronger materials. When added to cotton fabrics, CMC also increases how much water they can hold onto, making these treated fabrics particularly useful for applications that need high absorbency, such as wound dressings [51].

2.4.2 Applications in the Food Industry

CMC emerges as a frequently employed auxiliary agent within the food industry owing to its exceptional attributes, including being odorless, tasteless, non-caloric, physiologically inert, and capable of forming clear solutions without opacity [40]. Here is a clearer rephrasing:

CMC plays diverse and important roles in food applications, where it helps create better quality products with the right texture and mouthfeel while maintaining food safety standards. In the food industry, CMC serves multiple functions—it thickens products, keeps emulsions stable, binds moisture, suspends particles, improves texture, and helps retain water. Thanks to its pseudoplastic behavior, CMC can be used to adjust food products' flow properties, structure, flavor, and appearance. It is also used as a protective coating or packaging material that helps food products last longer on the shelf [65]. Research has documented many uses of CMC in food products. One notable example is its role in olive oil nano-emulsions, where it acts as a thickener. In these applications, CMC helps create better physical properties and more stable products by controlling factors like particle size, concentration, and texture [6]. The viscosity of CMC determines how it is used in food applications: When it has low viscosity, it works well for binding moisture, while high-viscosity CMC is effective as a gelling agent [66]. CMC serves as a thickening agent across a range of food products, including semi-solid dairy items, salad dressings, and fruit syrups [68]. CMC plays different roles depending on the food product. In fruit syrups and juices, it forms a hydrogel that helps ensure the food stays safe to consume. For a variety of other products—including dairy items, condiments, baked goods, acidic dairy products, and ice cream—CMC works as an emulsifier to keep ingredients properly mixed [66]. CMC helps maintain consistent texture in cocoa drinks and acidic milk beverages by preventing ingredients from settling to the bottom or separating into layers. This stabilizing effect works whether the drinks are cold or hot [89]. The food industry's use of CMC has evolved beyond traditional applications to more sophisticated uses, driven by its beneficial properties—it is hygienic, compatible with biological systems, and can help prevent certain diseases. For example, Dafe et al. [17] created a specialized food delivery system by combining CMC with k-carrageenan. This mixture was designed to carry probiotic foods directly to the colon, helping to prevent gastrointestinal infections.

2.4.3 Utilization in Water Treatment

CMC has garnered attention as an effective material for various water treatment applications over recent years. Innovative hybrid materials incorporating CMC have also been developed to remove diverse pollutants from wastewater. For example,

Salama et al. [77] synthesized a nano-adsorbent material, CMC/Fe_3O_4, via co-precipitation for methylene blue dye adsorption. This material exhibited strong adsorption, removing up to 64 mg/g of methylene blue at pH 7. Manzoor et al. [56] synthesized chitosan/CMC hybrid adsorbents, displaying excellent adsorption capacity against Cd^{2+} and Pb^{2+} ions in experimental conditions. Gasemloo et al. [30] demonstrated a sulfated-CMC-based nano-filter membrane for efficient Cr(VI) removal at optimum conditions. Zhang et al. [91] created an innovative hybrid material by combining CMC with ammonium phosphomolybdate. This cross-linked material works as an adsorbent that can effectively remove radioactive cesium-137 from wastewater produced by nuclear facilities. Cai et al. [14] engineered a specialized composite material that combines CMC with Fe_3O_4 in a core–shell structure. This material was designed to extract specific radioactive isotopes—europium-152 and europium-154—from wastewater.

2.4.4 Biomedical Applications

CMC and its hybrid materials have become increasingly important in biomedical applications thanks to their special surface characteristics. These materials are particularly valuable because they allow cells to attach easily, have low toxicity to cells, work well with biological systems, break down naturally in the body, and help keep cells healthy.

In the pharmaceutical field, CMC-based composites are especially useful for drug delivery. They can effectively transport powerful medical compounds through the body, and scientists have investigated many different ways to use these CMC-based biomaterials to deliver therapeutic treatments. For example, Oliveira et al. [64] developed a hydrogel composed of PVA/Na-CMC, demonstrating its ability to release flavonoids and phenols effectively to wound sites. This hydrogel exhibited excellent elasticity which can be quantified using parameters like Young's Modulus (E) or the elastic modulus. For example, we can measure the Young's Modulus of the hydrogel to determine its elasticity in pascals (Pa) or megapascals (MPa) [11], swelling properties which can be quantified using parameters like the swelling ratio or percentage. For instance, we can express the swelling capability as a percentage increase in volume when the hydrogel is fully hydrated. The formula for swelling ratio is [((Vh-Vd)/Vd)*100], where Vh is the volume when fully hydrated, and Vd is the initial dry volume [92], and antimicrobial capabilities which can be quantified using parameters, such as the Minimum Inhibitory Concentration (MIC) or zone of inhibition (ZOI). MIC measures the lowest concentration of the hydrogel required to inhibit the growth of microorganisms, typically expressed in micrograms per milliliter (μg/mL) or milligrams per milliliter (mg/mL). ZOI measures the area around the hydrogel where microbial growth is inhibited, usually expressed in millimeters (mm) [20]. Du et al. [22] developed a specialized microgel by combining two modified materials—carboxymethyl chitosan and oxidized CMC. They used this microgel to carry two different protein-based drugs: bovine serum albumin and silver sulfadiazine. The researchers incorporated these drug-containing microgels into a hydrogel using a

Schiff base reaction. One particularly important aspect of their work was that the resulting hydrogels could release drugs in response to changes in pH. When tested against Staphylococcus aureus bacteria, these hydrogels showed strong antibacterial properties. Furthermore, Joorabloo et al. [43] created a new type of hydrogel that combines nano-sized zinc oxide particles with polyvinyl alcohol and CMC. This hydrogel proved to be both biocompatible and non-toxic, successfully supporting cell life while also effectively fighting two types of bacteria—Staphylococcus aureus and Escherichia coli.

These research examples demonstrate how versatile CMC-based biomaterials can be for drug delivery, offering different solutions for various medical needs. The applications of CMC-based materials continue to expand, particularly in the innovative field of 3D bioprinting for tissue engineering. This cutting-edge process uses computer-controlled systems to build living tissue scaffolds by depositing layers of suitable biomaterials. For example, Habib et al. [35] demonstrated another innovative use of CMC by creating a hybrid hydrogel specifically for 3D bioprinting. They combined CMC with sodium alginate (another material known to work well with biological systems) to create a solution with better viscosity and printing qualities. Their hybrid hydrogel showed promise for drug delivery applications, particularly through the production of medicinal beads. Ji et al. [42] created two new versions of CMC by modifying it with amide and ester cross-linkers to serve as bio-inks for 3D printing. These cost-effective materials successfully allowed for the printing of complex three-dimensional structures.

2.4.5 Application of CMC in Pharmaceutical Industries

In recent years, CMC and its derivatives, including composites and copolymers, have garnered significant attention for their applications in pharmaceutical products. This heightened interest is due to their properties, including biocompatibility, extended in vivo half-life, exceptional stability, high drug-binding capacity and pH sensitivity. CMC serves various functions in pharmaceutical applications, for example, Sathasivam et al. [79] achieved controlled release of red palm oil's nutritional components in targeted intestinal fluid by encapsulating them in CMC beads derived from sago biomass. CMC also serves as a binder or a foundational component in drug formulations. For instance, anionic CMC is employed as a binding agent alongside lysozyme and curcumin: CMC forms a protective matrix around these bioactive molecules, hiding them from external elements that can cause destruction, such as heat and oxygen. This encapsulation allows for regulated release, extended efficacy, and enhanced absorption in the body. The application of CMC as a binding agent in pharmaceuticals helps to preserve the efficacy and integrity of curcumin and lysozyme, which eventually improves the quality of the finished product and the experience of the customer [53].

3 Comparison Between Carboxymethylcellulose from *Ziziphus lotus* and *Posidonia oceanica*

In this section, we will explore and compare the synthesis processes of CMC derived from two distinct sources: *Ziziphus lotus*, originating from arid regions, and *Posidonia*, sourced from marine environments. These two raw materials offer unique characteristics and challenges in CMC synthesis, making it essential to delve into their respective production methods and resulting properties.

3.1 Overview of Posidonia oceanica *and* Ziziphus lotus

Posidonia oceanica is a seagrass species endemic to the Mediterranean Sea and is one of the most ecologically important seagrass species in this region [83]. It forms extensive meadows, known as "*Posidonia* beds" in shallow coastal waters, creating complex habitats that support diverse marine ecosystems. This seagrass plays a crucial role in maintaining coastal biodiversity, as it provides a nursery ground and refuge for various marine species, including fish, crustaceans, and invertebrates [24]. The morphology of *Posidonia oceanica* consists of long, ribbon-like leaves that grow from rhizomes anchored to the seabed. These leaves can reach lengths of up to one meter and are an essential part of the plant's photosynthetic machinery, contributing to the seagrass's ability to capture sunlight and convert it into energy. Many scientific research works have utilized *Posidonia oceanica* as a raw material in various applications. For example, Khiari et al. [47] focused on extracting cellulose and CMC from *Posidonia oceanica* fibers. The research involved the use of alkali treatment and carboxymethylation processes to produce CMC from the raw material. Besides, Bettaieb et al. [10] explored the extraction and characterization of cellulose nanocrystals from *Posidonia oceanica* fibers. The research focused on obtaining nanostructures from the raw material and assessing their properties, such as morphology, crystallinity, and thermal stability. The study highlighted the potential of *Posidonia oceanica* as a sustainable and renewable source of cellulose nanocrystals for various nanotechnology applications.

Ziziphus lotus also referred to as *the lotus Jujube* or *lotus tree* has a long history and significant cultural value. *Ziziphus lotus* is a plant that is native to the Mediterranean region and some parts of Asia. It has been prized for its many traditional uses, including its palatable fruits and potential therapeutic benefits. The Rhamnaceae family-related plant is distinguished by its thorny branches, tiny yellow blooms, and tiny, date-like fruits. As a possible source of raw materials for the extraction of cellulose and CMC [76], *Ziziphus lotus* has come to light recently. Researchers have investigated how to separate cellulose fibers for use in numerous industries from various *Ziziphus lotus* plant components, including as stems and leaves (Fig. 3).

Fig. 3 *Ziziphus lotus* (**a**) and *Posidonia oceanica* (**b**)

3.2 Chemical Composition

Comparing the chemical compositions *of Ziziphus lotus* and *Posidonia oceanica* (Table 1) with other plant materials reveals distinct variations in key components. *Posidonia oceanica*, as reported by Khiari et al. [46], stands out with a notably higher cellulose content of 40%, compared to *Ziziphus lotus* (53.24%) from the study of [75] and *Opuntia ficus-indica* (53%) from the study of [55]. Moreover, *Ziziphus lotus* demonstrates a relatively low ash content of 1.49%, significantly lower than *Posidonia oceanica* (12.00%) and *Opuntia ficus-indica* (5.50%). In contrast, *Ziziphus lotus* exhibits a higher extractive content, especially in hot water and 1% NaOH extracts, showcasing its unique composition for potential applications but *Posidonia oceanica* possesses a higher lignin content (29.80%) compared to *Ziziphus lotus* (12.4%). Comparing these two with other plant materials, *Opuntia ficus-indica* stands out with high hot water extractives (36.30%) and cellulose content (53.60%).

Astragalus armatus, on the other hand, shows higher cellulose content (54.00%) but lower lignin content (16.70%) than *Posidonia oceanica.* Date Palm Rachis exhibits significant holocellulose content (74.80%), making it a potential candidate for various applications. These comparisons underline the diverse chemical compositions of these plant materials, offering insights into their suitability for various industrial processes, including pulping, biofuel production, and biorefinery applications. Each plant source presents its unique characteristics, making them valuable resources in the realm of sustainable materials and bioenergy production.

Table 1 Chemical composition comparison

Plant	C.W	H.W	NaOH 1%	A.B	Ash	Lign	Holo	Cell
Ziziphus lotus [75]	46.11	49.01	40.40	11.03	1.49	12.4	68.14	53.24
Posidonia oceanica [46]	7.30	12.20	16.50	10.70	12.00	29.80	61.80	40.00
Date Palm rachis [45]	5.00	8.10	20.80	6.30	5.00	27.20	74.80	45.00
Opuntia ficus-indica [55]	24.00	36.30	29.60	9.80	5.50	4.80	64.50	53.60
Retama raetam [26]	32.00	31.50	47.00	10.00	3.50	20.50	58.70	36.00
Nitraria retusa [26]	23.00	25.50	40.00	3.00	6.20	26.30	52.00	41.00
Astragalus armatus [62]	26.20	33.00	32.70	13.00	3.00	16.70	54.00	35.00
Stipagrostis pungens [27]	19.30	20.50	42.90	4.80	4.60	12.00	71.00	44.00
Tamarix Aphylla [58]	12.80	19.00	18.50	4.50	3.50	30.00	50.00	39.00
Schinus molle [74]	-	38.20	27.00	12.30	4.00	21.40	74.00	53.20

CW cold water solubility, *HW* hot water solubility, *AB* solubility in various organic solvents; 1% NaOH: 1% sodium hydroxide solubility, *Hol.* holocellulose, *Lign.* Klason lignin (%), *Cell.* cellulose

3.3 Methodology for Synthesis CMC from **Posidonia oceanica** *and* **Ziziphus lotus**

The synthesis of CMC from the celluloses extracted from *Posidonia oceanica* and *Ziziphus lotus*, respectively, are explained in the corresponding papers by Khiari et al. [46] and Saad et al. [76]. The extraction procedure for *Posidonia oceanica* required a novel optimization approach. *Posidonia oceanica* wastes were initially treated with a mild alkaline solution at a temperature of 100 °C [46]. The resulting unbleached material was then bleached at room temperature with a sodium hypochlorite solution. The bleached material was then used as the starting material for the synthesis of CMC. Saad et al. [76] outlined the synthesis technique for *Ziziphus lotus*-derived CMC in their work. To delignify the chips, an alkaline extraction stage employing sodium hydroxide solution (10% w/v) at 80 °C for 6 h was used. After that, the delignified chips were carefully rinsed until neutral pH. The obtained pulp was then bleached using sodium chlorite and acetic acid as part of the cellulose recovery process, which effectively removed impurities and enhancing the purity of the material. Alkaline hydrolysis, chloroacetylation, and etherification were the three steps in the carboxymethylation process that were used in each of these studies to transform the extracted cellulose into carboxymethylcellulose (CMC).

3.4 Comparison of Solubility Between the Two Types of CMC

Khiari et al.'s study [46] on *Posidonia*-derived CMC exhibited good water solubility, making it ideal for a variety of industrial uses. Saad's work on CMC from *Ziziphus lotus* [76], on the other hand, revealed equivalent or even greater water solubility when compared to commercially available CMC products. CMC's solubility is an important aspect in determining its adaptability and success in applications, such as food stabilizers, pharmaceutical binders, and cosmetic emulsifiers [87]. The distinct solubility profiles of CMC from *Posidonia* and *Ziziphus lotus* reveal their distinct possibilities in a variety of sectors, such as, in the food sector, CMC can be employed as a thickening agent or emulsifier, enhancing the texture and stability of products [12]. In the realm of water treatment, the distinct properties of these CMC variants offer potential advantages in improving water quality and treatment processes [29].

3.5 Comparison of Carboxymethylcelluloses' Substitution Degree

Direct comparisons between CMC extracted from *Posidonia oceanica* and *Ziziphus lotus*, as researched by Khiari et al. [46] and Saad et al. [76], respectively, extend to an assessment of their substitution degrees. In Khiari's study, CMC derived from *Posidonia oceanica* exhibits a range of substitution degrees spanning from 0.67 to 1.62. This variability signifies the varying levels of carboxymethyl group substitution along the cellulose chains, indicating the diverse chemical composition of the extracted CMC. Likewise, Saad's et al. [76] investigation into CMC obtained from *Ziziphus lotus* reveals substitution degrees of 0.74 and 0.85. This comparative analysis demonstrates the distinct ranges of substitution observed in both types of CMC. The variations in the degree of substitution are associated with the choice of the source of raw materials used.

3.6 Potential Applications of **Posidonia** *and* **Ziziphus lotus***-Derived CMC*

Carboxymethylcellulose (CMC) obtained from *Posidonia oceanica* and *Ziziphus lotus* have a vast range of prospective applications. Notably, Khiari's study [46] has been assessed for its potential as a high-performance adsorbent material. A comprehensive comparative analysis, involving two conventional adsorbent materials frequently utilized across several industries, revealed that the CMC preparation displayed significantly superior retention capabilities when contrasted with industrial-grade cellulose. This discovery underscores the exceptional adsorption properties of CMC, opening the way for its application in a diverse range of industries where efficient adsorbent materials are indispensable, such as in wastewater treatment which CMC's enhanced adsorption capacity can be operated to remove pollutants and impurities from water and in the pharmaceutical sector which can benefit from its properties as a super adsorbent, facilitating product formulations.

CMC is derived from *Ziziphus lotus*, on the other hand, as investigated by Saad et al. [76], revealing its substantial promise in the field of sustainable membrane materials for water treatment. This application is especially important in the context of dealing with water scarcity and the necessity for efficient purifying technologies. Furthermore, CMC derived from *Ziziphus lotus* emerges as an environmentally acceptable textile industry option. Its biodegradability, with the increased demand for environmentally friendly materials, displays a commitment to sustainable practices and lowering environmental impact in the textile production sector. These research findings highlight CMC's multifarious potential in a variety of industrial applications, fostering innovation and sustainability.

4 Conclusion

The comparative research on CMC extraction from *Posidonia oceanica* and *Ziziphus lotus* has provided vital insights into the potential of these varied sources for cellulose derivatives. The discrepancies in substitution degrees highlight the derived CMC variants' different chemical compositions and structural features. Additionally, there is potential for experimentation with better ways to make CMC, aiming to enhance how efficiently it is produced, its purity, and how environmentally friendly the process is. This research on CMC from *Posidonia oceanica* and *Ziziphus lotus* opens up exciting possibilities for more studies that link material science, practical applications, and eco-friendly practices.

References

1. Abraham J, Sharika T, Mishra R, Thomas S (2017) Rheological characteristics of nanomaterials and nanocomposites. In: Raghvendra KM, Sabu T, Nandakumar K (eds) Micro and nano fibrillar composites (MFCs and NFCs) from polymer blends. Elsevier, Amsterdam, The Netherlands, pp 327–350
2. Adinugraha MP, Marseno DW (2005) Synthesis and characterization of sodium carboxymethylcellulose from cavendish banana pseudo stem (Musa cavendishii LAMBERT). Carbohydr Polym 62:164–169
3. Akram F, ulHaq I, Raja SI, Mir AS, Qureshi SS, Aqeel A, Shah FI (2022) Current trends in biodiesel production technologies and future progressions: a possible displacement of the petro-diesel. J Clean Prod 370:133479
4. Amior A, Satha H, Laoutid F, Toncheva A, Dubois P (2022) Natural cellulose from ziziphus jujube fibers: extraction and characterization. Materials 16:385
5. An F, Fang K, Liu X, Li C, Liang, Y, Liu H (2020) Rheological properties of carboxymethylhydroxy propyl cellulose and its application in high quality reactive dye inkjet printing on wool fabrics. Int J Biol Macromol 164:4173–4182
6. Arancibia C, Navarro-Lisboa R, Zúñiga RN, Matiacevich S (2016) Application of CMC as thickener on nanoemulsions based on olive oil: physical properties and stability. Int J Polym Sci 2016

7. Bahlool SO, Kamel SA (2023) Utilization of Egyptian cotton waste fibers for production of Carboxymethyl cellulose (CMC). J Text Color Polym Sci 20:83–90
8. Bangar SP, Kajla P, Ghosh T (2022) Valorization of wheat straw in food packaging: a source of cellulose. Int J Biol Macromol 227:762–776
9. Barakat A, Kamoun EA, El-Moslamy SH, Ghazy MB, Fahmy A (2022) Photo-curable carboxymethylcellulose composite hydrogel as a promising biomaterial for biomedical applications. Int J Biol Macromol 207:1011–1021
10. Bettaieb F, Khiari R, Dufresne A, Mhenni MF, Belgacem MN (2015) Mechanical and thermal properties of *Posidonia oceanica* cellulose nanocrystal reinforced polymer. Carbohyd Polym 123:99–104
11. Bhat MA, Rather RA, Yaseen Z, Shalla AH (2022) Viscoelastic and smart swelling disposition of Carboxymethylcellulose based hydrogels substantiated by Gemini surfactant and in-vitro encapsulation and controlled release of Quercetin. Int J Biol Macromol 207:374–386
12. Bhavyasri K, Anila B, Sumakanth M (2023) Quantification of stabilizers in food products by a simple method development and validation. J Coast Life Med 11:1683–1687
13. Bidgoli H, Zamani A, Jeihanipour A, Taherzadeh M (2014) Preparation of carboxymethyl cellulose superabsorbents from waste textiles. Fibers and Polymers 15:431–436
14. Cai Y, Yuan F, Wang X, Sun Z, Chen Y, Liu Z, Wang X, Yang S, Wang S (2017) Synthesis of core–shell structured Fe_3O_4@carboxymethyl cellulose magnetic composite for highly efficient removal of Eu (III). Cellulose 24:175–190
15. Casaburi A, Rojo ÚM, Cerrutti P, Vázquez A, Foresti ML (2018) Carboxymethyl cellulose with tailored degree of substitution obtained from bacterial cellulose. Food Hydrocoll 75:147–156
16. Chen WQ, Lou DP (2014) Synthesis of sodium carboxymethyl cellulose based on pretreated bamboo shaving. Adv Mater Res 997:169–172
17. Dafe A, Etemadi H, Zarredar H, Mahdavinia GR (2017) Development of novel carboxymethyl cellulose/k-carrageenan blends as an enteric delivery vehicle for probiotic bacteria. Int J Biol Macromol 97:299–307
18. Dai H, Huang H (2017) Enhanced swelling and responsive properties of pineapple peel carboxymethyl cellulose-g-poly (acrylic acid-co-acrylamide) superabsorbent hydrogel by the introduction of carclazyte. J Agric Food Chem 65:565–574
19. Das SK, Vishakha K, Das S, Chakraborty D, Ganguli A (2022) Carboxymethyl cellulose and cardamom oil in a nanoemulsion edible coating inhibit the growth of foodborne pathogens and extend the shelf life of tomatoes. Biocatal Agric Biotechnol 42:102369
20. Das T, Ali F, Rahman MS (2022) Cellulase activity of a novel bacterial strain *Arthrobacter woluwensis* TDS9: its application on bioconversion of paper mill sludge. J Genet Eng Biotechnol 20:87
21. De Mattos NR, Colodette JL, de Oliveira CR (2019) Alkaline extraction and carboxymethylation of xylans from corn fiber. Cellulose 26:2177–2189
22. Du S, Chen X, Chen X, Li S, Yuan G, Zhou T, Li J, Jia Y, Xiong D, Tan H (2019) Covalent chitosan-cellulose hydrogels via schiff-base reaction containing macromolecular microgels for pH-sensitive drug delivery and wound dressing. Macromol Chem Phys 220:1900399
23. Edali M, Esmail MN, Vatistas GH (2001) Rheological properties of high concentrations of carboxymethyl cellulose solutions. J Appl Polym Sci 79:1787–1801
24. Fakiris E, Giannakopoulos V, Leftheriotis G, Dimas A, Papatheodorou G (2023) Predictive mapping of mediterranean seagrasses-exploring the influence of seafloor light and wave energy on their fine-scale spatial variability. Remote Sens 15(11):2943
25. Fakrul Alam A, Mondal MIH (2103) Utilization of cellulosic wastes in textile and garment industries. I. Synthesis and grafting characterization of carboxymethyl cellulose from knitted rag. J Appl Polym Sci 128:1206–1212
26. Ferhi F, Das S, Elaloui E, Moussaoui Y, Yanez JG (2014) Chemical characterisation and suitability for papermaking applications studied on four species naturally growing in Tunisia. Ind Crops Prod 61:180–185
27. Ferhi F, Das S, Moussaoui Y, Elaloui E, Yanez JG (2014) Paper from *Stipagrostis pungens*. Ind Crops Prod 59:109–114

28. Gahrooee TR, Moud AA, Danesh M, Hatzikiriakos SG (2021) Rheological characterization of CNC-CTAB network below and above critical micelle concentration (CMC). Carbohyd Polym 257:117552
29. Gao F (2019) An overview of surface functionalized magnetic nanoparticles: preparation and application for wastewater treatment. Chemistry Select 4:6805–6811
30. Gasemloo S, Khosravi M, Sohrabi MR, Dastmalchi S, Gharbani P (2019) Response surface methodology (RSM) modeling to improve removal of Cr (VI) ions from tannery wastewater using sulfated carboxymethyl cellulose nanofilter. J Clean Prod 208:736–742
31. Ghannam MT, Esmail MN (1997) Rheological properties of carboxymethylcellulose. J Appl Polym Sci 64:289–301
32. Gholamali I, Yadollahi M (2021) Bio-nanocomposite polymer hydrogels containing nanoparticles for drug delivery: a review. Regen Eng Transl Med 7:129–146
33. Golbaghi L, Khamforoush M, Hatami T (2017) Carboxymethyl cellulose production from sugarcane bagasse with steam explosion pulping: experimental, modeling, and optimization. Carbohyd Polym 174:780–788
34. Gutawa K, Willy A, Nikita G (2015) Synthesis and characterization of sodium carboxymethylcellulose from pod husk of Cacao (*Theobroma cacao* L.). Int J Food Microbiol 3:99–103
35. Habib A, Sathish V, Mallik S, Khoda B (2018) 3D printability of alginate-carboxymethyl cellulose hydrogel. Materials 11:454
36. Haleem N, Arshad M, Shahid M, Tahir MA (2014) Synthesis of carboxymethyl cellulose from waste of cotton ginning industry. Carbohyd Polym 113:249–255
37. Hivechi A, Bahrami SH, Arami M, Karimi A (2015) Ultrasonic mediated production of carboxymethyl cellulose: optimization of conditions using response surface methodology. Carbohyd Polym 134:278–284
38. Ibrahim AA, Adel AM, Abd El-Wahab ZH, Al-Shemy MT (2011) Utilization of carboxymethyl cellulose based on bean hulls as chelating agent. Synthesis, characterization and biological activity. Carbohyd Polym 83:94–115
39. Jahan IA, Sultana F, Islam MN, Hossain MA, Abedin J (2007) Studies on indigenous cotton linters for preparation of carboxymethylcellulose. Bangladesh J Sci Ind Res 42:29–36
40. Jayakody MM, Kaushani KG, Vanniarachchy MPG, Wijesekara I (2023) Hydrocolloid and water soluble polymers used in the food industry and their functional properties: a review. Polym Bull 80:3585–3610
41. Jeong DS, Chun TI (2013) Color fastness of digital textile printing on silk fabrics-the effect of the mixed pre-treatment agent. Fash TextE Res J 15:808–814
42. Ji S, Abaci A, Morrison T, Gramlich WM, Guvendiren M (2020) Novel bioinks from UV-responsive norbornene-functionalized carboxymethyl cellulose macromers. Bioprinting 18:e00083
43. Joorabloo A, Khorasani MT, Adeli H, Mansoori-Moghadam Z, Moghaddam A (2019) Fabrication of heparinized nanoZnO/poly (vinylalcohol)/carboxymethyl cellulose bionanocomposite hydrogels using artificial neural network for wound dressing application. J Ind Eng Chem 70:253–263
44. Joshi G, Naithani S, Varshney V, Bisht SS, Rana V, Gupta P (2015) Synthesis and characterization of carboxymethyl cellulose from office waste paper: a greener approach towards waste management. Waste Manage 38:33–40
45. Khiari R, Dridi-Dhaouadi S, Aguir C, Mhenni MF (2010) Experimental evaluation of eco-friendly flocculants prepared from date palm rachis. J Environ Sci 22:1539–1543
46. Khiari R, Mhenni MF, Belgacem MN, Mauret E (2010) Chemical composition and pulping of date palm rachis and *Posidonia oceanica*—a comparison with other wood and non-wood fibre sources. Biores Technol 101:775–780
47. Khiari R, Mhenni MF, Belgacem MN, Mauret E (2011) Valorisation of vegetal wastes as a source of cellulose and cellulose derivatives. J Polym Environ 19:80–89
48. Kolman K, Nechyporchuk O, Persson M, Holmberg K, Bordes R (2017) Preparation of silica/polyelectrolyte complexes for textile strengthening applied to painting canvas restoration. Colloids Surf, A 532:420–427

49. Koul B, Yakoob M, Shah MP (2022) Agricultural waste management strategies for environmental sustainability. Environ Res 206:112285
50. Kukrety A, Singh RK, Singh P, Ray SS (2018) Comprehension on the synthesis of carboxymethylcellulose (CMC) utilizing various cellulose rich waste biomass resources. Waste Biomass Valorization 9:1587–1595
51. Li P, Guo M, Sun F, Fu J, Gao W (2021) Study on gathering-and-twisting mechanism of fibers and CMC-Na/PAM/PVA solution optimization for enhancing cotton yarn performance by adhesive-aided ring spinning. Fibers Polym 22:3490–3500
52. Li S, Zhou G, Liu Z, Wang N, Wei Z, Liu W (2020) Synthesis and performance characteristics of a new ecofriendly crust-dust suppressant extracted from waste paper for surface mines. J Clean Prod 258:120620
53. Li Z, Wang Y, Pei Y, Xiong W, Zhang C, Xu W, Liu S, Li B (2015) Curcumin encapsulated in the complex of lysozyme/carboxymethylcellulose and implications for the antioxidant activity of curcumin. Food Res Int 75:98–105
54. Liu Y, Wei Q, Zhao Y, Yang G (2021) Sustainable extraction of carboxymethylcellulose from agricultural waste and its application in flexible piezoelectric nanogenerator. Carbohyd Polym 253:117221
55. Mannai F, Ammar M, Yanez JG, Elaloui E, Moussaoui Y (2016) Cellulose fiber from Tunisian Barbary Fig "*Opuntia ficus-indica*" for papermaking. Cellulose 23:2061–2072
56. Manzoor K, Ahmad M, Ahmad S, Ikram S (2019) Synthesis, characterization, kinetics, and thermodynamics of EDTA-modified chitosan-carboxymethyl cellulose as Cu (II) ion adsorbent. ACS Omega 4:17425–17437
57. Mazlan MM, Kian LK, Fouad H, Jawaid M, Karim Z, Saba N (2022) Synthesis and characterization of carboxymethyl cellulose from pineapple leaf and kenafcore biomass: a comparative study of new raw materials. Biomass Convers Biorefinery. https://doi.org/10.1007/s13399-022-03700-w
58. M'barek I, Slimi H, AlSukaibi AKD, Alimi F, Lajimi RH, Mechi L, Moussaoui Y (2022) Cellulose from *Tamarix aphylla*'s stem via acetocell for cadmium adsorption. Arab J Chem 15:103679
59. Mondal MIH, Yeasmin MS, Rahman MS (2015) Preparation of food grade carboxymethyl cellulose from corn husk agrowaste. Int J Biol Macromol 79:144–150
60. Mondal MIH, Yeasmin MS, Rahman MS, Sayeed MA (2015) Synthesis and characterization of high-purity food grade carboxymethyl celluloses from different parts of maize waste. In: Mondal MIH (ed) Cellulose and cellulose derivatives. Nova Science Publisher, New York, p 227
61. Moshood TD, Nawanir G, Mahmud F, Mohamad F, Ahmad MH, Abdul Ghani A (2022) Biodegradable plastic applications towards sustainability: a recent innovations in the green product. Clean Eng Technol 6:100404
62. Moussaoui Y, Ferhi F, Elaloui E, ben Salem R, Belgacem MN (2011) Utilisation of *Astragalus armatus* roots in papermaking. Bio Resources 6:4969–4978
63. Obele CM, Ibenta ME, Chukwuneke JL, Nwanonenyi SC (2021) Carboxymethyl cellulose and cellulose nanocrystals from cassava stem as thickeners in reactive printing of cotton. Cellulose 28:2615–2633
64. Oliveira RN, Moreira APD, Thiré RMDSM, Quilty B, Passos TM, Simon P, Mancini MC, McGuinness GB (2017) Absorbent polyvinyl alcohol–sodium carboxymethyl cellulose hydrogels for propolis delivery in wound healing applications. Polym Eng Sci 57:1224–1233
65. Oztuna Taner O, Ekici L, Akyuz L (2023) CMC-based edible coating composite films from Brewer's spent grain waste: a novel approach for the fresh strawberry package. Polym Bull 80:9033–9058
66. Paria A, Rai VK (2022) The fate of carboxymethyl cellulose as a polymer of pharmaceutical importance. Biol Sci 2:204–215
67. Peighambardoust SJ, Aghamohammadi-Bavil O, Foroutan R, Arsalani N (2020) Removal of malachite green using carboxymethyl cellulose-g-polyacrylamide/montmorillonite nanocomposite hydrogel. Int J Biol Macromol 159:1122–1131

68. Pirsa S, Hafezi K (2023) Hydrocolloids: structure, preparation method, and application in food industry. Food Chem 399:133967
69. Prigioniero A, Scarano P, Ruggieri V, Marziano M, Tartaglia M, Sciarrillo R, Guarino C (2020) Plants named "Lotus" in antiquity: historiography, biogeography, and ethnobotany. Harv Pap Bot 25:59–71
70. Pushpamalar V, Langford SJ, Ahmad M, Lim YY (2006) Optimization of reaction conditions for preparing carboxymethyl cellulose from sago waste. Carbohyd Polym 64:312–318
71. Putro PA, Sumardi T, Sulaeman AS, Roza L, Ramza H, Anugrah DSB (2023) Synthesizing cellulose and its derivatives from pineapple peel: a systematic literature review. Biointerface Res Appl Chem 13:575
72. Rahaman MS, Nahar SS, Islam JM, Samadder R, Rahaman MS, Chadni RK, Saghor FR, Rahman ML, Ruhane TK, Khan MA (2022) Utilization of waste textile cotton by synthesizing sodium carboxymethyl cellulose: an approach to minimize textile solid waste. Adv Polym Technol 2022:4255409
73. Ramzi K, Ikhlass F, Farouk MHM (2010) Study of liquids absorption and retention capacities of new cellulosic materials and sodium cellulose carboxylmethylate prepared from *Posidonia*. Fibers Polym 11:593–597
74. Razzak A, Khiari R, Moussaoui Y, Belgacem MN (2022) Cellulose nanofibers from *Schinus molle*: preparation and characterization. Molecules 27:6738
75. Saad S, Dávila I, Mannai F, Labidi J, Moussaoui Y (2022) Effect of the autohydrolysis treatment on the integral revalorisation of *Ziziphus lotus*. Biomass Convers Biorefinery. https://doi.org/10.1007/s13399-022-02457-6
76. Saad S, Dávila I, Morales A, Labidi J, Moussaoui Y (2022) Cross-linked carboxymethylcellulose adsorbtion membranes from *Ziziphus lotus* for the removal of organic dye pollutants. Materials 15:8760
77. Salama A, Etri S, Mohamed SA, El-Sakhawy M (2018) Carboxymethyl cellulose prepared from mesquite tree: new source for promising nanocomposite materials. Carbohyd Polym 189:138–144
78. Saputra AH, Qadhayna L, Pitaloka AB (2014) Synthesis and characterization of carboxymethyl cellulose (CMC) from water hyacinth using ethanol-isobutyl alcohol mixture as the solvents. Int J Chem Eng Appl 5:36–40
79. Sathasivam T, Muniyandy S, Chuah LH, Janarthanan P (2018) Encapsulation of red palm oil in carboxymethyl sago cellulose beads by emulsification and vibration technology: physicochemical characterization and in vitro digestion. J Food Eng 231:10–21
80. Seddiqi H, Oliaei E, Honarkar H, Jin J, Geonzon LC, Bacabac RG, Klein-Nulend J (2021) Cellulose and its derivatives: towards biomedical applications. Cellulose 28:1893–1931
81. Shui T, Feng S, Chen G, Li A, Yuan Z, Shui H, Kuboki T, Xu C (2017) Synthesis of sodium carboxymethyl cellulose using bleached crude cellulose fractionated from cornstalk. Biomass Bioenerg 105:51–58. https://doi.org/10.1016/j.biombioe.2017.06.016
82. Singh RK, Singh AK (2013) Optimization of reaction conditions for preparing carboxymethyl cellulose from corn cobic agricultural waste. Waste Biomass Valorization 4:129–137
83. Stipcich P, Resaikos V, Ceccherelli G (2023) Experimental thermocline deepening highlights the resilience of the seagrass Posidonia oceanica: an opportunity to investigate shoot adaptability. Mar Pollut Bull 189:114824
84. Tardy BL, Mattos BD, Otoni CG, Beaumont M, Majoinen J, Kämäräinen T, Rojas OJ (2021) Deconstruction and reassembly of renewable polymers and biocolloids into next generation structured materials. Chem Rev 121:14088–14188
85. Tasaso P (2015) Optimization of reaction conditions for synthesis of carboxymethyl cellulose from oil palm fronds. Int J Chem Eng Appl 6:101
86. Teixeira MA, Paiva MC, Amorim MTP, Felgueiras HP (2020) Electrospun nanocomposites containing cellulose and its derivatives modified with specialized biomolecules for an enhanced wound healing. Nanomaterials 10:557
87. Teo SH, Chee CY, Fahmi MZ, Wibawa Sakti SC, Lee HV (2022) Review of functional aspects of nanocellulose-based pickering emulsifier for non-toxic application and its colloid stabilization mechanism. Molecules 27:7170

88. Wongvitvichot W, Pithakratanayothin S, Wongkasemjit S, Chaisuwan T (2021) Fast and practical synthesis of carboxymethyl cellulose from office paper waste by ultrasonic-assisted technique at ambient temperature. Polym Degrad Stab 184:109473
89. Xie L, Shen M, Wang Z, Xie J (2021) Structure, function and food applications of carboxymethylated polysaccharides: a comprehensive review. Trends Food Sci Technol 118:539–557
90. Zainith P, Mishra NK (2021) Experimental investigations on stability and viscosity of carboxymethyl cellulose (CMC)-based non-newtoniannanofluids with different nanoparticles with the combination of distilled water. Int J Thermophys 42:137
91. Zhang N, Chen S, Hu J, Shi J, Guo Y, Deng T (2020) Robust and recyclable sodium carboxymethyl cellulose–ammonium phosphomolybdate composites for cesium removal from wastewater. RSC Adv 10:6139–6145
92. Zhang Y, He Q, Kobayashi K, Kusumi R, Wada M (2023) Hydrogels from dextran/carboxymethyl cellulose exhibiting high post-drying swelling ratios and recovery. Cellulose 30:263–276

Preparation and Characterization of Cellulosic Derivatives

Fedia Bettaieb, Noureddine Baaka, and Mohamed Taha

Abstract Different types of plant waste are available in Tunisia, and their exploitation and possible recovery represent an economic and environmental interest. Among them, *Posidonia oceanica* is a marine plant abundant on the Tunisian coast. The *Posidonia oceanica* wastes are rejected on the beaches in huge quantities every year, and these beaches must therefore be cleaned each summer season. The wastes of Posidonia, consisting of a large percentage of cellulose, represent an important source of cellulose fibers at low cost. This chapter presents the derivative cellulosic products obtained from *Posidonia oceanica* and its applications. These studies serve two important purposes: First, they enable the efficient use of waste materials, and second, they lead to the production of derivatives that are significantly more cost-effective than those currently on the market. This is particularly significant since these derivatives are currently imported from developed nations.

Keywords *Posidonia oceanica* · Valorization · Cellulosic derivatives · Preparation

---

F. Bettaieb · M. Taha (✉)
Université Grenoble Alpes, LGP2, Grenoble, France
e-mail: mohamed_taha@aast.edu

N. Baaka
Higher Institute of Fashion of Monastir, Monastir, Tunisia

M. Taha
Mechanical Engineering Department, College of Engineering and Technology, Arab Academy for Science, Technology and Maritime Transport, Aswan, Egypt

F. Bettaieb
CNRS, LGP2, Grenoble, France

R. Khiari and M. Jawaid (eds.), *Current Status and Opportunity in Fibre and Composites*, Sustainable Materials and Technology,
https://doi.org/10.1007/978-981-96-4546-6_6

1 Introduction

The increasing global consumption of fossil fuels has given rise to significant social and environmental challenges, prompting research efforts to identify more sustainable solutions. The bioeconomy concept aims to utilize renewable resources efficiently to produce chemicals, materials, and energy in a sustainable manner. This has led to a growing interest in the development of various cellulose-based products.

Cellulose is the most abundant natural polymer on Earth, found in the cells of plants, some marine animals, algae, fungi, and bacterial biofilms. The unique properties of cellulose, including its renewable source, biodegradability, biocompatibility, and ease of modification or functionalization, make it a promising material for a wide range of applications. While cellulose is currently utilized in the pulp, paper, textile, and pharmaceutical sectors, its chemical and structural versatility offer potential for expansion into other fields, such as packaging, construction, separation membranes, bioplastics, bioengineering, and optoelectronics.

Modifying polysaccharides through chemical processes is crucial for enhancing natural biopolymers' properties and utilizing this renewable resource sustainably. Virtually any chemical reaction from organic chemistry can be applied to these modifications. Current research focuses on two main areas: improving existing products and synthesis methods, while also developing new derivatives and alternative approaches to synthesis [33].

Carboxymethylation is one of the most studied methods for converting polysaccharides, both for its simplicity and for the versatile properties of its end products. The process follows the Williamson ether synthesis: The polysaccharide is first activated using an alkali hydroxide (typically sodium hydroxide) and then reacted with either monochloroacetic acid or its sodium salt to produce carboxymethyl (CM) polysaccharide derivatives. This process is not limited to cellulose—it can be applied to polysaccharides from various sources: animal-derived chitin and chitosan, bacterial-sourced curdlan and dextran, and plant-based materials like starch, amylase, and amylopectin [33].

Various solvent-soluble cellulose derivatives can be produced, including carboxymethyl cellulose (CMC), methylcellulose, hydroxypropyl cellulose, and hydroxypropyl methylcellulose. Traditionally, these derivatives are manufactured using cellulose from two main sources: wood pulp (which contains 40–50% cellulose) and cotton linters (which contains 90% cellulose). However, using these traditional cellulose sources is becoming less favorable due to high production costs and growing environmental concerns, particularly regarding deforestation and its contribution to greenhouse effects [18, 62].

As fossil resources become increasingly scarce, renewable materials are growing in importance. Cellulose-rich biomass—including bamboo, weeds, fibers, and agricultural waste—offers a promising alternative source of chemical raw materials. These sources are particularly valuable because they contain cellulose, hemicellulose, and lignin, which have various functional groups that can be chemically modified for different applications [66]. CMC (carboxymethyl cellulose) can be synthesized using cellulose extracted from various types of agricultural waste, with wheat straw being one notable example [17], barley straw [17], rice hull [17], sago waste [59], sugarcane straw [23], sugarcane bagasse [42], palm kernel cake [18], corncob agricultural waste [62], and date palm rachis [41]. Many researchers have documented the successful synthesis of CMC from agricultural waste. Additionally, CMC has been synthesized from invasive weeds like Lantana camara and Parthenium hysterophorus [66], marine waste, such as *Posidonia Oceanica* [4], and industrial waste, such as paper industries.

Many cellulosic sources are available in Tunisia in very large quantities. There are more than 58,748 million vines (vitis vinifera L.) which cover more than 32,600 hectares. The agricultural residues coming from the exploitation of these vines are to date left on the spot in the vineyards (the fields of vines) and are not the object of a particular treatment. In addition, the Posidonia (*Posidonia oceanica*) is, as for her, a marine plant abundant in the Tunisian seas. The debris of this plant are rejected by the sea in the form of balls and accumulate in large quantities on the coast each summer season. Given the importance of the tourist activity in the Tunisian economy, the beaches must be cleared of this debris. The recovery of this waste can therefore present an interesting economic activity allowing the production of products and biosourced materials, i.e., produced from plant biomass. During this chapter, this chapter addresses the synthesis and various applications of CMCs in various fields, with particular attention to CMCs derived from *Posidonia oceanica*.

2 Posidonia oceanica

P. oceanica is a long-lived marine Magnoliophyte endemic to the Mediterranean belonging to the Posidoniaceae family. It was described accurately for the first time in 1623 by Bauhin [30] under the name Alga marina. It is only in 1813 that Delile, in the flora of Egypt, gave it its current name of *P. oceanica* (Linnaeus) Delile. The structure of Posidonia shoots has been widely described in numerous works [15, 25, 30, 32, 61]. They have the same cell structure [28] and same morphology as the other marine Magnoliophytes. Like all marine Magnoliophytes, Posidonia oceanica is organized in creeping or erect stems of woody consistency and slightly flattened, generally buried in the sediment, called rhizomes. The flattened leaves have at their base a short sheath corresponding to the petiole of terrestrial plants. This is separated from the rest of the leaf by a small woody tongue called a ligule [25].

When the senescent leaf falls, the ligule (sheath) remains attached to the rhizome and forms the scales [43] which can persist for thousands of years (more than 4000 years) according to [21, 61]. Leaf renewal is continuous in Posidonia oceanica,throughout the year, new leaves appear in the center of the bundle while old leaves fall off at its periphery [54, 60]. Posidonia oceanica represents an excellent substratum for colonization by a sessile fauna and a flora forming the epiphytic community of Posidonia leaves.

The epiphytic flora is composed on the one hand of an epiphytic felting including diatoms and cyanobacteria and on the other hand of macroalgae. In the macroalgae, we find mainly rhodophyceae (red algae), phaeophyceae (brown algae), and chlorophyceae (green algae). Encrusting algae, such as melobesias and corallinaceae, are also very present. The biomass of epiphytes can represent up to 30% of the biomass of the canopy [19, 32, 46, 52].

Because of their abundance, epiphytic algae represent an important compartment in the Posidonia oceanica meadow. Calcareous algae are responsible for a deposition of calcium and magnesium carbonate that can reach more than 100 g of mixed carbonate m-2 year^{-1} [22]. The different contributions to total primary production in seagrass beds are generally 24–38% for leaf production, 33–42% for benthic macroalgae, and 20–60% for epiphytic flora [31, 46, 48].

It should be noted that a significant part of the epiphytes are sessile animals. The epiphytic fauna is dominated by filter feeders, ascidians, hydrozoans, bryozoans, foraminifera, and polychaetes. Several species are attached to Posidonia oceanica, including the bryozoan Electra posidoniae [19, 45, 60]. These suspension-feeding animals play an active role in the sedimentation of particles in seagrass beds [3].

The leaves are ribboned and have a basal growth. Depending on the season, they can measure from a few centimeters to more than a meter long. Their width on the other hand is rather constant throughout the year, about 0.9 cm [32]. The leaves are arranged in a bundle, and the new leaves appear in its center. The oldest leaves are thus outside the said bundle. If the growth is done throughout the year, it is more important in spring [32]. In autumn and winter, the old leaves more or less necrosed are easily torn off by the numerous storms raging in the Mediterranean at these times. The base of the fallen dead leaves remains attached to the rhizome and forms the scales. The average life time of a leaf is between 7 and 12 months, depending on the period of its appearance. The complete replacement of the canopy, at 10 m, takes place on average every 55–90 days. The biomass of the canopy at 10 m over a year is about 600 g ms m^{-2} [15, 32]. The lifetime of a leaf and the time of canopy replacement for Posidonia oceanica can be considered relatively long compared to those corresponding to other marine phanerogams [69]. The stems carrying the leaf bundles are rhizomes.

The rhizomes show either a horizontal growth (plagiotropic) allowing the plant to conquer the uncovered surfaces and thus extend the meadow, or a vertical growth (orthotropic) allowing it to fight against burial by the sediment and the continuous development of the canopy [25, 57]. The plagiotropic rhizomes have adventitious roots that fix the plant to the sediment, with a diameter of more than 2 mm, with a rough appearance and woody texture that can descend up to 70 cm into the sediment [21]. These roots have two roles: a role of anchoring the plant and a role of taking nutrients from the substrate [20]. It is this dimorphism linked to the growth of rhizomes that is the basis of the formation of the matte [25]. The latter is made up of a network of rhizomes, dead and/or alive, roots, and heterogeneous sediment (of autochthonous and/or allochthonous origin) which fills the interstices [21, 57]. The rhizome has roots that attach the plant to the sediment and allow it to take the nutrients necessary for its growth and development [36, 63]. Rhizomes and roots are real sediment traps. These mattes can reach a thickness of 8 m at a rate of 1 m per century, depending on the rate of sedimentation and the erosion-construction cycles, in the Hyères roadstead, a minimum thickness of 10 m has been observed. As a result, in the bays, they can reach the surface or come close to it [21, 57]. It can lead, after a certain time, in the sheltered zones, to the outcrop of the meadow at the surface of the water, forming a barrier reef, which isolates behind it a lagoon occupied by the biocenosis of the "Sables Vaseux en Mode Calme" [57]. The accidental degradation of these mats can mean, locally, the death of the meadow by collapse, erosion, and breakage of the plants [10].

Concerning the geographical distribution of *Posidonia oceanica*, in spite of doubtful records in the last century in Portugal and on the Basque coast [30], the specie is absent from the Black Sea and the Atlantic. The report of P. oceanica in Texas by American authors would be due to a confusion with Thalassia testudinum [68]. Posidonia oceanica is a strict Mediterranean endemic. In favorable biotopes, P. oceanica develops throughout the infralittoral stage, from near the surface to around 20–40 m depth, depending on the transparency of the water, most often on a soft substratum (whatever the granulometry), more rarely on rocks,it forms meadows that are usually called meadows [57].

Considering that the meadow occupies 1–2% of the Mediterranean sea bed, its surface area can be estimated at 35–37,000 km^2 (3.5–3.7 million hectares [55, 56], and it is the main climatic population. The distribution of P. oceanica is continuous all around the Mediterranean basin (Fig. 5), with the exception of the Gibraltar area (Atlantic influence) [57], the Sea of Marmara, the coasts of Israel [27], the mouths of rivers, as well as in the northern Adriatic Sea [32]. On the Syrian-Lebanese coasts, the Posidonia oceanica meadow has only been found in two localities (north-west of the island of Rouad and near Ras Ibn Hani), where it appears to be very threatened [50].

Herbariums are rare on the Languedoc coast, from the Camargue to the Pyrenees (French) and in front of the Nile delta [7], probably due to excessive sedimentary movements and desalination. Indeed, Posidonia oceanica is extremely stenohaline and disappears when the salinity is lower than 33%, hence, its total absence in the brackish ponds of Languedoc, the eastern coast of Corsica and Tunisia. On the other hand, the species supports relatively important thermal variations (from 9 to 29 °C). Numerous studies have addressed the different aspects of its distribution [26, 55], ecology, physiology [6, 38, 39, 45, 46], and more recently, genetics [24, 51] (Fig. 1).

The life cycle of *Posidonia oceanica* is annual (Fig. 2). Reproduction can be asexual or more rarely sexual. In case of unfavorable conditions for germination, the seeds enter into dormancy. Sexual reproduction: Sexual reproduction of *Posidonia oceanica* is done by the production of flowers, fruits, and seeds. However, the flowering of *Posidonia oceanica* is infrequent, occurring in the autumn (September–November). The flowers are hermaphroditic, and 4–10 flowers are grouped in an inflorescence at the top of a floral peduncle from 10 to 30 cm long [54]. Flowering does not occur every year, especially in the relatively cold waters of the north of the western Mediterranean. Some years have been marked by a particularly intense

Fig. 1 Distribution of *Posidonia Oceanica* in Mediterranean Sea; green dots correspond to limits of species' range; white dot to "relict site" in Sea of Marmara

flowering, on the scale of the whole Mediterranean, for example, the years 1971, 1982, 1993, 1997, and 2003 [1, 21, 25, 26, 32, 50, 52, 54, 57].

They can give rise to fruits called "sea olives" which contain a single seed [20, 21] (Fig. 2). It takes 6–9 months for the fruits of *Posidonia oceanica* to mature. Between May and July, they detach and float for some time, then fall to the bottom and, if the nature of the substratum and the physico-chemical factors are favorable, the germination of an embryo released by the dehiscence of the fruit can take place [58].

Fig. 2 Fruits of *Posidonia Oceanica* ("sea olive"): **a** Beginning of fruiting, **b** Ripe fruits, and **c** Fruits at surface of water [44]

Asexual reproduction: Seed maturation seems to be relatively rare and *Posidonia oceanica* reproduction is essentially asexual [25, 49, 56] by vegetative multiplication, consisting of a natural fragmentation of the plant rhizomes, terminated by a living bundle, following storms or marine currents [49]. However, this mode of propagation requires important constraints for the implantation of cuttings and thus remains infrequent. The vegetative propagation of Posidonia oceanica occurs essentially by multiplication and increase of the axes [25]. This reproduction is carried out by stolonization. It is done by the multiplication and the growth of the orthotropic and plagiotropic rhizomes. This process is particularly slow,the orthotropic rhizome grows about one centimeter in one year, and the plagiotropic rhizome increases by 3.5–7.5 cm per year [11]. Another form of vegetative reproduction by pseudo viviparity was recently observed in May 2004 in the Balearic Islands [12]. Vegetative seedlings are formed directly on inflorescences and replace the organs of sexual reproduction. This strategy contributes to a short distance dispersal. It is not known, for the moment, if this is a very local mode of reproduction, or if it concerns other regions of the Mediterranean.

3 Carboxymethylcellulose

Sodium carboxymethylcellulose (CMCNa) is one of the most widely used cellulose ethers available in the market. It is produced through the etherification of the hydroxyl groups present in cellulose, resulting in a long-chain, anionic, linear polymer that is water soluble. There is a substantial body of research on the synthesis of CMC from various cellulose sources, including studies by [2, 13, 16, 29, 40, 65].

CMC has a broad range of applications across multiple industries. It is extensively used in the food, paper, and textile sectors [41, 67] and is also found in cosmetic and pharmaceutical formulations as well as in adhesive products [14]. The production of CMC involves two main steps: First, the cellulose is transformed into an alkaline derivative, and then, monochloroacetic acid (MCA) is introduced into the reaction medium to react with the cellulose, resulting in the formation of CMC.

The degree of substitution (DS) of commercial CMCNa can range from 0.4 to 1.2. When the DS is between 0.1 and 0.4, CMCNa is only soluble in alkaline solutions. Above a DS of 0.4, carboxymethylcellulose becomes soluble in water, which is typical of most commercial water-soluble CMCNa products that usually have DS values between 0.5 and 0.9. These aqueous solutions remain stable even when diluted with 50% alcohol. Generally, CMCNa is insoluble in organic solvents, and due to its anionic nature, it forms insoluble salts with divalent and trivalent metals in solution.

Carboxymethylcellulose can exhibit varying degrees of polymerization (DP), typically ranging from 100 to 600, which influences the viscosity of the resulting solutions. The viscosities of CMCNa solutions in water can vary significantly, ranging from 20 centipoise (cp) to 5000 cp in a 2% solution. These solutions are often thixotropic, a property that is particularly pronounced in high-viscosity CMCNa, meaning their viscosity decreases under shear stress but returns to its original state when the stress is removed.

The formation of carboxymethylcellulose is carried out in several steps: The first corresponds to the formation of cellulose alcoholate in the presence of soda (Eq. 1) and the second corresponds to the formation of the final product (CMCNa) in the presence of MAC (Eq. 2 and 3). The CMC formation reaction is therefore a simple etherification reaction. However, several side reactions (Eq. 4–6) can occur. Unlike reaction 4, reactions 5 and 6 have a very low probability of occurring because the formation of alcoholate from a primary alcohol in the presence of soda is not easy (the departure of the acid proton is difficult).

$$\text{Cell-----OH} + \text{NaOH} \rightarrow \text{Cell-----NaOH} + H_2O \tag{1}$$

$$ClCH_2CO_2H + NaOH \rightarrow ClCH_2CO_2Na + H_2O \tag{2}$$

$$\text{Cell-----ONa} + ClCH_2CO_2Na \rightarrow \text{Cell-----OCH}_2CO_2Na + NaCl \tag{3}$$

$$ClCH_2CO_2Na + ClCH_2CO_2H \rightarrow ClCH_2CO_2CH_2CO_2H + NaCl \tag{4}$$

$$ROH + NaOH + ClCH_2CO_2H \rightarrow RONa + H_2O \tag{5}$$

$$RONa + ClCH_2CO_2Na \rightarrow ROCH_2CO_2Na + NaCl \tag{6}$$

Treating cellulose with soda (sodium hydroxide) is considered an extremely effective method for activating cellulose, as described in Eq. 1. When cellulose is exposed to increasing concentrations of soda, the cellulose polymer begins to swell and gradually loses its crystalline structure, eventually becoming completely amorphous. At a sufficiently high concentration of soda, complete activation of the material occurs, allowing reagents full access to all hydroxyl (OH) groups on the cellulose. This change in the cellulose structure and accessibility can be observed through analytical techniques, such as X-ray Photoelectron Spectroscopy (XPS) and Carbon-13 Nuclear Magnetic Resonance (13C NMR) specifically on alkali-celluloses, as documented by [34, 35].

In contrast, alcohols like butanol and ethanol behave differently due to the high polarity of their OH bonds, which reduces the lability of the functional hydrogen atom. Although the hydrogen in these alcohols can be replaced directly to form an alcoholate, this substitution requires the presence of a very strong base capable of releasing the labile hydrogen. As a result, ionization in solution is almost negligible, with dissociation constants (K_a) on the order of 10^{-16} to 10^{-19}. This low ionization occurs because the action of soda (OH^-) establishes an unfavorable equilibrium for the formation of alcoholates, making their formation in solution highly unlikely.

$$\mathrm{ROH} + \mathrm{Na^+OH^-} \rightleftarrows \mathrm{RONa} + \mathrm{H_2O}$$

Achieving the desired reaction equilibrium in carboxymethylation processes often requires artificially shifting the equilibrium, such as by removing water formed during the reaction. However, when an alcoholate is dissolved in water, it undergoes almost complete hydrolysis [8]. To optimize the carboxymethylation reaction, several key factors must be controlled:

Formation of Monochloroacetic Acid Salt: It is crucial to carefully control the reaction outlined in Eq. 2, which involves the formation of the monochloroacetic acid (MAC) salt.

Regulating Reagent Quantities and Contact Time: It is essential to precisely control the amounts of monochloroacetic acid (MAC) and sodium hydroxide (NaOH), as well as optimize the contact time of MAC with soda, to prevent the side reaction described in Eq. 4, which leads to the formation of an adduct of two MAC molecules.

Promoting CMCNa Synthesis: Simultaneously, conditions must be adjusted to favor the reaction in Eq. 3, which leads to the synthesis of sodium carboxymethylcellulose (CMCNa).

Cellulose derivatives, including carboxymethylcelluloses (CMCs), are typically produced from bleached cellulose pulp derived from wood. However, there is increasing interest in synthesizing these derivatives from alternative plant materials, such as annual plants or agricultural waste. Numerous studies have explored the use of non-wood sources for cellulose derivatives, including the work of [4, 9, 13, 35, 37, 40, 47, 53].

Carboxymethylcelluloses derived from these alternative sources differ primarily in their degree of substitution (DS) and degree of polymerization (DP), which vary depending on the cellulose origin and the synthesis process used. Typically, for all grades of CMC produced, the DS does not exceed 1. Achieving a high DS is challenging mainly due to insufficient understanding of the reaction kinetics, making it difficult to optimize the addition rates and concentrations of each reagent to favor carboxymethylcellulose.

To enhance the yield of CMC with a higher DS, a stepwise modification technique can be employed, as developed by Ahmad et al. [5] and further utilized by Barba et al. [14]. This approach involves repeating the carboxymethylation reaction multiple times, with each successive reaction performed on the product from the previous step. For example, in the study by Aguir and M'Henni [4], this method was used to achieve a CMC with a DS of 2.75 from delignified and bleached seagrass through a synthesis conducted in three stages. This stepwise approach allows for a more controlled and iterative enhancement of the DS in CMC products.

4 Applications of CMCNa from Posidonia: *Retention Agent*

Single-use hygiene products are designed for one-time use and are not intended to be reused after serving their purpose. These items are pre-packaged, are ready for use, and help minimize the risk of contamination for the user and their surroundings. Many hygiene products, such as baby diapers, sanitary pads, and adult incontinence items, are predominantly available in disposable formats. In the healthcare sector, single-use products are particularly critical as they significantly reduce the transmission of germs and bacteria in hospital environments.

A disposable baby diaper, for instance, is constructed with an absorbent core that comprises two main components: cellulose fibers and super-absorbent polymers (SAPs), encased between two layers of synthetic material. The absorbent core is engineered to efficiently absorb and retain fluids, while the non-woven outer layer provides a comfortable feel and helps prevent leakage. The production of these diapers involves a multi-step process where adhesive labels and repositioning strips are attached using mechanical systems, and other components are bonded with glue applied at high temperatures (130–155 °C) or through ultrasonic vibrations. Elastic elements are carefully positioned and secured to the sides of the diaper to ensure a snug fit around the baby's legs.

Diapers primarily consist of two essential materials: super-absorbent polymers (SAPs) and cellulose pulp. The SAPs typically used in diapers are sodium polyacrylate salts, known for their remarkable ability to absorb and retain large volumes of liquid relative to their weight, forming a gel-like substance. These SAPs have been used in various consumer products for decades, particularly in baby diapers, adult incontinence products, and sanitary pads, significantly enhancing their absorbency and fluid retention properties. SAPs are well regarded for their skin compatibility, low potential for irritation, and safety for use near mucous membranes.

The cellulose pulp used in these products is typically a bleached fluff pulp, valued for its absorption capabilities. There is ongoing research into developing new absorbent materials based on cellulosic fibers, which offer promising improvements in absorption and retention characteristics. Chemical modification of these fibers is a common approach to enhancing their absorbent properties. Since the 1960s, various chemical modifications, including cross-linking, substitution, grafting, and combinations of these techniques (such as cross-linking with substitution or grafting with substitution), have been explored to optimize the performance of cellulosic fibers in absorbent products [64].

In this study, different lignocellulosic materials and a series of low degree of substitution (DS) sodium carboxymethylcellulose (CMCNa), derived from seagrass waste, such as Posidonia, were tested as absorbent materials. The absorption capacity of these materials is defined as the ratio of the weight of the absorbed liquid, taken over a 30-min absorption period followed by a 5-min draining period, to the weight of the dry material, and is expressed in grams of liquid per gram of material (g/g). The retention capacity measures the amount of liquid retained after the absorbent capacity test, followed by centrifugation of the sample for 16 min.

Tests were conducted on CMCNa samples derived from seagrass with degrees of substitution ranging from 0.18 to 0.39. The results, depicted in Figs. 3 and 4, illustrate the variations in absorption and retention capacities as a function of the degree of substitution. This evaluation helps in understanding how modifications to the chemical structure of CMCNa influence its performance as an absorbent material, offering insights into the potential of using seagrass-derived CMCNa in disposable hygiene products.

The results indicated that for all types of CMCNa tested, deionized water was the liquid that showed the highest absorption and retention, demonstrating superior performance compared to other tested liquids. In contrast, the absorption and retention of synthetic urine were the lowest, which is a commonly observed outcome in such studies. For each type of liquid tested, both absorption and retention capacities of the materials increased with the degree of substitution (DS) of the CMCNa.

Specifically, the absorption capacity of the various CMCNa samples ranged from 4.69 to 14.37 g/g, depending on the DS. Similarly, the retention capacities varied from 0.97 to 5.47 g/g. These variations highlight the influence of DS on the performance of CMCNa in terms of fluid management.

The evaluation of the prepared cellulosic materials, detailed in Figs. 5 and 6, demonstrated a clear trend: as the DS of CMCNa increased, so did the absorption

Fig. 3 Evolution of absorption capacity function DS

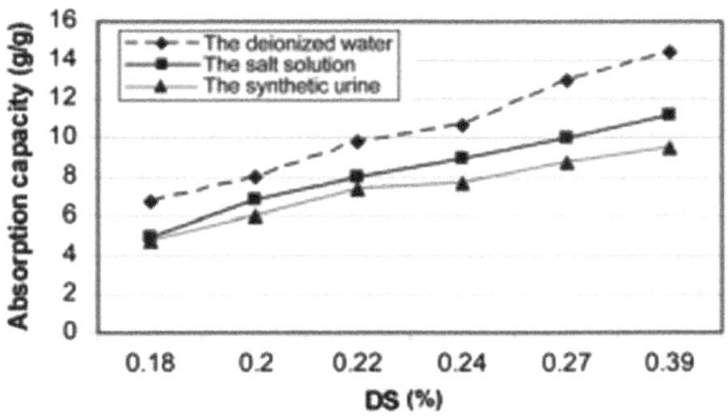

Fig. 4 Evolution of retention capacity function DS

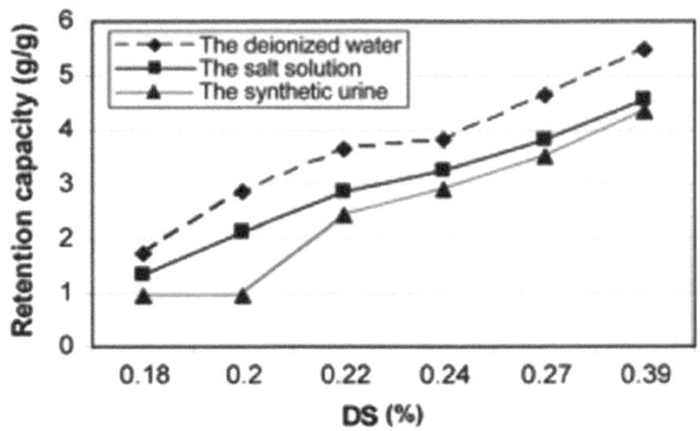

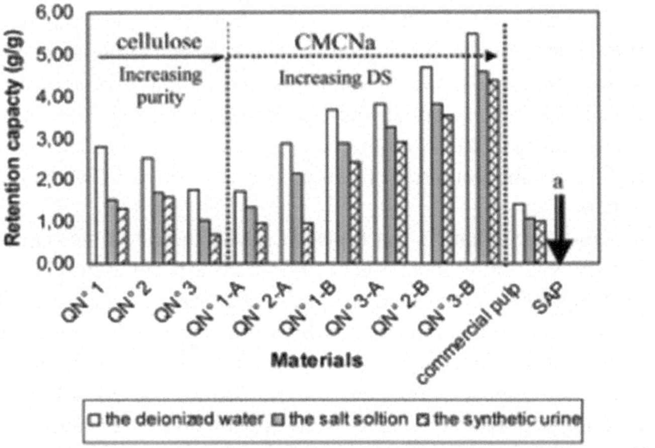

a:liquids tested	Deionized water	Salt water	Synthetic urine
retention Capacity of SAP :	172	32	24

Fig. 5 Evolution of retention capacity

and retention capacities across all liquids tested. This suggests that enhancing the DS is a key factor in improving the efficacy of CMCNa as an absorbent material, making it more suitable for applications where high fluid retention and absorption are critical, such as in disposable hygiene products.

As the cellulose content increased in the three grades of the prepared materials (QN°1, QN°2, and QN°3), a decrease in both absorption and retention capacities was

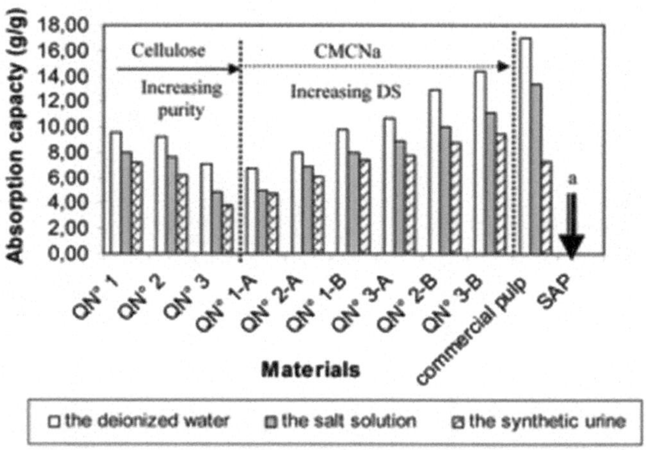

a: liquids tested	Deionized water	Salt water	Synthetic urine
Capacity of absorption SAP :	440	241	155

Fig. 6 Evolution of absorption capacity

observed. Conversely, when the degree of substitution (DS) of the CMCNa increased, there was a notable improvement in both absorption and retention capacities.

In a comparative study with the two standard absorbent materials used in the industry—super-absorbent polymers (SAP) and industrial cellulose pulp—it was found that the CMCNa prepared in this study exhibited significantly better retention capacity than industrial cellulose pulp for all tested liquids. This difference was particularly pronounced with synthetic urine, where CMCNa's retention capacity was up to three times higher than that of industrial cellulose pulp, despite synthetic urine being the liquid with the lowest retention.

However, when it came to absorption capacity, CMCNa's performance was slightly lower than that of commercial cellulose pulp, except in the case of synthetic urine. For synthetic urine, CMCNa outperformed, showing an absorption capacity that was 130% higher than that of commercial cellulose pulp. This substantial advantage highlights the potential of CMCNa, especially in applications where the retention of challenging fluids like synthetic urine is critical.

5 Conclusion

In an effort to explore the potential applications of various qualities of sodium carboxymethyl cellulose (CMCNa) derived from agricultural waste, such as date palm rachis and seagrass (Posidonia), three primary applications were investigated: as absorbent and retention materials for hygiene products, as flocculating agents for reducing water turbidity, and as sizing agents for cotton threads.

The study evaluated the absorption and retention capacities of different liquids (deionized water, saline water, and synthetic urine) by various grades of CMCNa prepared from seagrass. These results were then compared with those of commercial materials commonly used in the hygiene products industry: super-absorbent polymers (SAP) and cellulose pulp (paper pulp). The findings revealed that all grades of CMCNa demonstrated significantly higher retention capacities compared to cellulose pulp, with some grades showing up to four times the retention capacity. In terms of absorption, CMCNa generally performed better for synthetic urine, while the reference cellulose pulp absorbed deionized water and saline solution more effectively.

Additionally, certain CMCNa samples (with DS ranging from 0.98 to 1.86) were tested as sizing agents and flocculating agents. The performance of these samples was compared with commercial products. CMCNa derived from date palm waste was used as a sizing agent to improve the mechanical properties of textile yarns, such as hairiness, elongation, and breaking strength. The desizing process was also evaluated. The results indicated that the sizing performance of CMCNa from date palm waste was comparable to that of commercial CMCNa.

The same samples were also assessed for their effectiveness as flocculating agents. Optimal coagulation-flocculation conditions—including dosage, stirring speed, and pH—were determined. The best results were achieved at a pH of 8, a flocculant concentration of 100 mg/L, and a flocculation speed of 30 rpm using CMCNa QR2, which was obtained from raw date palm rachis. These results were comparable to those of a commercial anionic flocculant (A100 PWC: polyacrylamide), which is widely used in water treatment. Notably, in some cases, the performance of CMCNa with a DS of 1.17 was 10% higher than that of the commercial product tested.

Overall, these findings demonstrate the versatility and effectiveness of CMCNa from agricultural waste, highlighting its potential as a sustainable alternative in various industrial applications.

References

1. Acunto S et al (2000) Characteristics of the mediterranean seagrass Posidonia oceanica contributing to high-frequency acoustic scattering
2. Adinugraha MP, Marseno DW (2005) Synthesis and characterization of sodium carboxymethylcellulose from cavendish banana pseudo stem (Musa cavendishii LAMBERT). Carbohyd Polym 62(2):164–169
3. Agawin NS, Duarte CM (2002) Evidence of direct particle trapping by a tropical seagrass meadow. Estuaries 25:1205–1209
4. Aguir C, M'Henni MF (2006) Experimental study on carboxymethylation of cellulose extracted from Posidonia oceanica. J Appl Polym Sci 99(4):1808–1816
5. Ahmad A et al (2021) Screening of soil fungi for cellulase production using carboxymethyl cellulose media. Path of Science 7(12):1006–1011
6. Alcoverro T et al (1997) Spatial and temporal variations in nutrient limitation of seagrass Posidonia oceanica growth in the NW Mediterranean. Mar Ecol Prog Ser 146:155–161
7. Aleem A (1984) Distribution and ecology of seagrass communities in the Western Indian Ocean. Deep-Sea Res 31:919–933
8. Arnaud-Neu F et al (1997) Complexation of Pr 3+, Eu 3+, Yb 3+ and Th 4+ ions by calixarene carboxylates. J Chem Soc, Dalton Trans 3:329–334
9. Baar A et al (1994) Nuclear magnetic resonance spectroscopic characterization of carboxymethylcellulose. Macromol Chem Phys 195(5):1483–1492
10. Badalamenti F et al (2011) The impact of dredge-fill on Posidonia oceanica seagrass meadows: regression and patterns of recovery. Mar Pollut Bull 62(3):483–489
11. Balestri E et al (1998) Survival and growth of transplanted and natural seedlings of Posidonia oceanica (L.) Delile in a damaged coastal area. J Exp Mar Biol Ecol 228(2):209–225
12. Ballesteros E et al (2005) Pseudovivipary, a new form of asexual reproduction in the seagrass Posidonia oceanica. Walter de Gruyter
13. Barai BK et al (1997) Optimization of a process for preparing car□ ymethyl cellulose from water hyacinth (Eichornia crassipes). Carbohyd Polym 32(3–4):229–231
14. Barba C et al (2002) Synthesis and characterization of carboxymethylcelluloses from non-wood pulps II. Rheological behavior of CMC in aqueous solution. Cellulose 9:327–335
15. Bay D (1984) A field study of the growth dynamics and productivity of Posidonia oceanica (L.) Delile in Calvi Bay, Corsica. Aquatic Botany 20(1–2):43–64
16. Bettaieb F et al (2015) Mechanical and thermal properties of Posidonia oceanica cellulose nanocrystal reinforced polymer. Carbohyd Polym 123:99–104
17. Biswas A et al (2014) Conversion of agricultural residues to carboxymethylcellulose and carboxymethylcellulose acetate. Ind Crops Prod 60:259–265
18. Bono A et al (2009) Synthesis and characterization of carboxymethyl cellulose from palm kernel cake. Adv Nat Appl Sci 3(1):5–12
19. Borowitzka MA et al (2006) Epiphytes of seagrasses. In: Seagrasses: biology, ecology and conservation, pp 441–461

20. Boudouresque C et al (2006) The outstanding traits of the functioning of the Posidonia oceanica seagrass ecosystem. Biol Mar Medit 13(4):109–113
21. Boudouresque CF et al (1983) Jeudy De grissac. Lab. Ecol. du Benthos et de Biol
22. Canals M, Ballesteros E (1997) Production of carbonate particles by phytobenthic communities on the Mallorca-Menorca shelf, northwestern Mediterranean Sea. Deep Sea Res Part II 44(3–4):611–629
23. Candido R, Gonçalves A (2016) Synthesis of cellulose acetate and carboxymethylcellulose from sugarcane straw. Carbohyd Polym 152:679–686
24. Capiomont A et al (1996) Enzyme polymorphism in Posidonia oceanica, a seagrass endemic to the Mediterranean. Aquat Bot 54(4):265–277
25. Caye G, Meinesz A (1985) Observations on the vegetative development, flowering and seeding of Cymodocea nodosa (Ucria) Ascherson on the Mediterranean coasts of France. Aquat Bot 22(3–4):277–289
26. Ceccherelli G et al (2000) Response of the non-indigenous Caulerpa racemosa (Forsskål) J. Agardh to the native seagrass Posidonia oceanica (L.) Delile: effect of density of shoots and orientation of edges of meadows. J Exp Mar Biol Ecol 243(2):227–240
27. Çelebi B (2007) A study on Posidonia oceanica (L.) delile, 1813 seagrass meadows in the levant sea, Middle East Technical University
28. Colombo PM et al (1983) Posidonia oceanica (L.) Delile: a structural study of the photosynthetic apparatus. Marine Ecol 4(2):133–145
29. Coulter A et al (2015) Potassium bitartrate crystallisation in wine and its inhibition. Aust J Grape Wine Res 21:627–641
30. Den Hartog C (1970) The sea-grasses of the world. North-Holland, Amsterdam
31. Duarte CM, Chiscano CL (1999) Seagrass biomass and production: a reassessment. Aquat Bot 65(1–4):159–174
32. Gobert JD (2005) The effects of different learning tasks on model-building in plate tectonics: diagramming versus explaining. J Geosci Educ 53(4):444–455
33. Heinze T, Koschella A (2005) Solvents applied in the field of cellulose chemistry: a mini review. Polímeros 15:84–90
34. Heinze T, Liebert T (2001) Unconventional methods in cellulose functionalization. Prog Polym Sci 26(9):1689–1762
35. Heinze T, Pfeiffer K (1999) Studies on the synthesis and characterization of carboxymethylcellulose. Die Angewandte Makromolekulare Chemie 266(1):37–45
36. Hemminga M (1998) The root/rhizome system of seagrasses: an asset and a burden. J Sea Res 39(3–4):183–196
37. Hidayat A et al (2000) Cyanogenic content of cassava root of 179 cultivars grown in Indonesia. J Food Compos Anal 13(1):71–82
38. Invers O et al (2004) Effects of nitrogen addition on nitrogen metabolism and carbon reserves in the temperate seagrass Posidonia oceanica. J Exp Mar Biol Ecol 303(1):97–114
39. Invers O et al (2002) Seasonal nitrogen speciation in temperate seagrass Posidonia oceanica (L.) Delile. J Exp Mar Biol Ecol 273(2):219–240
40. Käuper P et al (1998) Development and evaluation of methods for determining the pattern of functionalization in sodium carboxymethylcelluloses. Die Angewandte Makromolekulare Chemie 260(1):53–63
41. Khiari R et al (2011) Valorisation of vegetal wastes as a source of cellulose and cellulose derivatives. J Polym Environ 19:80–89
42. Koh MH (2013) Preparation and characterization of carboxymethyl cellulose from sugarcane bagasse, UTAR
43. Kuo J (1978) Morphology, anatomy and histochemistry of the australian seagrasses of the genus Posidonia könig (posidoniaceae). I. Leaf blade and leaf sheath of Posidonia australis hook F. Aquat Bot 5:171–190
44. Lafabrie C et al (2007) Trace metals assessment in water, sediment, mussel and seagrass species–validation of the use of Posidonia oceanica as a metal biomonitor. Chemosphere 68(11):2033–2039

45. Larkum AW et al (2006) Seagrass beds and coastal biogeochemistry. In: Seagrasses: biology, ecologyand conservation, pp 135–157
46. Lepoint G et al (1999) Fauna vs flora contribution to the leaf epiphytes biomass in a Posidonia oceanica seagrass bed (Revellata Bay, Corsica). Hydrobiologia 394:63–67
47. Makhanya FM (2020) Conversion of sugarcane bagasse into carboxymethl cellulose
48. Mateo M et al (2006) Carbon flux in seagrass ecosystems. In: Seagrasses: biology, ecology and conservation, pp 159–192
49. Meinesz A et al (1993) Polymorphism and development of Posidonia oceanica transplanted from different parts of the Mediterranean into the National Park of Port-Cros
50. Meinesz A, Hesse B (1991) Introduction et invasion de l'algue tropicale Caulerpa taxifolia en Méditerranée nord-occidentale. Oceanol Acta 14(4):415–426
51. Micheli C et al (2005) Genetic variability of Posidonia oceanica (L.) Delile in relation to local factors and biogeographic patterns. Aquatic Botany 82(3):210–221
52. Modigh M et al (1998) Carbon assimilation in Posidonia oceanica: biotic determinants
53. Olaru N et al (1998) Carboxymethylcellulose synthesis in organic media containing ethanol and/or acetone. J Appl Polym Sci 67(3):481–486
54. Panayotidis P, Giraud G (1981) Sur un cycle de renouvellement des feuilles de Posidonia oceanica (Linnaeus) Delile dans le golfe de Marseille. Vie et Milieu/Life Environ 31:129–136
55. Pasqualini V et al (1998) Mapping ofposidonia oceanicausing aerial photographs and side scan sonar: application off the Island of Corsica (France). Estuar Coast Shelf Sci 47(3):359–367
56. Pergent-Martini C et al (1995) Nutrient impact on Posidonia oceanica seagrass meadows: preliminary data. Marine life (Marseille) 5(2):3–9
57. Pergent G, Pergent-Martini C (1995) Dynamique et évolution de l'herbier à Posidonia oceanica en Méditerranée. Plant Biosyst 129(1):303–317
58. Piazzi L et al (1999) In situ survival and development of Posidonia oceanica (L.) Delile seedlings. Aquatic Botany 63(2):103–112
59. Pushpamalar V et al (2006) Optimization of reaction conditions for preparing carboxymethyl cellulose from sago waste. Carbohyd Polym 64(2):312–318
60. Romero J et al (1994) The belowground organs of the Mediterranean seagrass Posidonia oceanica as a biogeochemical sink. Aquat Bot 47(1):13–19
61. Semroud R et al (1992) On a broad-leaved form of the seagrass Posidonia oceanica (Posidoniaceae) from Algiers (Algeria). Aquat Bot 43(2):181–198
62. Singh RK, Singh AK (2013) Optimization of reaction conditions for preparing carboxymethyl cellulose from corn cobic agricultural waste. Waste Biomass Valorization 4:129–137
63. Smart RM, Barko JW (1985) Laboratory culture of submersed freshwater macrophytes on natural sediments. Aquat Bot 21(3):251–263
64. Specklin P (1994) □ e super absorbents. Industrie Textile 1257(9):48–52
65. Toğrul H, Arslan N (2003) Production of carboxymethyl cellulose from sugar beet pulp cellulose and rheological behaviour of carboxymethyl cellulose. Carbohyd Polym 54(1):73–82
66. Varshney V, Naithani S (2011) Chemical functionalization of cellulose derived from nonconventional sources. Cellul Fibers Bio Nano-Polym Compos Green Chem Technol 43–60
67. Wakelyn PJ et al (2006) Cotton fiber chemistry and technology. CRC Press
68. Walker D, McComb A (1988) Seasonal variation in the production, biomass and nutrient status of Amphibolis Antarctica (Labill.) Sonder ex Aschers. and Posidonia australis Hook. f. in Shark Bay, Western Australia. Aquatic Botany 31(3–4):259–275
69. Waycott M (2000) Mating systems and population genetics of marine angiosperms (seagrasses). In: Monocots: systematics and evolution, vol 1. CSIRO Publishing, Sydney, pp 277–285

The Use of Nanocellulose from Three Potential Mediterranean Lignocellulosic Residues—A Review for Novel Applications

Jihene Belhaj, Araceli García, and Ramzi Khiari

Abstract Lignocellulosic residues represent a potential source of renewable and economic cellulosic fibers for the production of high value-added biomaterials. In this chapter, we present various news related to agricultural and marine wastes. Special attention is given to the plant *Posidonia Oceanica*, the almond tree (*Prunus amygdalus L.*) and the fig stem (*Ficus carica L.*). First, a detailed description of these materials is discussed in terms of their geographical distribution, botanical description and morphology, as well as the scope of their application. The description of the chemical composition of the materials has also been taken into account. Finally, an illustration of nanocelluloses and their terminological composition has been studied using the selected materials as an example.

Keywords Posidonia Oceanica · Almond tree (*Prunus amygdalus L.*) · Fig stem (*Ficus carica L.*) · Fiber · Cellulose · Nanoscale · Nanocellulose

1 Introduction

The worldwide production of textiles is mainly based on the use of natural fibers that can be produced from the shells, stems, leaves and fruits of annual plants (cotton, jute, etc.), from livestock activities with a wide range of animals (sheep, goat, etc.) or

J. Belhaj
National Engineering School of Monastir, University of Monastir, Monastir, Tunisia

J. Belhaj · A. García (✉)
FQM-383 NANOVAL Group, Organic Chemistry Department, University of Córdoba, Córdoba, Spain
e-mail: qo2ganua@uco.es

R. Khiari
Department of Textile, Higher Institute of Technological Studies of Ksar Hellal, 5070 Ksar Hellal, Tunisia

CNRS, Grenoble INP, LGP2, Université Grenoble Alpes, F-3800 Grenoble, France

R. Khiari and M. Jawaid (eds.), *Current Status and Opportunity in Fibre and Composites*, Sustainable Materials and Technology,
https://doi.org/10.1007/978-981-96-4546-6_7

from marine resources (Posidonia, seaweed, etc.). Lignocellulose is considered the most useful and important source of textile fibers, but in the last century, it was being systematically replaced by synthetic fibers, high-yield and low-cost produced from petrochemical industries. These oil-derived fibers have excellent mechanical and chemical properties, whereas natural fibers have a wide dispersion and heterogeneity of properties and a very high associated implementation cost.

The rise of petrochemicals prices and the increasing social interest toward the reduction of our environmental impact in recent years have plunged the market of the textile industry. Since our society is becoming more and more of a pollution generator, directly related to an exponential growth of industrial activity, new strategies and novel vision of the current production system must be reconsidered.

In the twenty-first century, researchers and industrialists have moved toward the valorization of natural biomass, as well as become aware of the future of renewable energies, sustainable and biodegradable fibrous materials.

Nature, in its abundance, offers plenty of materials that can be considered as textile. Several parts of plants, such as stems, leaves, straws, grasses and seeds are lignocellulosic resources, available in large quantities and inexpensively, and they constitute alternatives to be applied in several fields not only in textiles, but also in buildings, as a source of biopolymers, source of food/feed as well as chemicals, natural cosmetics and energy [14, 73]. Moreover, lignocellulosic fibers are considered non-toxic, biodegradability, low density and high recyclability and of good mechanical performance [91]. For all this, researchers have investigated the use of this noble material in several areas of textile application.

In order to reduce the toxicity of wastewater discharges, manufacturers have equipped themselves with physicochemical treatment and detoxification stations for their effluents, mainly by a physical or/and chemical modification of polluting species. In this sense, a higher demand of more ecological alternatives has been related to the exploitation of natural lignocellulosic residues in applications of water discoloration and raw or modified lignocellulosic fibers have been used as adsorbents, allowing the retention of toxic products. [21] treated hazelnut and almond shells in aqueous solutions for the reduction and elimination of Cd (II), Ni (II) and Pb (II). Also in waste treatment plants, agricultural waste has been used for the reduction of basic and acid dyes [61], pesticides [83], as well as the static adsorption examination of anthracene [42]. In the context of the elimination and reduction of contaminants from wastewater and residual waters, several efforts have been made for the preparation of activated carbons from agricultural residues such as hulls and bagasse [118], succeeding in preparing activated carbon from almond shells for the adsorption of cationic dyes [8].

Specific properties of natural fibers, such as hygroscopicity, no rejection of substances harmful to health, non-allergic effect, low cost, biosourced origin, biodegradable and safer behavior in the event compared to synthetic or artificial fibers [71, 89, 91] became very interesting in the field of composites. As a result, several studies have proved that the incorporation of fibers of natural origin (flax, hemp, sisal, among others) in thermoplastic or thermosetting resins can lead in products that are competitive with those containing fiberglass, Kevlar or carbon fibers, for various applications and especially in the automotive industry.

However, cellulosic hydrophilic reinforcements can have drawbacks, including the lack of poor adhesion or compatibility with most polymers, especially polyolefins that are hydrophobic matrices. The surface modification of the fibers results, then mandatory to optimize the adhesion between them and the polymer. There are currently several surface treatment methods such as mercerization, isocyanate grafting, acetylation and acylation. In this case, the surface treatment makes it possible to improve the resistance of the fibers and their roughness in addition to fiber/matrix adhesion [52]. Besides the use of agricultural residues, also marine waste is being widely investigated. Thus, Posidonia oceanica was used [69] to produce new lignocellulosic particles in order to reinforce a commercial biodegradable thermoplastic matrix, and more recently, in 2021, a study of the biodegradability of composites reinforced with lignocellulosic fibers was carried out [31] to clearly show the speed of degradation of lignocellulosic fibers compared to synthetic polymers.

Also, the paper industry has suffered a major economic crisis due to different factors influencing that strike the base of this industry. Among these factors, the increase in the price of raw materials and energy result key in this subject, and thus the impressive growth of the Asian paper industry with its competitive low production costs and numerous exports. So, it becomes necessary for pulp and paper manufacturers to advocate for recycling by developing new alternative scenarios for their pulp production, their co-products and forest resources. Researchers are finding new ways of developing, through chemical modifications of paper pulp fibers, which are essentially made up of cellulose, to develop new methods of recovery. In an effort to develop new paper products, in 2014 a one-step cellulose fiber crosslinking approach was developed for ultra-strength packaging papers made from Kraft paper pulp [45]. Also, [29] made a paperboard using an enzymatically treated sample from the agricultural waste of tomato plants.

The nonwovens industry is organized differently from the textile industry, aiming above all at the rapid and low-cost production of large volumes. According to the International Nonwovens & Disposables Association (INDA), in 2011, the global industry produced approximately 7.6 million tons of nonwovens products, with a growth rate of nearly 8% per year. The growth rate started in the 1970s as soon as they entered the filtration field. Today, the global filtration industry is growing by 4 to 8% every 6 years, with development that should further accelerate.

Since the trend is the exploitation and use of natural waste for textile applications, researchers [101] have been manufacturing 100% kenaf paper for special applications such as cigarette filters. Also, [47] used fiber felts to make air filters for particles removal. Essentially, these fibers can be modified to remove specifically certain types of contaminants. In the same context of filtration, scientists and researchers from the Department of Science and Technology have developed in 2019 a new non-woven fabric from natural abaca fibers and synthetic polymers capable of filtering heavy metals dissolved in a liquid.

Finally, in order to replace typical expensive synthetic polymers and petroleum derivatives, researchers are considering renewable resources derived from plant biomass as a potential economic candidate. The use of renewable biomass as a raw material to prepare cellulose nanofibers (CNF) has become a very important approach to expand the applications of carbonaceous materials [127]. In the concept of green chemistry, which is a concept that is gaining recognition in several industries, [70] have established the development of new derivatives based on renewable natural resources. Their work focused on the preparation and characterization of cellulose carbonate using dimethyl carbonate as a green reagent in the ethanoic solution of KOH. Electrospinning technology too is an easy and cost-effective approach by electrostatic forces to produce long continuous nanofibers [77].

In the present work, with the aim of finding new applications for agricultural and marine wastes in the textile field, a study of the morphology and characteristics was elaborated for three natural biomasses: the almond tree (*Prunus amygdalus* L.), the fig tree (*Ficus carica* L.) and posidonia (*Posidonia oceanica*).

In Tunisia, the cultivation of the almond tree, is well distributed throughout the territory. The total area of almond trees in 2021 increased by 213.471 ha for local and new varieties of almond (FAOSTAT 2023). Central Tunisia, covering 87.000 ha in Sfax, occupies 80% of the total area. Fig cultivation plays an important role in agricultural development in Tunisia, and fig production in some governorates is a major factor and an important source of income for citizens. In 2021 the average total production is estimated at 20.000 tons for all Tunisia (FAOSTAT 2023). The disponibility of Posidonia oceanica is localized around the Mediterranean basin; in 2017, the total areas covered by seagrass beds have been estimated at between 25.000 and 50.000 km^2 ("Posidonia, the lung of the Mediterranean," [98] and in Tunisia the Posidonia covering in 2016 a total area of 156.8 ha [50].

This agricultural and marine waste has actually any industrial application, being normally incinerated or dumped. The burning of agricultural or marine residues causes environmental problems such as air and water pollution and the degradation of fauna and flora, the reduction of biological activity and soil erosion. In textile industry, global fiber production reached 107 million MT in 2018, and it is projected to increase significantly to 145 million MT by 2030 (Subash and Muthiah [114]. In addition, the valorization of this residues represents a very interesting economic activity making it possible to produce new bioresourced products. The research that we have started in this thesis work is above all a study aimed at the reuse of cellulosic waste to obtain nanomaterials with high added value.

2 Exploring Novel Cellulose Sources from Mediterranean Area

In the present work, three Mediterranean plants (Fig. 1) are featured as potential source of waste-derived cellulose: almond tree (*Prunus amygdalus* L.), fig tree (*Ficus carica* L.) and posidonia (*Posidonia oceanica*) also named Neptune grass.

2.1 History of the Species

Almond tree has its origin in the western regions of Asia and China and was probably first cultivated there. Also, it is mentioned in the Bible and even in Egypt has been found in the tomb of Tutankhamun (around 1325 years before our era). Experts believe that almond trees were one of the first domesticated trees as cultivated almond trees appeared in the early Bronze Age (around 3000–2000 years before our era) in Numeira, Jordan. Two types of almonds are different: sweet (edible) almonds and bitter almonds, which are naturally toxic since they contain high amounts of cyanide. While both almond cultivars exist, historians believe that since the sweet almond mutation is relatively common among bitter almonds, early farmers realized that these trees provided safe and delicious treats. The ancients who ate wild almonds most likely roasted and washed them to remove any lingering toxicity. Since then, people have enjoyed roasted sweet almonds, but they are delicious and can be safely eaten raw.

The fig, a very old fruit, is known all over the world and its history begins since antiquity, it was cultivated from 5000 years BC in the Middle East. It is the fruit of the common fig tree known as *Ficus carica.* Its French name comes from the Occitan "figa", a dialect from the south of France [62]. The fig tree is an easy-to-grow fruit tree and can reach up to 5 m in height. Originally from Syria and Afghanistan, it can, however, acclimatize in most regions, provided that the temperatures are not lower than -15 °C in winter. 32 to 37 °C are favorable temperatures for fruit development

Fig. 1 Three potential Mediterranean sources of cellulose: almond tree (left) fig tree (middle) and Neptune grass (right). *Source* https://unsplash.com/

and ripening. When the temperature reaches 43 °C, the fruit hardens. The fig tree adapts to a wide range of soils, from heavy clay soils to sandy soils, but prefers loamy-clay soils. There are two varieties of fig tree: The yellow drop of gold fig tree gives garnet red figs in June–July, then from August to October. Due to its size, it is particularly interesting for cultivation in small spaces (2.5–4 m). Uniferous variety. It also finds the black carob fig tree and produces purple-brown figs in July and then in September–October. During the second harvest, its very fragrant fruits can be dried. Biferous variety.

Regarding posidonia first finding, Carl von Linné (1707–1778), a Swedish naturalist known as the father of modern taxonomy and ecology that named this underwater plant Posidonia in honor of Poseidon, since he was the god of turbulent oceans and seas. This plant was also known in Greek mythology as the "Earth Shaker" God of springs and earthquakes. Posidonia is an aquatic plant and although it lives underwater it is not an alga. Five species of Posidonia are initially recorded: oceanica, australis, angustifolia, sinuosa and ostenfeldii. Then, four other species were mentioned from what Kuo and Cambridge 1984 called the "Posidonia ostenfeldii Complex": denhartogii, kirkmanii, coriacea, Robersonelle. All the species mentioned are probably descended from three fossil species which flourished about 65 million years ago in the Tethys.

2.2 Geographical Distribution

In Tunisia, the cultivation of the almond tree is well distributed throughout all the territory (Fig. 2). The total area of almond trees increased by 213.471 ha intended for local and new varieties of almond tree (FAOSTAT 2023). The center of Tunisia, which extends over 87.000 ha in Sfax, occupies 80% of the total area, being the main almond-growing region and containing more than half of the plantations. In this area, almond trees are interspersed with olive trees or grown as the main fruit. The North of Tunisia, precisely the North-East region of Bizerte (RafRaf, Ghar ElMelh, Metline, Ras-Jebel), is distinguished by small orchards located near the sea (a hundred meters away) where the almond tree and the potato have for several centuries been a joyous association, typical of this region. The other almond plantations are located in the provinces of Mahdia, Tunisia, Sousse, Kairouan, Kasserine, Monastir, Sidi Bouzid, Medenine and Bizerte as shown in Fig. 2.

The cultivation of the fig tree is part of the historical and cultural heritage of the country, and it is very old in Tunisia. From one extreme to another of the Tunisian territory, there are fig trees with an important varietal richness. This species is present in various bioclimatic zones of the country. Surveys carried out in different regions have made it possible to describe and list several dozen cultivars [78–79]. In Tunisia, the average total production of figs is estimated at 20.000 tons (FAOSTAT 2023), mainly in the governorates of Beja Djebba (3630 tons), Siliana kesra (3500 tons), Gafsa (2600 tons) and Gabes (2800 tons) [24]. The fig trees cultivation plays a key role in agricultural development of Beja and Siliana. The production of figs occupies

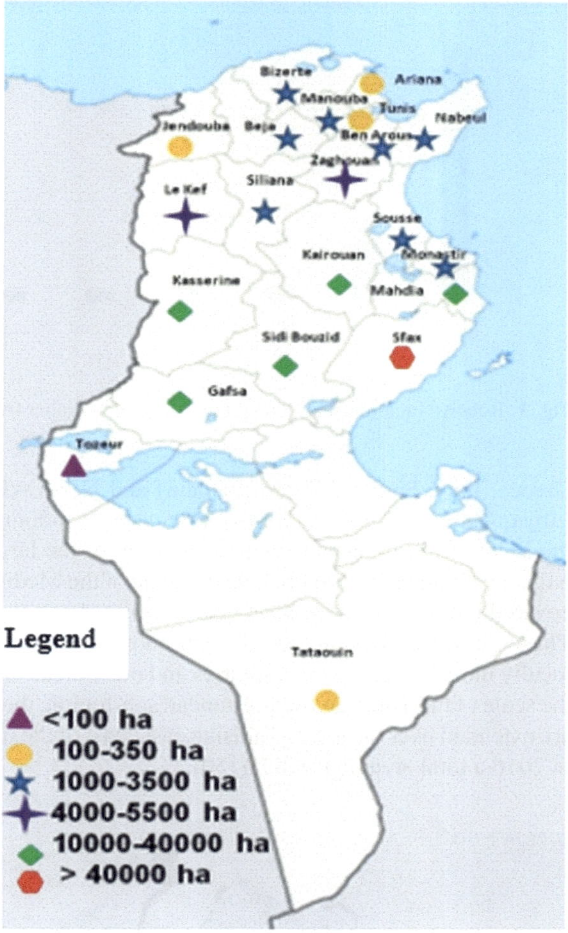

Fig. 2 Geographical distribution and areas of the almond tree in Tunisia

a prominent place and represents an important source of income for the citizens. In the region of Djebba, the fig tree represents more than 62% of the total fruit trees even exceeding the olive tree (Fig. 3), and the total area it occupies, which is about 140 ha. The dispersion of plantations is very accentuated in the region. The fig tree is almost always associated with other fruit or vegetable crops. A few small plots of fig trees are found in Djebba downstream.

Today Posidonia oceanica is only found on the Mediterranean coast, at a depth of up to 40 m depending on the clarity of the water. The other 8 species of Posidonia are only found on certain Australian coasts. Unfortunately, along the French coast the meadows formed by this plant have been reduced, the most important is indicated on the map as shown in Fig. 4. This map represents the distribution of Posidonia oceanica, which is found along European coasts (France, Corsica, Yugoslavia,

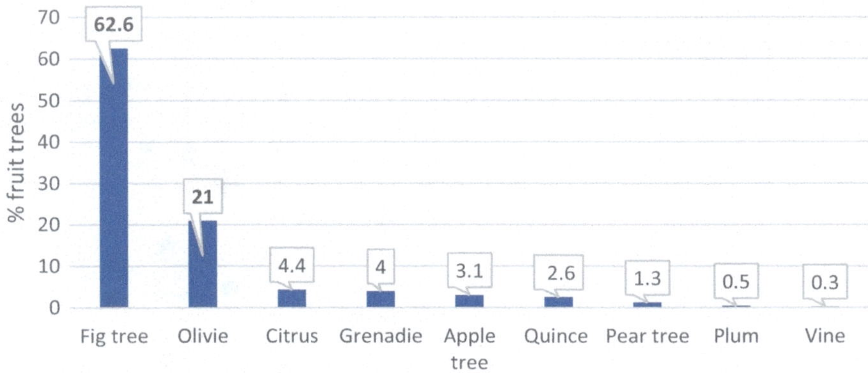

Fig. 3 Relative importance of the fig tree compared to other fruit species in Djebba

Greece, Italy, Sardinia, Turkey, Spain) and North African coasts (Egypt, Tunisia, Libya, Algeria, Morocco). More particularly, Posidonia oceanica is localized around the Mediterranean basin except at the level of the large estuaries of the Rhône, the Nile, the Po and the Ebro [96]. At the scale of the Mediterranean basin, the total areas covered by seagrass beds have been estimated at between 30.000 and 40.000 km^2. These meadows form an almost continuous belt all around the Mediterranean Sea, locally interrupted at ports, estuaries and other areas of high anthropization [109]. At the scale of the Tunisian Mediterranean sub-region, the Posidonia oceanica meadows are extended over the entire Tunisian east coast of the order of 1300 km [18] covering in 2016 a total area of 156.8 ha [50].

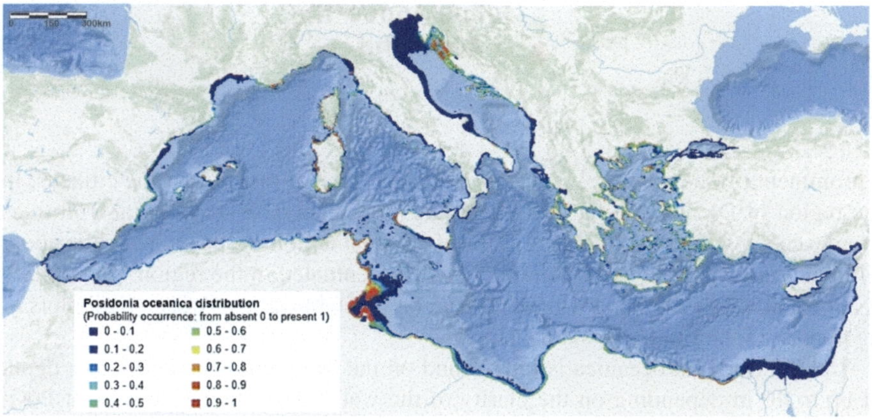

Fig. 4 Geographic distribution of posidonia along the Mediterranean coasts. *Source* https://emodnet.ec.europa.eu/en/map-week-%E2%80%93-posidonia-oceanica-distribution, visited 11 July 2023

2.3 Botanical Description and Morphology

The almond trees are deciduous with a rustic dormancy and carry pale pink to white flowers at the beginning of spring before being leafy. In cooler climates, the growing fruit (a drupe) resembles a peach until it approaches maturity, as it ripens, the leathery outer covering, or hull, splits open, curls outward and discharges the pit. Despite their common label, almonds are not true nuts (a type of dry fruit) but rather seeds enclosed in a hard fruit covering. A botanical description of almond tree is displayed in Table 1.

Secondly, the fig tree (*Ficus carica* L.) is a dicot of the Moraceae family. From a systematic point of view, the botanical classification of the fig tree, as described by [44], is shown in Table 2. The fig tree has a great capacity to adapt to different climates, which has made it possible to generate several types of varieties with different aspects and morphologies. For example, it can be found in its shrub form, as it can also be found in the form of a tree that could reach 12–15 m in height in humid regions. In our conditions, the fig tree is a deciduous tree which falls in winter, a period of vegetative rest. It has twigs covered with fine non-rough bark, which turns with age from green to white or pale gray [48, 94], the tree contains a milky, gummy and sour juice called the latex which contains proteolytic enzymes such as Ficin and which cause irritation on contact with the skin. The root system of the plant is generally shallow and spreading [13].

Finally, an aquatic plant is considered in this work. *Posidonia oceanica* is the botanical name for Posidonia. It is a perennial aquatic plant, which can be 30 to 80

Table 1 Botanical description of the different components of the almond tree

Component	Taxonomy	Description
Tree	• Family: Rosaceae • Typical habitat: Mediterranean shrubs • Area of distribution: Mediterranean • Exposure: Sunny • Ground: Ordinary	The almond tree, a tree whose cultivation is economically important, is mainly established in areas with a Mediterranean climate between 28° and 48° north latitude and 20° and 40° south latitude. They reach a height of 6–9 m
Flower	• Type of inflorescence: corymb • Sex distribution: hermaphrodite • Type of pollination: entomogamous • Mode of dissemination: endozoochore • Flowering: spring	When flowering, fragrant, pale pink to white, five-petaled flowers appear from late January to early April in the northern hemisphere. The flowers are self-incompatible, implying that pollinating insects facilitate cross-pollination with other cultivars
Fruit	• Type of fruit: drupe • Harvest: end of summer	4 to 6 cm long, they are green, velvety and contain 1 or 2 seeds When it ripens, the leather outer husk tears open, curls up and releases the kernel. In fact, almonds are not true nuts (type of dried fruit), but rather seeds enclosed in a coating of hard fruits

Table 2 Systematic position of the fig tree [44]

Reign	Vegetal
Branch	Phanerogams
Sub-phylum	Angiosperms
Class	Dicotyledons
Subclass	Hamamelidae
Series	Unisexual apetalous
Order	Urticales
Family	Moraceae
Kind	Ficus
Species	Ficus carica L

cm high, with a thick, brown and fibrous rhizome. Rosette leaves are linear, flat, with parallel veins. Flowering is extremely rare and when it occurs it produces a loose inflorescence, in the shape of an umbel, made up of a few short spikes provided at their base with a bract. Flowering occurs in the fall and gives fruits called sea olives which, after being detached from the plant, float and wash up on the beaches between May and June.

As shown in Fig. 5, Posidonia is made up of a vegetative apparatus comprising a stem or rhizome (1) buried in the sand. The rhizome bears roots (2) and rootlets upwards and downwards from the leaves (3) dark green, banded, narrow (1 cm) and elongated (they can reach 1 m long) grouped in bundles of 6–7 leaves with parallel veins: Posidonia is an aquatic geophyte with a rhizome. (The geophyte is a type of perennial plant with organs that allow it to spend the bad season buried in the ground.) The rhizome is covered with scars left by falling leaves, and it is also covered with fibrous remains (4) which torn off, agglomerate under the effect of currents and waves to form pellets or ægagropiles (5) which can be harvested on the shore; moreover, it grows slowly horizontally (about 6 cm per year) as it settles into its new environment and grows vertically, and the plant is buried in sediment trapped by the leaves.

The tangled rhizome gradually dies in its lower part, while in its upper part live Posidonia leading to the formation of mats. The rhizomes produce leaves throughout the year, and their lifespan is about 5 to 8 months. When they die, the leaves break off their scales and are swept away by the currents, depositing them on the shore. The accumulation of these leaves is called the sea line. These plants decompose very slowly and pile up in places to form veritable shoals.

Posidonia can reproduce in two ways:

- Asexual reproduction is a simple vegetative multiplication, and it is ensured by the plagiotropic rhizomes (horizontally) which allow the transformation of the leaves as they grow.
- Sexual reproduction is when the Posidonia flower, at the beginning of autumn. The color becomes very light green, and the flowers merge with the leaves and are difficult to see. They form an inflorescence carried by a peduncle (stem 10–30 cm long) from a bundle of leaves.

Fig. 5 Posidonia plant components. *Source* https://www.animateur-nature.com/connaissances_flore/zoom_posidonie.php, visited 24 July 2023

2.4 Fields of Application

For many years, almonds have been used in different fields (medicine, pharmaceuticals, para-cosmetics, etc.). Indeed, it has several health benefits since it is particularly rich in lipids, especially in unsaturated fatty acids such as omega 6 and 9, it contains several active components and a significant amount of carbohydrates and proteins. Among other things, it is an excellent source of minerals (magnesium, manganese, copper, phosphorus and calcium) and trace elements such as vitamins E, B9 and B2. In addition, for several decades, a new spirit has emerged to use biodegradable products and to try to eliminate as many toxic products as possible for the environment for everyone in the textile field. Several researchers have chosen the agricultural wastes derived from almond tree cultivars for its low cost and availability.

According to the research, almond tree waste has been used for very diverse textile applications (almond stem, almond shell, almond leaf, etc.). Almond shells have been used as an adsorbent of heavy metals in textile applications, researchers, [21], treat the shells of hazelnuts and almonds, which are cheap and economical materials available for reduction. And the elimination of Cd (II), Ni (II) and Pb (II) in aqueous solutions, and not only an adsorbent of metals but also an adsorbent of several ranges of dyes, the dye Direct "Red 80 (DR 80)" [37], the acid dye "acid blue 25 (AB25)" and the basic dye "methylene blue (MB)", etc. Almond shells were also used by [34] to evaluate the biosorption potential of the methyl orange dye.

Several efforts have been made for the preparation of activated carbon from agricultural residues such as hulls and sugarcane bagasse [118]. As well as activated carbon was prepared from the agricultural biomass of almond shells, for the adsorption of cationic dyes [8], And of course, this shows that the shells of almonds can be used as a potential source for the production of activated carbons. For almond leaves, some authors (H. S. [49] used leaves for the extraction of cellulosic fibers and then incorporated them into sustainable composites with recycled polypropylene

(PPr) as a matrix. [60] exploited African almond (AL) leaves for the preparation of a durable and environmentally friendly adsorbent by chemical activation followed by pyrolysis at 700 °C for the adsorption of blue dye of methylene (MB) from an aqueous solution.

Almond tree waste has also been valorized for the biogenic synthesis of selenium nanoparticles with tunable morphologies such as rods and brooms, from almond skin extract [103]. As well as an ecological approach was presented by the valorization of the external green shell of the almond fruit as a new natural colorant [43].

The fig tree is widely used in traditional medicine, with the pulp of the fruit that relieves pain, gingivitis and aphtes; it contains a nitrogenous substance (0.7–2.5%) and a coloring matter. It is very rich in vitamins A and C (similar to quercetol). Omega-3 and omega-6 essential fatty acids and the phytosterol in dried figs play an important role in reducing cholesterol levels, as well as in the proper functioning of the brain, nervous system and heart. In addition, the fig is used in treatments for lung ailments, cough, flu, tuberculosis, varicose veins, blood circulation disorders, intestinal disorders (tired and irritated intestines, constipation), asthma, and irritation of the trachea and throat.

Fig leaves are also used against insect bites and warts since they are rich in tyrosine and glutamic acid and are two amino acids with many benefits for humans, such as prevention of stress, improved mood or alleviation of depressive states. In the fruit latex, enzymes peptonize the buminoid substances (in alkaline or acid medium), acting like amylase on the starch and coagulating the milk. This latex is also used to manufacture rubber. But, despite the availability of production, the exploitation of the fig tree is very limited and is not valued in other areas. In addition, [81] valorized the stems of fig trees by two methods: (1) extraction of the cellulose pulp and isolation of lignins for the valorization of delignification baths in the dyeing of wool, silk and polyamide fabric and (2) production of cellulosic derivatives, as the synthesis of carboxymethylcellulose (CMC) from cellulose from the stems of the fig tree.

The different ways of recovering Posidonia debris are the basis of traditional applications. The dead leaves of Posidonia oceanica have the first components that have been tested since antiquity by the human being for various applications; for example, the Venetians used the leaves of Posidonia to transport and pack their delicate and famous glassware. According to [20], these organs were known as the "straw of Venice". Also, in France in the Lazaret cave (Alpes-Maritimes, France), the leaves were used to make bedding for sleeping [33], but this type of recovery was interrupted in 1995 [95]. Were discovered that these leaves are harmful to the skin and contain phenolic acids.

The Egyptians also used the balls of Posidonia to build felts for use in the manufacture of shoes [22, 115]. The use of Posidonia leaves remained until the beginning of the twentieth century, the populations of southern Europe (France, Corsica, Sardinia, Italy, Yugoslavia, Greece, Turkey and Spain) and the coastal populations of South Africa. North (Egypt, Libya and Tunisia) used dry Posidonia leaves, which provide good thermal insulation [19], for the construction of roofs Le Floc'h, [75].

Over the past fifty years, several tests for producing a compost based on dead Posidonia leaves [104] have been regularly carried out in Greece, Italy and Tunisia.

In addition, the Posidonia fruits (Figure 8) washed up on the beaches were eaten by cattle (in Tunisia), pigs (in Corsica) and even by humans during periods of deficiency [19].

Over the years, with current environmental awareness and growing interest in advanced and sustainable materials, the use of natural fibers has become a common practice due to their attractive characteristics. Therefore, the valorization from this plant becomes more and more innovative, and researches have been carried out for the possibility of exploitation and reuse of Posidonia in new ways of applications. Nevertheless, studies have been devoted to the delignification of Posidonia to produce cellulosic derivatives or cellulose [7, 67–69, 116]. Two research groups led by Mhenni, M, F., (Research Unit LCE2P, Monastir) and Seffen, M., (URCA, Sousse) have investigated the exploitation of posidonia as an adsorbent dye material [7, 23, 36, 41, 82, 84].

In literature, [74] the hardness of gypsum plaster composed of Mediterranean seagrass (Posidonia oceanica) and pine (Pinus sylvestris) fibers was studied by evaluating the mechanical properties in this with respect to bending and compression, resistance to bending to shocks [51]. Also focused on the thermal and mechanical properties of hardened cement paste reinforced with Posidonia-Oceanica natural fibers.

Hydrolytic degradation of green PLA/Posidonia Oceanica composites showed that these biocomposites degraded much faster than pure PLA under different conditions [107]. Another important use has been investigated, dealing with the removal of Cr (VI), one of the most toxic heavy metals in biological systems, from wastewater using activated carbon produced by pyrolysis and chemical activation of Posidonia oceanica [9].

3 The Lignocellulosic Biomass: A Challenge for a Green Future

Plant biomass is made up of several macromolecules closely linked together within the plant cell wall. These lignocellulosic materials are abundant on earth and are mainly constituted by three components: 40–60% cellulose, 20–40% hemicelluloses and 10–50% lignin [91].

3.1 The Cellulose

Cellulose was discovered as a chemical substance by Anselme Payen in 1838. He suggested that plant cells are almost all composed of the same fibrous substance that resists treatments with ammonia or acid. Cellulose structure, morphology and chemical composition have remained unknown for a very long time, even though

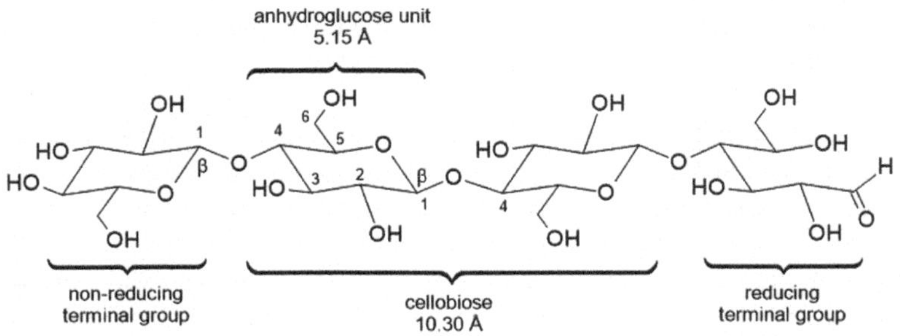

Fig. 6 Cellulose macromolecule [80]

this material has been used for centuries in very diverse applications. The main constituent of photosynthetic organisms is this long polysaccharide, which protects plant organisms [11].

The elemental composition of the cellulose macromolecules is distributed as 49.4% oxygen, 44.4% carbon and 6.2% hydrogen [133]. It is built of linear macromolecules formed by the sequence of cellobiose units. The latter is composed of a sequence of D-glucopyranose units having free OH groups and linked together by β-(1–4) glycosidic bonds (Fig. 6), being proposed its general formula as $(C_6H_{10}O_5)_n$ [4]. Certain chemical polarity is conferred on native cellulose because of the regularity of the assembly of cellobiose molecules. Once the two ends of cellulose have different chemical functionality, this molecule is provided with a relative polarity with a non-reducing end at the C4 position and a reducing end at the C1 position due to a hemiacetal function of the terminal secondary alcohol.

The antiparallel or parallel arrangement of the chains inside the cellulose crystal is because of the polarity of the cellulose chain. Two arrangements in the crystal are therefore possible: either the reducing ends are alternated (organization in antiparallel chains) or they are all placed on the same side (organization in parallel chains). The numerous hydroxyl groups on the chain allow the formation of several intra and intermolecular hydrogen bonds.

Intramolecular hydrogen bonds are created between the hydrogen atom carried by the hydroxyl group of the C3 carbon of a ring and the oxygen atom of the adjacent ring. The interaction between the hydrogen carried by the primary OH group of carbon C6 and the oxygen of the hydroxyl of C2 carbon of the adjacent ring. The intermolecular bonds are between the hydrogen of the primary hydroxyl OH-6 and the oxygen in the O-3 position of a neighboring unit ring.

The creation of a large number of inter and intramolecular hydrogen bonds gives cellulose resistance to solvents [30], and high-temperature stability which inhibits melting. The degree of polymerization (DP) in the native state of cellulose depends on the nature of the plant species (Table 3).

The existence of numerous intra/intermolecular hydrogen bonds is responsible for the formation of so-called supramolecular structures. This natural structure of

Table 3 Range the degree of polymerization with different sources of cellulose [46]

Cellulose sources	Degree of polymerization (DP)
Alga valonia	20,000–26,000
Cotton	2500–3000
Bacterial cellulose	2000–2700
Wood	600–1200
Regenerated cellulose	250–500
Cellulose powder	100–250

cellulose consists of microfibrils composed of 30–40 cellulose aggregated chains which forms fibrils and then rigid fibers (Fig. 7).

Several polymorphic varieties denote the crystal structure of cellulose. X-ray diffraction technique distinguishes up to four different crystal structures named I to IV [28]. Cellulose in its natural or native form is named cellulose I, its crystallographic structure determined by X-ray diffraction and infrared spectrometry results in a

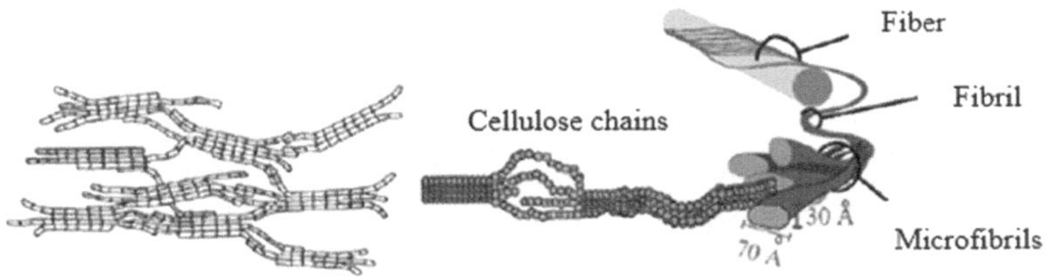

Fig. 7 Organization of the fibrillar structure of cellulose [71]

Fig. 8 Cellulose crystal lattice [28]

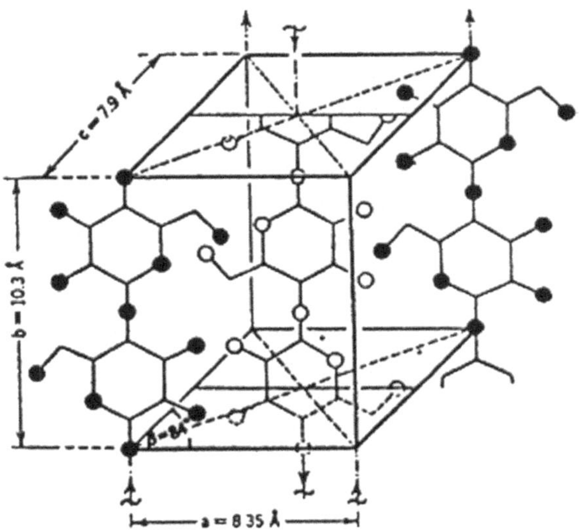

monoclinic pattern which has two cellulose chains arranged in parallel to the b axis, in opposite directions along the fiber axis (Fig. 8). Glucose cycles are located in the ab plane. The macromolecular chains are linked laterally by hydrogen bonds. In 1984, researchers demonstrated, by carbon nuclear magnetic resonance, that native cellulose had two allomorphs noted as Iα and Iβ [123]. The Iα crystal form is dominant in species with a simple and less developed architecture of other species such as algae, as opposed to plants with a more complex architecture where the Iβ crystal structure is more likely [10, 131].

Previous research projects have shown that the Iα crystal lattice results from the repetition of a primitive triclinic lattice and the more thermodynamically stable Iβ phase [128], from the repetition of a monoclinic cell (Fig. 8).

3.2 The Hemicelluloses

The hemicelluloses at the base are copolymers of saccharides. The chains are much more branched and shorter than cellulose. The monomeric units consist of uronic acids, pentoses, hexoses and deoxyoses. The main chain of a polyose consists of a homopolymer (single unit), such as that of hardwood xylans (Fig. 9A), and they are the most abundant hemicelluloses in hardwoods. The xylose units are linked together by β-1,4 glycosidic bonds. The xylan chains are linked, at irregular intervals, to 4-O-methyl-glucuronic acid structures by α-1,2 glycosidic bonds. The OH groups carried by the C2 and C3 carbons can also be substituted by O-acetyl groups. The xylans extracted from deciduous trees have a xylose to 4-O-methylglucuronic acid ratio of around 10:1. The degree of polymerization (DP) of xylans depends on the nature of the species, and it can reach values ranging from 100 to 200 in hardwoods.

However, the xylans of conifers do not contain acetyl groups on the main chain. Figure 9B must be also noted the presence of arabinofuranose units linked to the main xylan chain through α-1,3 glycosidic bonds. Conifer xylan molecules are shorter than hardwood polysaccharides. They have a DP between 70 and 130 (Jodin [64]. Other research show that in the absence of uronic acids or acetyl groups, xylans are able to crystallize as a hexagonal unit cell. As well as, the ability of xylans to form hydrogen bonds between them or to adsorb polyvalent cationic metals or water molecules. These properties are strongly conditioned by the acetyl groups carried by the mannose and glucose units of the main chain [39].

Hemicelluloses are generally amorphous polymers. They can, if devoid of uronic or deacetylated acids, adopt hexagonal crystalline forms [119]. FT-IR spectroscopy is a technique commonly used for the characterization of hemicelluloses. It makes it possible to control the degree of acetylation of hemicelluloses through the intensity of the peak at 1740–1717 cm^{-1}, characteristic of the $C = O$ groups of carboxylic acids and xylan esters [119]. In addition, it allows the qualitative analysis of the nature of the constituent monosaccharides of the polymer chain. It is possible to distinguish, respectively, the presence of xylans and glucomannans from the bands

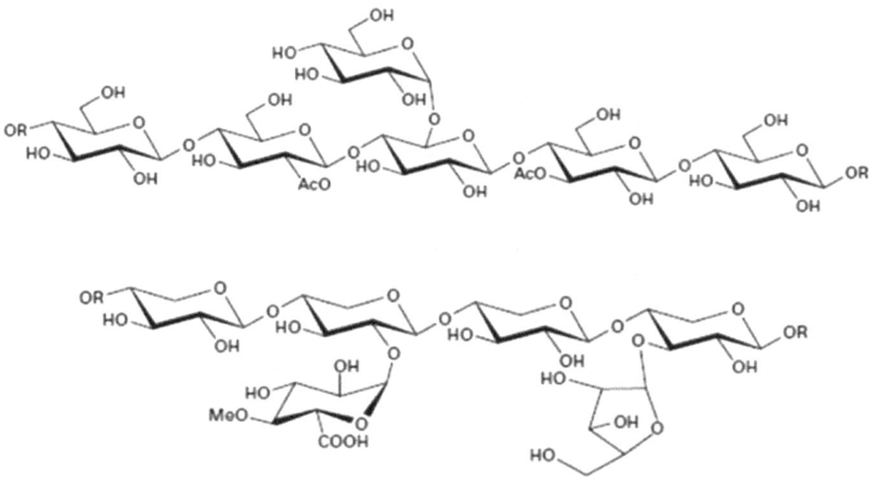

Fig. 9 Chemical structure of a 4-O-methylglucunoxylan in **a** α- hardwoods and **b** β- conifers [93]

at 808–807 and at 760–774 cm^{-1} [55]. It is also possible to study hemicellulose-water interactions based on the FT-IR technique [87].

3.3 The Lignin

Lignin is the second major organic biopolymer in plants after cellulose. This amorphous three-dimensional polymer expels the water contained in the cell wall during its formation. It increases the rigidity of cell walls and makes them hydrophobic. Lignin influences the rheological characteristics of plant tissues, but this influence is not always favorable. In the manufacture of paper pulp, for example, lignin has to be eliminated during the extraction of the cellulose since it makes the process more difficult. The digestibility of fodder for livestock depends on the lignin content. However, thanks to a high percentage of carbon, lignin strongly contributes to the calorific value of wood, thus a high proportion of lignin is an advantage when wood is used as fuel [16].

In order to control the proportion of lignin in plants, several investigations in biotechnology have been developed based on genetic modifications. The chemical structure of lignin is not well defined because it is not easy to extract it without subjecting it to degradation; moreover, the term "lignin" represents a group of polymers whose structure is influenced by the polymerization conditions. In this way, as shown in Fig. 10, the three monomers of lignin can assemble in multiple ways forming a highly branched three-dimensional structure [65].

The exact chemical composition of lignin cannot be determined precisely unlike the other polymers that make up the plant wall Drzal [38]. The origin of this

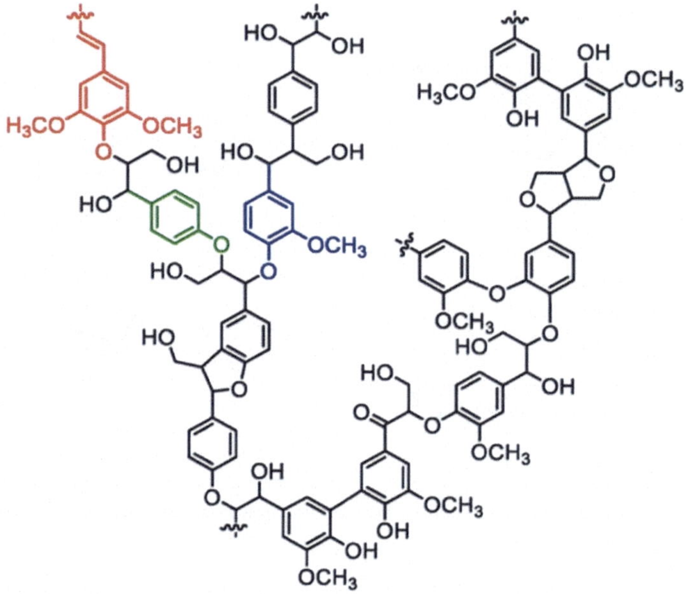

Fig. 10 Schematic representation for the macromolecular structure of lignin (with major monolignol units are colored as sinapyl alcohol-red, guaiacyl alcohol-blue, p-coumaryl alcohol-green) proposed by [65]

complexity is attributed to the conditions of lignin polymerization and the inevitable natural diversity. The different factors that influence polymerization and lignin formation were highlighted by [117]:

- Lignin monomers vary with cell age and type.
- The sequential aspect of lignification follows the stages of formation of the primary and secondary walls. In time and space, the synthesized lignins can differ even within the wall of a cell
- Polymerization conditions can be influenced by the activity of enzymes, the relative amounts of monomers, the presence of inorganic compounds such as calcium and the pH of the reaction site.

As said above, lignins are obtained from the radical polymerization of three phenylpropanoid alcohols or more commonly called monolignols (p-coumaryl (H), coniferyl (G) and sinapyl (S)). These three units differ in the degree of substitution of the hydroxyl groups, respectively, a hydroxyl or methoxyl group, two or three for the H, G or S structures. These different alcohols lead, after polymerization, respectively, to the p-hydroxyphenyl, guaiacyl and syringyl units, respectively, constituting the reticular structure of lignin.

4 Nanocelluloses and Their Terminologies

The nanocellulose term is used to describe cellulosic materials isolated from lignocellulosic materials with a dimension on the nanometer scale. In recent years, the synthesis and application of nanocellulose have achieved remarkable results as a polymer reinforcement to create high-performance biomaterials. Cellulose molecules of at least one dimension in the nanoscale (1–100 nm) are referred to as nanocellulose [57]. The main reason for the growing interest in nanosize material is because the highly uniform material with improved mechanical fasteners can be obtained by reducing the size of the cellulose fiber [59]. Nanocellulose is considered as a possible emerging solution to curb all these environmental problems, due to their biocompatible, biodegradable and highly physicochemical properties [110]. Characteristic properties of nanocellulose such as crystallinities, surface area and mechanical properties vary with extraction methods and processing techniques (Abitbol et al., [6]. Depending on the technique and conditions of synthesis of nanocellulose, which determine its dimensions, composition and properties, it can be divided into three classifications (Fig. 11):

- (CNC) Cellulose nanocrystals, also known as cellulose whiskers.
- (CNF) Cellulose nanofibrils, also known as (MFC) microfibrillated cellulose, (NFC) Nanofibrillated cellulose, or cellulose nanofibers.
- (BNC) Bacterial nanocellulose.

The two first categories, CNC and CNF, are produced by the disintegration of cellulose fibers into nanoscale particles; however, in the case of BNC, the dissolved cellulose is generated by bacteria [85].

4.1 Cellulose Nanocrystals (CNC)

CNC are the most widely used and renewable nanocellulose subfamily of nanomaterials obtained from amorphous sections of wood pulp by acid hydrolysis [121].

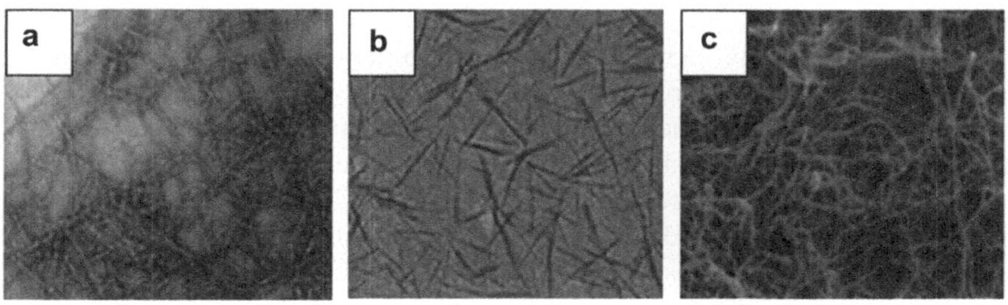

Fig. 11 Transition electron microscopy images (A: CNF/500 nm, B: CNC/600 nm, C: BNC/1000 nm) [18]

The acid hydrolysis process consists of two steps: the alkaline treatment of cellulosic pulp to remove polysaccharide, followed by heat treatment in acid medium for several hours to destroy the amorphous zone more easily and cellulose fiber regions become accessible. At the appropriate point in this step, the acid mixture is spent, and residual acids and impurities can be completely removed by repeated centrifugation and dialysis to obtain purified and highly crystalline CNC [72].

Morphologically, CNC exist in rod-like or needle-like structures when viewed under a scanning electron microscopy (SEM). The structure, properties, phase separation behavior and dimensions, including length and diameter, of these CNC vary depending on the type of raw material, acid, hydrolysis temperature, time and the intensity of ultrasonic irradiation [27]. For example, CNC with a diameter of 200–300 nm and a length of 10–30 nm were obtained when cotton fibers were hydrolyzed by sulfuric acid (H_2SO_4) at 65 °C.

On the other hand, changing the source to CNC produced by ascidian with a diameter of less than 1 μm under the same experimental conditions Additionally, hydrolysis of cotton with HCl at 65 °C produced CNC with a length of 100–200 nm and 10 nm in diameter [54, 92]. Besides acid hydrolysis, CNC are also produced by treating wood pulp with other reagents such as 2,2,6,6-tetramethylpiperidin-1-oxyl (TEMPO) [105], ammonium persulfate (APS) [124] and some bio-based enzymes, TEMPO-oxidized CNCs are produced by oxidizing cellulose fibers with TEMPO, NaBr and NaClO, where the reaction is carried out in an alkaline condition by constantly adding NaOH solution to the reaction mixture until a constant pH is obtained [105].

Enzymatic hydrolysis of microcrystalline cellulose (MCC) by the use of cellulolytic fungi yielded 20% CNC; however, the high cost of this process limited its commercial application [106]. CNC show considerable dispersion in aqueous mixtures as well as organic solvents with dielectric constants, such as ethylene glycol and dimethyl sulfoxide (DMSO). However, these form agglomerates in highly hydrophobic solutions. Agglomerated CNC can be redispersed by sonication, for example in N, N-dimethylformamide (DMF) [12]. CNC act as fillers with different materials to form functional materials for various applications due to their biocompatibility, renewability, biodegradability, low density, high mechanical properties and high crystallinity. As a reinforcement material, CNC are mixed with different polymer materials, such as polyacrylamide, poly (N,N-dimethylacrylamide) [129], poly (vinyl alcohol) [5], poly (acrylic acid-coacrylamide) [132], alginate [125], poly (acrylic acid) [132], BC [35] and gelatin [88], to improve their physicochemical, mechanical and biological properties, such as rigidity, crystallinity, mechanical resistance, swelling and biocompatibility. The addition of different ions such as Na + , K + , Mg + and Ca + in CNC-based suspensions forms nematic-ordered hydrogels by aggregating the nanocellulose particles [17]. Using CNC as reinforcement materials, researchers have developed different functional materials such as double membrane hydrogels [76], macroporous hydrogels [77] and self-healing hydrogels [108, 130]. These CNC-reinforced composite hydrogels have applications in photonics, filters, sensors and models in different fields [66].

4.2 Cellulose Nanofibrils (CNF)

The first production process of the cellulose nanofibers CNF is with ITT the Rayonier company developed in the 80s in the USA. CNF are long-tangled fibrils with a nanometer in diameter. This type of nanocellulose was produced by high-pressure grinding of the cellulosic pulp suspension, a process from which highly entangled networks of nanofibrils are formed [120]. Unlike CNC, which have nearly perfect crystallinity (about 90%), CNF contain both amorphous and crystalline cellulosic domains in individual fibers [63].

CNF extraction from cellulosic fibers can be achieved by three types of processes: (1) mechanical treatments (e.g., homogenization, milling and milling), (2) chemical treatments (e.g., TEMPO oxidation) and (3) a combination of chemical and mechanical treatments [1]. The preparation of CNF is very simple, involving the physical separation of cellulose fibers by grinding, homogenization, ultrasonication, etc., and requires essentially no modification of the chemical structure of the cellulosic chain [25, 122]. Besides the physical process, CNF can also be produced by chemical methods like oxidation of the raw material by TEMPO under magnetic stirring [105]. Additionally, physical and chemical processes can be used together for the preparation of the individual CNF, for example, carboxymethylation and high-pressure homogenization produce uniformly distributed CNF [124]. Compared to the formation of hydrogel-based CNCs by self-aggregation, CNF-based hydrogels are formed by the mechanical treatment of their aqueous suspensions with alkalis followed by neutralization [2, 3]. Studies have shown improved 3D network compactness and mechanical properties of polyacrylamide [86] and gelatin-containing nanocellulose hydrogels [126].

CNF can be prepared from different materials such as wood and cotton fibers. The diameter and length of CNF vary depending on the source material; however, these typically range from -50 nm in diameter and a few micrometers in length [26]. CNF possess a high aspect ratio, specific surface area, high mechanical properties, high crystallinity and precoating thresholds. These characteristics allow the formation of a very rigid network of CNF in different matrices and improve their mechanical properties.

4.3 Nanocelluloses Applications

Nanocellulose is believed that can to replace synthetic materials for more environmentally friendly products and is an addition to brand-new types of biomaterials, i.e., cellulose nanocomposites. Nowadays, cellulose nanocomposites are used in different fields such as medical, automotive, electronics, packaging, building, wastewater, etc.

4.3.1 Nanocellulose in the Paper Industry

About 100 million tons per year of commercially harvested cellulose are used for the production of paper and board [90]. The papermaking process involves steps including preparing paper components, wet refining, forming a wet sheet, pressing, drying, calendering and finishing. The refining of cellulose fibers in an aqueous medium is a mandatory step in the papermaking process to obtain a strong paper. [56, 58] showed the probability of increasing paper strength with an additive of nanocellulose particles to paper compositions. Such sheets exhibit admirable mechanical properties. According to [53], these properties are at least 2 to 5 times higher than those of common papers formed from conventional refining processes. According to [32], the use of nanocellulose, especially TEMPO-oxidized cellulose nanofibers (TOCNF), improves the staining efficiency of the colored paper preparation. Moreover, the non-toxicity, rheological property, biodegradability and biocompatibility of nanocellulose allow its use as food packaging material [15].

4.3.2 Nanocellulose in the Textile Industry

In recent years, the revolutionary potential of nanotechnology has provided enormous development opportunities in the textile industry. The quality of textiles can be improved for special and general applications by incorporating nanomaterials as part of the polymer matrix or coated on the surface of the fibers. The properties of nanocellulose include good stiffness and strength as well as a large surface area. For example, the large number of hydroxyl groups in the structure of nanocellulose particles makes it possible to reform the textile surface.Nanocellulose fibers are cheaper, electrically conductive and offer less weight and high strength, indicating their ability for a wide range of applications. Nanocellulose fibers are also insensitive to environmental changes Sreeraj et al. [113]. The good adsorption yield of the fibers can also be considered for the treatment of colored water pollutants, by surface modification of nanocellulose-based adsorbents [100].

4.3.3 Nanocellulose in the Composites Industry

In recent years, there has been a remarkable increase in interest in using nanocellulose as a polymeric reinforcement to create high-performance biomaterials. The main reason for the attractiveness of nanoscale cellulose is that it is a material with greater uniformity and fewer defects with improved mechanical properties compared to macroscale cellulose [52, 112]. It can be utilized as a reinforcing filler to prepare composites with solutions of water-soluble polymers to modify the viscosity and improve the mechanical properties of dry composites. The addition of nanocellulose to biodegradable polymers is of great environmental importance, as it also accelerates the rate of biodegradation [102].

4.3.4 Nanocellulose in the Biomedical Industry

Nanocellulose is a biodegradable natural material that is ideal for the biomedical industry. Pure nanocellulose is non-toxic to humans and biocompatible. It can therefore be used for healthcare applications such as personal care products, cosmetics and biopharmaceuticals. One of the more modest applications of nanocellulose is the stabilization of medical suspensions against phase separation and sedimentation of heavy constituents. Chemically modified cellulose can be a promising support for the immobilization of enzymes and other drugs [40]. Due to its nanometre size, such a carrier-drug complex can penetrate through skin pores and treat skin diseases. Likewise, it can be used as a gentle yet active exfoliating agent in cosmetics. Nanocellulose is also used in many pharmaceutical applications as polymer in hydrogels, transdermal drug delivery systems, aerogels, healing dressing materials, as superdisintegrants in fast dissolving tablets, microparticles, gels, foams, thickening agents, stabilizers, medical implants, tissue engineering and liposomes [99].

References

1. Abdul Khalil HPS, Bhat AH, Ireana Yusra AF (2012) Green composites from sustainable cellulose nanofibrils: a review. Carbohyd Polym 87:963–979. https://doi.org/10.1016/j.carbpol.2011.08.078
2. Abe K, Yano H (2012) Cellulose nanofiber-based hydrogels with high mechanical strength. Cellulose 19:1907–1912. https://doi.org/10.1007/s10570-012-9784-3
3. Abe K, Yano H (2011) Formation of hydrogels from cellulose nanofibers. Carbohyd Polym 85:733–737. https://doi.org/10.1016/j.carbpol.2011.03.028
4. Abidi N (2021) Cellulose macromolecule as a source for advanced materials preparation. Mater Today: Proc, Fourth Ed Int Conf Mater Environ Sci 45:7473–7476. https://doi.org/10.1016/j.matpr.2021.02.051
5. Abitbol T, Johnstone T, Quinn TM, Gray DG (2011) Reinforcement with cellulose nanocrystals of poly(vinyl alcohol) hydrogels prepared by cyclic freezing and thawing. Soft Matter 7:2373–2379. https://doi.org/10.1039/C0SM01172J
6. Abitbol T, Rivkin A, Cao Y, Nevo Y, Abraham E, Ben-Shalom T, Lapidot S, Shoseyov O (2016) Nanocellulose, a tiny fiber with huge applications. Current Opinion in Biotechnology, Systems biology, Nanobiotechnology 39:76–88. https://doi.org/10.1016/j.copbio.2016.01.002
7. Aguir C, M'Henni MF (2006) Experimental study on carboxymethylation of cellulose extracted from Posidonia oceanica. J Appl Polym Sci 99:1808–1816. https://doi.org/10.1002/app.22713
8. Ait Ahsaine H, Zbair M, Anfar Z, Naciri Y, El Haouti R, El Alem N, Ezahri M (2018) Cationic dyes adsorption onto high surface area 'almond shell' activated carbon: Kinetics, equilibrium isotherms and surface statistical modeling. Mater Today Chem 8:121–132. https://doi.org/10.1016/j.mtchem.2018.03.004
9. Asimakopoulos G, Baikousi M, Salmas C, Bourlinos AB, Zboril R, Karakassides MA (2021) Advanced Cr(VI) sorption properties of activated carbon produced via pyrolysis of the "Posidonia oceanica" seagrass. J Hazard Mater 405:124274. https://doi.org/10.1016/j.jhazmat.2020.124274
10. Atalla RH, Vanderhart DL (1984) Native cellulose: a composite of two distinct crystalline forms. Science 223:283–285. https://doi.org/10.1126/science.223.4633.283

11. Azizi Samir MAS, Alloin F, Dufresne A (2005) Review of recent research into cellulosic whiskers, their properties and their application in nanocomposite field. Biomacromol 6:612–626. https://doi.org/10.1021/bm0493685
12. Azizi Samir MAS, Alloin F, Sanchez J-Y, Dufresne A (2004) Cellulose nanocrystals reinforced poly(oxyethylene). Polymer 45:4149–4157. https://doi.org/10.1016/j.polymer.2004.03.094
13. Badgujar SB, Patel VV, Bandivdekar AH, Mahajan RT (2014) Traditional uses, phytochemistry and pharmacology of Ficus carica: a review. Pharm Biol 52:1487–1503. https://doi.org/10.3109/13880209.2014.892515
14. Baranwal J, Barse B, Fais A, Delogu GL, Kumar A (2022) Biopolymer: a sustainable material for food and medical applications. Polymers 14:983. https://doi.org/10.3390/polym14050983
15. Bashir O, Rashid S, Masoodi N, Khan SA, Majid I, Malik M, Beenish (2022) Nanocellulose: chemistry, preparation, and applications in the food industry, In: Industrial applications of nanocellulose and its nanocomposites. Elsevier, pp 155–177. https://doi.org/10.1016/B978-0-323-89909-3.00008-0
16. Baucher M, Christensen JH, Meyermans H, Chen C, Van Doorsselaere J, Leplé J-C, Pilate G, Petit-Conil M, Jouanin L, Chabbert B, Monties B, Van Montagu M, Boerjan W (1998) Applications of molecular genetics for biosynthesis of novel lignins. Polym Degrad Stab, Biodegrad Polym Macromol 59:47–52. https://doi.org/10.1016/S0141-3910(97)00201-2
17. Bertsch P, Isabettini S, Fischer P (2017) Ion-induced hydrogel formation and nematic ordering of nanocrystalline cellulose suspensions. Biomacromol 18:4060–4066. https://doi.org/10.1021/acs.biomac.7b01119
18. Bettaieb F (2015) Valorisation des déchets cellulosiques tunisiens (phdthesis). Université Grenoble Alpes; Faculté des Sciences de Monastir (Tunisie)
19. Boudouresque and Meinesz (1982) Découverte de l'herbier de Posidonie. Cah. Parc nation. Port-Cros, Fr., 4, 1–79.—References–Scientific research publishing [WWW Document]. URL https://www.scirp.org/(S(czeh2tfqw2orz553k1w0r45))/reference/referencespapers.aspx?referenceid=1915305. Accessed 23 July 23
20. Boudouresque CHF, Meinesz A (1983) Découverte de l'herbier de posidonie. GIS POSIDONIE
21. Bulut Y, Tez Z (2007) Adsorption studies on ground shells of hazelnut and almond. J Hazard Mater 149:35–41. https://doi.org/10.1016/j.jhazmat.2007.03.044
22. Cannon JFM (1979) An experimental investigation of Posidonia balls. Aquat Bot 6:407–410. https://doi.org/10.1016/0304-3770(79)90080-9
23. Cengiz S, Tanrikulu F, Aksu S (2012) An alternative source of adsorbent for the removal of dyes from textile waters: Posidonia oceanica (L.). Chem Eng J 189–190:32–40. https://doi.org/10.1016/j.cej.2012.02.015
24. Chebbi HE, Pellissier J-P, Khechimi W, Rolland J-P (2019) Rapport de synthèse sur l'agriculture en Tunisie (Research Report). CIHEAM-IAMM
25. Chen W, Li Q, Wang Y, Yi X, Zeng J, Yu H, Liu Y, Li J (2014) Comparative study of aerogels obtained from differently prepared nanocellulose fibers. Chemsuschem 7:154–161. https://doi.org/10.1002/cssc.201300950
26. Chen W, Yu H, Lee S-Y, Wei T, Li J, Fan Z (2018) Nanocellulose: a promising nanomaterial for advanced electrochemical energy storage. Chem Soc Rev 47:2837–2872. https://doi.org/10.1039/C7CS00790F
27. Chen Y, Liu C, Chang PR, Cao X, Anderson DP (2009) Bionanocomposites based on pea starch and cellulose nanowhiskers hydrolyzed from pea hull fibre: Effect of hydrolysis time. Carbohyd Polym 76:607–615. https://doi.org/10.1016/j.carbpol.2008.11.030
28. Chikkol Venkateshappa S (2014) Natural areca fiber: surface modification and spectral studies. J Adv Chem 10:3263–3273
29. Covino C, Sorrentino A, Di Pierro P, Roscigno G, Vece AP, Masi P (2020) Lignocellulosic fibres from enzyme-treated tomato plants: Characterisation and application in paperboard manufacturing. Int J Biol Macromol 161:787–796. https://doi.org/10.1016/j.ijbiomac.2020.06.077

30. Cuissinat C, Navard P (2006) Swelling and dissolution of cellulose part 1: free floating cotton and wood fibres in N-Methylmorpholine-N-oxide–water mixtures. Macromol Symp 244:1–18. https://doi.org/10.1002/masy.200651201
31. Da Silva AMB, Martins AB, Santana RMC (2021) Biodegradability studies of lignocellulosic fiber reinforced composites, In: Fiber reinforced composites. Elsevier, pp 241–271. https://doi.org/10.1016/B978-0-12-821090-1.00006-5
32. Dai L, Wang X, Jiang X, Han Q, Jiang F, Zhu X, Xiong C, Ni Y (2022) Role of nanocellulose in colored paper preparation. Int J Biol Macromol 206:355–362. https://doi.org/10.1016/j.ijbiomac.2022.02.151
33. De Lumley H (1969) M07–(1969) Une cabane acheuléenne dans la grotte du Lazaret, Nice - H. de LUMLEY - Société Préhistorique française [WWW Document]. URL https://www.prehistoire.org/shop_515-17330-0-0/m07-1969-une-cabane-acheuleenne-dans-la-grotte-du-lazaret-nice-h.-de-lumley.html. Accessed 23 July 23
34. Deniz F (2013) Dye removal by almond shell residues: Studies on biosorption performance and process design. Mater Sci Eng, C 33:2821–2826. https://doi.org/10.1016/j.msec.2013.03.009
35. Di Z, Shi Z, Ullah MW, Li S, Yang G (2017) A transparent wound dressing based on bacterial cellulose whisker and poly(2-hydroxyethyl methacrylate). Int J Biol Macromol 105:638–644. https://doi.org/10.1016/j.ijbiomac.2017.07.075
36. Douissa NB, Bergaoui L, Mansouri S, Khiari R, Mhenni MF (2013) Macroscopic and microscopic studies of methylene blue sorption onto extracted celluloses from Posidonia oceanica. Ind Crops Prod 45:106–113. https://doi.org/10.1016/j.indcrop.2012.12.007
37. Doulati Ardejani F, Badii Kh, Limaee NY, Shafaei SZ, Mirhabibi AR (2008) Adsorption of direct red 80 dye from aqueous solution onto almond shells: effect of pH, initial concentration and shell type. J Hazard Mater 151:730–737. https://doi.org/10.1016/j.jhazmat.2007.06.048
38. Drzal AKM, Manjusri Misra, Lawrence T (Ed.) (2005) Natural fibers, biopolymers, and biocomposites. CRC Press, Boca Raton. https://doi.org/10.1201/9780203508206
39. Dudkin MS, Danilova EI, Shchelkunov LF (2002) Isolation and characterization of holocellulose from Alfalfao. Appl Biochem Microbiol 38:72–77. https://doi.org/10.1023/A:1013260823802
40. Dufresne A (2013) Nanocellulose: a new ageless bionanomaterial. Mater Today 16:220–227. https://doi.org/10.1016/j.mattod.2013.06.004
41. Dural MU, Cavas L, Papageorgiou SK, Katsaros FK (2011) Methylene blue adsorption on activated carbon prepared from Posidonia oceanica (L.) dead leaves: Kinetics and equilibrium studies. Chem Eng J 168:77–85. https://doi.org/10.1016/j.cej.2010.12.038
42. El Khames Saad M, Khiari R, Elaloui E, Moussaoui Y (2014) Adsorption of anthracene using activated carbon and Posidonia oceanica. Arab J Chem, Spec Issue: Environ Chem 7:109–113. https://doi.org/10.1016/j.arabjc.2013.11.002
43. Erdem İşmal Ö, Yıldırım L, Özdoğan E (2014) Use of almond shell extracts plus biomordants as effective textile dye. J Clean Prod 70:61–67. https://doi.org/10.1016/j.jclepro.2014.01.055
44. Falistocco E (2020) The millenary history of the fig tree (*Ficus carica L.*). Adv Agric, Hortic Entomol 2020. https://doi.org/10.37722/AAHAE.202051
45. Faugeras P, Zerrouki R, Brouillette F (2014) Selective crosslinking of kraft pulp fibres by click chemistry–characterisation of the properties of reaction intermediates and final product. Nord Pulp Pap Res J 29:673–678. https://doi.org/10.3183/npprj-2014-29-04-p673-678
46. French AD et al, B, N, Brown RM, Chanzy H, Glasser W, Hattori K, Gray D (2004) Publication : USDA ARS [WWW Document]. URL https://www.ars.usda.gov/research/publications/publication/?seqNo115=140738. Accessed 23 July 23
47. Gabriele B, Cerchiara T, Salerno G, Chidichimo G, Vetere MV, Alampi C, Gallucci MC, Conidi C, Cassano A (2010) A new physical–chemical process for the efficient production of cellulose fibers from Spanish broom (*Spartium junceum L.*). Biores Technol 101:724–729. https://doi.org/10.1016/j.biortech.2009.08.014
48. Giraldo E, López-Corrales M, Hormaza JI (2010) Selection of the most discriminating morphological qualitative variables for characterization of fig germplasm. J Am Soc Hortic Sci 135:240–249. https://doi.org/10.21273/JASHS.135.3.240

49. Souza Rosa HS, C, Gonçalves Mothé M, Vieira Marques MF, Gonçalves Mothé C, Neves Monteiro S (2020) Steam-exploded fibers of almond tree leaves as reinforcement of novel recycled polypropylene composites. J Mater Res Technol 9:11791–11800. https://doi.org/10.1016/j.jmrt.2020.08.069
50. Hachani MA, Ziadi B, Langar H, Sami DA, Turki S, Aleya L (2016) The mapping of the Posidonia oceanica (L.) Delile barrier reef meadow in the southeastern Gulf of Tunis (Tunisia). J Afr Earth Sc 121:358–364. https://doi.org/10.1016/j.jafrearsci.2016.05.030
51. Hamdaoui O, Limam O, Ibos L, Mazioud A (2021) Thermal and mechanical properties of hardened cement paste reinforced with Posidonia-Oceanica natural fibers. Constr Build Mater 269:121339. https://doi.org/10.1016/j.conbuildmat.2020.121339
52. Hasan A, Rabbi MS, Maruf Billah Md (2022) Making the lignocellulosic fibers chemically compatible for composite: a comprehensive review. Clean Mater 4:100078. https://doi.org/10.1016/j.clema.2022.100078
53. Henriksson M, Berglund LA, Isaksson P, Lindström T, Nishino T (2008) Cellulose nanopaper structures of high toughness. Biomacromol 9:1579–1585. https://doi.org/10.1021/bm800038n
54. Heux L, Chauve G, Bonini C (2000) Nonflocculating and chiral-nematic self-ordering of cellulose microcrystals suspensions in nonpolar solvents. Langmuir 16:8210–8212. https://doi.org/10.1021/la9913957
55. Hult E-L, Iversen T, Sugiyama J (2003) Characterization of the supermolecular structure of cellulose in wood pulp fibres. Cellulose 10:103–110. https://doi.org/10.1023/A:1024080700873
56. Ioelovich M (2022) Nanocelluloses and their potential applications 2022
57. Ioelovich M (2008) Cellulose as a nanostructured polymer: a short review. BioRes 3:1403–1418. https://doi.org/10.15376/biores.3.4.Ioelovich
58. Ioelovich M, Leykin A (2004) Nano-cellulose and its application 6:17–25
59. Isogai A (2013) Wood nanocelluloses: fundamentals and applications as new bio-based nanomaterials. J Wood Sci 59:449–459. https://doi.org/10.1007/s10086-013-1365-z
60. Jabar JM, Odusote YA, Ayinde YT, Yılmaz M (2022) African almond (*Terminalia catappa L*) leaves biochar prepared through pyrolysis using H_3PO_4 as chemical activator for sequestration of methylene blue dye. Results Eng 14:100385. https://doi.org/10.1016/j.rineng.2022.100385
61. Jabli M, Gamha E, Sebeia N, Hamdaoui M (2017) Almond shell waste (*Prunus dulcis*): Functionalization with [dimethy-diallyl-ammonium-chloride-diallylamin- co-polymer] and chitosan polymer and its investigation in dye adsorption. J Mol Liq 240:35–44. https://doi.org/10.1016/j.molliq.2017.05.041
62. Jeddi L (2009) Valorisation des figues de Taounate: potentiel, mode et stratégies proposées–Sécheresse info [WWW Document]. URL http://www.secheresse.info/spip.php?article71937. Accessed 23 July 23
63. Jiang F, Hsieh Y-L (2013) Chemically and mechanically isolated nanocellulose and their self-assembled structures. Carbohyd Polym 95:32–40. https://doi.org/10.1016/j.carbpol.2013.02.022
64. Philippe J (1994) Le bois matériau d'ingénierie/textes rassemblés par Philippe Jodin; [remerciements Philippe Jodin,...] [préface Jacques Militon,...]. association pour la recherche sur le bois en Lorraine école nationale du génie rural, des eaux et des forêts, Nancy
65. Karunarathna MS, Smith RC (2020) Valorization of lignin as a sustainable component of structural materials and composites: advances from 2011 to 2019. Sustainability 12:734. https://doi.org/10.3390/su12020734
66. Kelly JA, Shukaliak AM, Cheung CCY, Shopsowitz KE, Hamad WY, MacLachlan MJ (2013) Responsive photonic hydrogels based on nanocrystalline cellulose. Angew Chem Int Ed 52:8912–8916. https://doi.org/10.1002/anie.201302687
67. Khiari R (2010) Valorisation des déchets d'origine agricole et marine: application dans les domaines textiles, papetiers et des composites (These de doctorat). Grenoble INPG
68. Khiari R, Krouit M, Belgacem N, Mauret E, Mhenni MF (2012) Préparation et caractérisation d'un matériau composite à base de Posidonia Oceanica. Matériaux Tech 100:369–375. https://doi.org/10.1051/mattech/2011130

69. Khiari R, Marrakchi Z, Belgacem MN, Mauret E, Mhenni F (2011) New lignocellulosic fibres-reinforced composite materials: a stepforward in the valorisation of the Posidonia oceanica balls. Compos Sci Technol 71:1867–1872. https://doi.org/10.1016/j.compscitech.2011.08.022
70. Khiari R, Salon M-CB, Mhenni MF, Mauret E, Belgacem MN (2017) Synthesis and characterization of cellulose carbonate using greenchemistry: Surface modification of Avicel. Carbohyd Polym 163:254–260. https://doi.org/10.1016/j.carbpol.2017.01.037
71. Klemm D, Heublein B, Fink H-P, Bohn A (2005) Cellulose: fascinating biopolymer and sustainable raw material. Angew Chem Int Ed 44:3358–3393. https://doi.org/10.1002/anie.200460587
72. Klemm D, Kramer F, Moritz S, Lindström T, Ankerfors M, Gray D, Dorris A (2011) Nanocelluloses: a new family of nature-based materials. Angew Chem Int Ed 50:5438–5466. https://doi.org/10.1002/anie.201001273
73. Kozlowski R, Mackiewicz-Talarczyk M (2000) Inventory of world fibres and involvement of FAO in fibre research. molecular crystals and liquid crystals science and technology. section A. Mol Cryst Liq Cryst 353:133–148. https://doi.org/10.1080/10587250008025654
74. Kuqo A, Mai C (2021) Mechanical properties of lightweight gypsum composites comprised of seagrass Posidonia oceanica and pine (*Pinus sylvestris*) wood fibers. Constr Build Mater 282:122714. https://doi.org/10.1016/j.conbuildmat.2021.122714
75. Le Floc'h É (19-) A (1983) Contribution à une étude ethnobotanique de la flore tunisienne. Imprimerie officielle de la République Tunisienne. Tunis
76. Lin N, Gèze A, Wouessidjewe D, Huang J, Dufresne A (2016) Biocompatible double-membrane hydrogels from cationic cellulose nanocrystals and anionic alginate as complexing drugs codelivery. ACS Appl Mater Interfaces 8:6880–6889. https://doi.org/10.1021/acsami.6b00555
77. Liu S, Jin M, Chen Y, Gao H, Shi X, Cheng W, Ren L, Wang Y (2017) High internal phase emulsions stabilised by supramolecular cellulose nanocrystals and their application as cell-adhesive macroporous hydrogel monoliths. J Mater Chem B 5:2671–2678. https://doi.org/10.1039/C7TB00145B
78. Mars M, Chatti K, Saddoud O, Salhi Hannachi A, Trifi M, Marrakchi M (2008) Fig cultivation and genetic resources in Tunisia, an overview. Acta Horticulturae 798:27–32. https://doi.org/10.17660/ActaHortic.2008.798.2
79. Mars M, Chebli T, Marrakchi M (1998) Multivariate analysis of fig (*FICUS CARICA L.*) germplasm in southern Tunisia, In: Acta Horticulturae. Int Soc Hortic Sci (ISHS), Leuven, Belgium, pp 75–82. https://doi.org/10.17660/ActaHortic.1998.480.10
80. Montet E (2021) Investigation of the consequences of the use of ozone in the bleaching of cellulosic fibres
81. Moussa I, Khiari R, Moussa A, Belgacem MN, Mhenni MF (2019) Preparation and characterization of carboxymethyl cellulose with a high degree of substitution from agricultural wastes. Fibers Polym 20:933–943. https://doi.org/10.1007/s12221-019-8665-x
82. Ncibi MC, Altenor S, Seffen M, Brouers F, Gaspard S (2008) Modeling single compound adsorption onto porous and non-porous sorbents using a deformed Weibull exponential isotherm. Chem Eng J 145:196–202. https://doi.org/10.1016/j.cej.2008.04.001
83. Ncibi MC, Mahjoub B, Seffen M (2007) Kinetic and equilibrium studies of methylene blue biosorption by *Posidonia oceanica* (*L.*) fibres. J Hazard Mater 139:280–285. https://doi.org/10.1016/j.jhazmat.2006.06.029
84. Ncibi MC, Mahjoub B, Seffen M (2006) Studies on the biosorption of textile dyes from aqueous solutions using *Posidonia Oceanica* (*L.*) leaf sheath fibres. Adsorpt Sci Technol 24:461–474. https://doi.org/10.1260/026361706780154383
85. Nechyporchuk O, Belgacem MN, Bras J (2016) Production of cellulose nanofibrils: a review of recent advances. Ind Crop Prod, Nanocellulose: Prod, Funct Appl 93:2–25. https://doi.org/10.1016/j.indcrop.2016.02.016
86. Niu J, Wang J, Dai X, Shao Z, Huang X (2018) Dual physically crosslinked healable polyacrylamide/cellulose nanofibers nanocomposite hydrogels with excellent mechanical properties. Carbohydr Polym 193:73–81. https://doi.org/10.1016/j.carbpol.2018.03.086

87. Olsson A-M, Salmén L (2004) The association of water to cellulose and hemicellulose in paper examined by FTIR spectroscopy. Carbohydr Res 339:813–818. https://doi.org/10.1016/j.carres.2004.01.005
88. Ooi SY, Ahmad I, Amin Mohd CIM (2016) Cellulose nanocrystals extracted from rice husks as a reinforcing material in gelatin hydrogels for use in controlled drug delivery systems. Ind Crop Prod, Nanocellulose: Prod, Funct Appl 93:227–234. https://doi.org/10.1016/j.indcrop.2015.11.082
89. Orts WJ, Shey J, Imam SH, Glenn GM, Guttman ME, Revol J-F (2005) Application of cellulose microfibrils in polymer nanocomposites. J Polym Environ 13:301–306. https://doi.org/10.1007/s10924-005-5514-3
90. Osong SH, Norgren S, Engstrand P (2016) Processing of wood-based microfibrillated cellulose and nanofibrillated cellulose, and applications relating to papermaking: a review. Cellulose 23. https://doi.org/10.1007/s10570-015-0798-5
91. Oushabi A (2019) The pull-out behavior of chemically treated lignocellulosic fibers/polymeric matrix interface (LF/PM): A review. Compos B Eng 174:107059. https://doi.org/10.1016/j.compositesb.2019.107059
92. Pakzad A, Simonsen J, Yassar RS (2012) Gradient of nanomechanical properties in the interphase of cellulose nanocrystal composites. Compos Sci Technol 72:314–319. https://doi.org/10.1016/j.compscitech.2011.11.020
93. Pérez J, Muñoz-Dorado J, de la Rubia T, Martínez J (2002) Biodegradation and biological treatments of cellulose, hemicellulose and lignin: an overview. Int Microbiol 5:53–63. https://doi.org/10.1007/s10123-002-0062-3
94. Perez-Jiménez M, López B, Dorado G, Pujadas-Salvá A, Guzmán G, Hernandez P (2012) Analysis of genetic diversity of southern Spain fig tree (*Ficus carica L.*) and reference materials as a tool for breeding and conservation. Hereditas 149:108–113. https://doi.org/10.1111/j.1601-5223.2012.02154.x
95. Pergent G, Pergent-Martini C, Boudouresque C (1995) Utilisation de l'herbier à Posidonia oceanica comme indicateur biologique de la qualité du milieu littoral en Méditerranée: État des connaissances. Mésogée 54:3–27
96. Phillips RC, Meñez EG (1988) Seagrasses
97. Phuong NT, Sollogoub C, Guinault A (2010) Relationship between fiber chemical treatment and properties of recycled pp/bamboo fiber composites. J Reinf Plast Compos 29:3244–3256. https://doi.org/10.1177/0731684410370905
98. Posidonia, the lung of the Mediterranean [WWW Document], 2017. URL https://medwet.org/2017/10/mediterranean-posidonia/. Accessed 23 July 23
99. Pradeep HK, Patel DH, Onkarappa HS, Pratiksha CC, Prasanna GD (2022) Role of nanocellulose in industrial and pharmaceutical sectors–a review. Int J Biol Macromol 207:1038–1047. https://doi.org/10.1016/j.ijbiomac.2022.03.171
100. Qiao A, Cui M, Huang R, Ding G, Qi W, He Z, Klemeš JJ, Su R (2021) Advances in nanocellulose-based materials as adsorbents of heavy metals and dyes. Carbohyd Polym 272:118471. https://doi.org/10.1016/j.carbpol.2021.118471
101. Ramaswamy GN, Boyd CR, Bel-Burger P, Kimmel L (1995) Kenaf/cotton blends for textiles. Fam Consum Sci Res J 24:180–190. https://doi.org/10.1177/1077727X950242005
102. Saba N et al (2015) Preparation and characterization of fire retardant nano-filler from oil palm empty fruit bunch fibers: BioResources [WWW Document]. URL https://bioresources.cnr.ncsu.edu/. Accessed 23 July 23
103. Sadalage PS, Nimbalkar MS, Sharma KKK, Patil PS, Pawar KD (2020) Sustainable approach to almond skin mediated synthesis of tunable selenium microstructures for coating cotton fabric to impart specific antibacterial activity. J Colloid Interface Sci 569:346–357. https://doi.org/10.1016/j.jcis.2020.02.094
104. Saidane A, De Waele N, Van De Velde R (1979) Contribution à l'étude du compostage de plantes marines en vue de la préparation d'un amendement organique et d'un substrat horticole. Contribution to the study of a compost of marine plants to be used as an organic amendment and as a horticultural substratum

105. Saito T, Kimura S, Nishiyama Y, Isogai A (2007) Cellulose nanofibers prepared by TEMPO-mediated oxidation of native cellulose. Biomacromol 8:2485–2491. https://doi.org/10.1021/bm0703970
106. Satyamurthy P, Jain P, Balasubramanya R, Nadanathangam V (2011) Preparation and characterization of cellulose nanowhiskers from cotton fibres by controlled microbial hydrolysis. Carbohyd Polym 83:122–129. https://doi.org/10.1016/j.carbpol.2010.07.029
107. Scaffaro R, Maio A, Gulino EF (2021) Hydrolytic degradation of PLA/Posidonia Oceanica green composites: a simple model based on starting morpho-chemical properties. Compos Sci Technol 213:108930. https://doi.org/10.1016/j.compscitech.2021.108930
108. Shao C, Wang M, Chang H, Xu F, Yang J (2017) A self-healing cellulose Nanocrystal-Poly(ethylene glycol) nanocomposite hydrogel via diels-alder click reaction. ACS Sustain Chem Eng 5. https://doi.org/10.1021/acssuschemeng.7b01060
109. Short F, Green E (2003) World atlas of seagrasses
110. Singh G, Saquib S, Gupta A, Swati (2022) Environmental, legal, health, and safety issue of nanocellulose, In: Nanocellulose materials. Elsevier, pp 265–288. https://doi.org/10.1016/B978-0-12-823963-6.00008-9
111. Souii A. n.d. Biovalorization of posidonia oceanica waste biomass as a green approach toward a sustainable bioeconomy
112. Spence K, Habibi Y, Dufresne A (2011) Nanocellulose-based composites, In: Cellulose fibers: bio-and nano-polymer composites. pp 179–213. https://doi.org/10.1007/978-3-642-17370-7_7
113. Sreeraj PR, Mishra SKr, Singh PK (2022) Characteristic features and functions of nanocellulose for its feasible application in textile industry, In: Nanocellulose Materials. Elsevier, pp 105–122. https://doi.org/10.1016/B978-0-12-823963-6.00013-2
114. Subash MC, Muthiah P (2021) Eco-friendly degumming of natural fibers for textile applications: a comprehensive review. Clean Eng Technol 5:100304. https://doi.org/10.1016/j.clet.2021.100304
115. Tackholm VD et al (1954) Flore d'Egypte. Volume III : Angiospermae, part Monocotyledones (Liliaceae - Musaceae) par Tackholm, Vivi & Drar, Mohammed: Good Paperback (1954) | Parc Mike Ltée [WWW Document]. URL https://www.iberlibro.com/Flora-Egypt-Volume-III-Angiospermae-part/30203655537/bd. Accessed 18 July 23
116. Tarchoun AF, Trache D, Klapötke TM (2019) Microcrystalline cellulose from Posidonia oceanica brown algae: extraction and characterization. Int J Biol Macromol 138:837–845. https://doi.org/10.1016/j.ijbiomac.2019.07.176
117. Terashima N (1990) A new mechanism for formation of a structurally ordered protolignin macromolecule in the cell wall of tree xylem
118. Thitame P, Shukla S (2015) Porosity development of activated carbons prepared from wild almond shells and coir pith using phosphoric acid. Chem Eng Commun 203. https://doi.org/10.1080/00986445.2015.1104503
119. Timell TE (1967) Recent progress in the chemistry of wood hemicelluloses. Wood Sci Technol 1:45–70. https://doi.org/10.1007/BF00592255
120. Tonoli GHD, Teixeira EM, Corrêa AC, Marconcini JM, Caixeta LA, Pereira-da-Silva MA, Mattoso LHC (2012) Cellulose micro/nanofibres from Eucalyptus kraft pulp: preparation and properties. Carbohyd Polym 89:80–88. https://doi.org/10.1016/j.carbpol.2012.02.052
121. Trache D, Hussin MH, Haafiz MKM, Thakur VK (2017) Recent progress in cellulose nanocrystals: sources and production. Nanoscale 9:1763–1786. https://doi.org/10.1039/C6NR09494E
122. Uetani K, Yano H (2011) Nanofibrillation of wood pulp using a high-speed blender. Biomacromol 12:348–353. https://doi.org/10.1021/bm101103p
123. VanderHart DL, Atalla RH (1984) Studies of microstructure in native celluloses using solid-state carbon-13 NMR. Macromolecules 17:1465–1472. https://doi.org/10.1021/ma00138a009
124. Wågberg L, Decher G, Norgren M, Lindström T, Ankerfors M, Axnäs K (2008) The build-up of polyelectrolyte multilayers of microfibrillated cellulose and cationic polyelectrolytes. Langmuir 24:784–795. https://doi.org/10.1021/la702481v

125. Wang K, Nune KC, Misra RDK (2016) The functional response of alginate-gelatin-nanocrystalline cellulose injectable hydrogels toward delivery of cells and bioactive molecules. Acta Biomaterialia 36. https://doi.org/10.1016/j.actbio.2016.03.016
126. Wang W, Zhang X, Teng A, Liu A (2017) Mechanical reinforcement of gelatin hydrogel with nanofiber cellulose as a function of percolation concentration. Int J Bio Macro 103. https://doi.org/10.1016/j.ijbiomac.2017.05.027
127. Wang Y, Qu Q, Cui J, Lu T, Li F, Zhang M, Liu K, Zhang Q, He S, Huang C (2021) Design and fabrication of cellulose derived free-standing carbon nanofiber membranes for high performance supercapacitors. Carbohydr Polym Technol Appl 2:100117. https://doi.org/10.1016/j.carpta.2021.100117
128. Yamamoto H, Horii F, Odani H (1989) Structural changes of native cellulose crystals induced by annealing in aqueous alkaline and acidic solutions at high temperatures. Macromolecules 22:4130–4132. https://doi.org/10.1021/ma00200a058
129. Yang J, Han C, Xu F, Sun R (2014) Simple approach to reinforce hydrogels with cellulose nanocrystals. Nanoscale 6:5934. https://doi.org/10.1039/c4nr01214c
130. Yang X, Liu G, Peng L, Guo J, Tao L, Yuan J, Chang C, Wei Y, Zhang L (2017) Highly Efficient self-healable and dual responsive cellulose-based hydrogels for controlled release and 3D cell culture. Adv Funct Mater 27:1703174. https://doi.org/10.1002/adfm.201703174
131. Yildiz S, Gümüşkaya E (2007) The effects of thermal modification on crystalline structure of cellulose in soft and hardwood. Build Environ 42:62–67. https://doi.org/10.1016/j.buildenv.2005.07.009
132. Zhang T, Zuo T, Hu D, Chang C (2017) Dual physically cross-linked nanocomposite hydrogels reinforced by tunicate cellulose nanocrystals with high toughness and good self-recoverability. ACS Appl Mater Interfaces 9:24230–24237. https://doi.org/10.1021/acsami.7b06219
133. Zhou X, Hao Y, Zhang X, He X, Zhang C (2021) Cellulose-based polymers. Phys Sci Rev. https://doi.org/10.1515/psr-2020-0067

Posidonia Oceanica at Multiscale from Fibre to Nanocellulose

Malek Khadraoui, Ramzi Khiari, Latifa Bergaoui, and Evelyne Mauret

Abstract *Posidonia ocean*ica is an abundant marine plant in the Tunisian seas. The waste of this plant is rejected by the sea in the form of balls or leaves and accumulates in significant quantities on the coasts each summer season. The valorisation of this waste is therefore an interesting avenue of research for the production of new bio-based products or materials. This chapter provides an overview of the overall of *Posidonia* as an annual marine plant, its chemical composition, its multi-scale structure (inspired by the organisation of plants and wood) and its valorisation into nanomaterials. General information on the subject will be presented (for non-expert readers), as well as specific data from recent scientific literature (for expert readers).

Keywords *Posidonica oceanica* · Cellulose · Fibres · Nanocellulose · Cellulose nanocrystal · Cellulose nanofibre

1 *Posidonia Oceanica* Fibres: A Multi-Scale Material

In Greek mythology, the term *Posidonia* comes from Poseidon, God of the seas and oceans. *Posidonia* is an underwater flowering plant, growing at the bottom of the sea. According to the literature, nine species of *Posidonia* exist: on the Australian coasts are eight species, while the ninth is located on the Mediterranean coasts [33]. The

M. Khadraoui · L. Bergaoui
Department of Biological and Chemical Engineering, National Institute of Applied Sciences and Technology, Carthage University, Carthage, Tunisia

R. Khiari
Department of Textile, Higher Institute of Technological Studies of Ksar Hellal, 5070 Ksar Hellal, Tunisia

R. Khiari (✉)
CNRS, Grenoble INP, LGP2, Université Grenoble Alpes, F-3800 Grenoble, France
e-mail: khiari_ramzi2000@yahoo.fr

M. Khadraoui · E. Mauret
University Grenoble Alpes, CNRS, Grenoble INP, LGP2, Grenoble, France

R. Khiari and M. Jawaid (eds.), *Current Status and Opportunity in Fibre and Composites*, Sustainable Materials and Technology,
https://doi.org/10.1007/978-981-96-4546-6_8

Table 1 List of marine phanerogam species based on [33]

Genus	Species	References
Posidonia	Angustifolia	Kuo [47]
Posidonia	Australis	Hooker [37]
Posidonia	Coriacea	Kuo and Cambridge [48]
Posidonia	Denhartogii	Kuo and Cambridge [48]
Posidonia	Kirkmanii	Kuo and Cambridge [48]
Posidonia	Ostendfeldii	Den Hartog [22]
Posidonia	Sinuosa	Kuo and Cambridge [48]
Posidonia	Robertsoniae	Kuo and Cambridge [48]
Posidonia	Oceanica	Ott [65]

nine species of marine phanerogams are as follows: Angustifolia, Australis, Coriácea, Denhartogii, Kirkmanii, Ostendfeldii, Sinuosa, Robertsoniae and Oceanica. They are listed in Table 1. Only *Posidonia oceanica* has been studied in our literature review. At the beginning of the nineteenth century, studies on the seagrasses of the Mediterranean basin started to describe the most widespread species, *P. oceanica* for the first time. Since then, the distribution, ecology, physiology, faunal and algal assemblages have been detailed in several papers.

In fact, *P. oceanica* covers about 50,000 km of sandy or rocky coastline [30]. Duarte [23] reported that *P. oceanica* is distributed according to the penetration of light into the underwater environment. Indeed, this plant forms meadows from the surface to a depth up to 40 m. The total primary production of *Posidonia* in the Mediterranean basin per year has been evaluated to be in the range of 5×10^6–5×10^7 tonne [66]. Depending on year, Green et al. [33] reviewed that meadows seems to be very dense with over 1000 shoots/m^2. *P. oceanica* forms a biogenic structure called "matte" that grows from the bottom up to a few metres and can be thousands of years old (this matte is slowly rising, around 1 m per century). This is due to the horizontal and vertical growth of rhizomes and its slow decomposition. Indeed, as shown in Fig. 1, *P. oceanica* consists of creeping or upright stems, usually buried in the sediment, called rhizomes. This rhizome (1) carries roots (2) and rootlets to the bottom, ribbed, thin (1 cm) but potentially long (up to 1 m length), dark green leaves (3) to the top, grouped in bundles of 6–7 parallel veined leaves [17].

Lefebvre et al. [50] reported that the annual sexual reproduction cycle of *P. oceanica* starts with an autumnal hydrophilic pollination. After six to nine months (May–July), the plant releases mature fruits, the pericarp rots, leaving the seed to sink to the bottom of the sea, after floating for several weeks. Under favourable conditions, roots develop, and a plant is produced from the seed. Due to the decrease in light and stormy weather during the autumn season, plant leaves fall off while splitting at the ligule (between the blade and the petiole) keeping the petiole (also called leaf sheath) on the rhizome (Fig. 2b). Some of the dead leaves accumulate at the foot of the meadows and form a litter (Fig. 2c) mixed with other plant debris (algae, rhizomes, roots) and epiphytes.

Fig. 1 Schematic presentation of *P. oceanica* from [82]

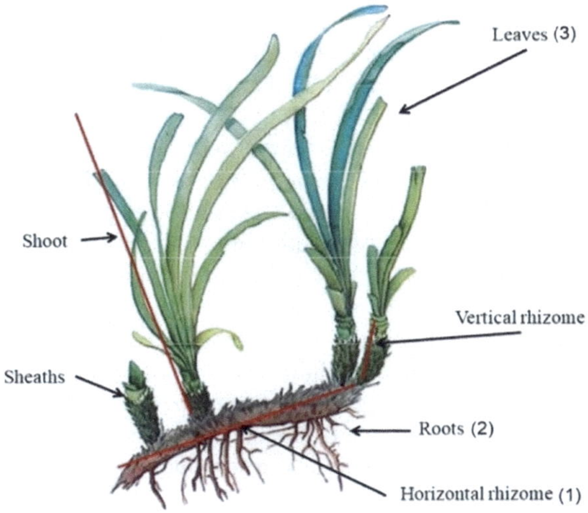

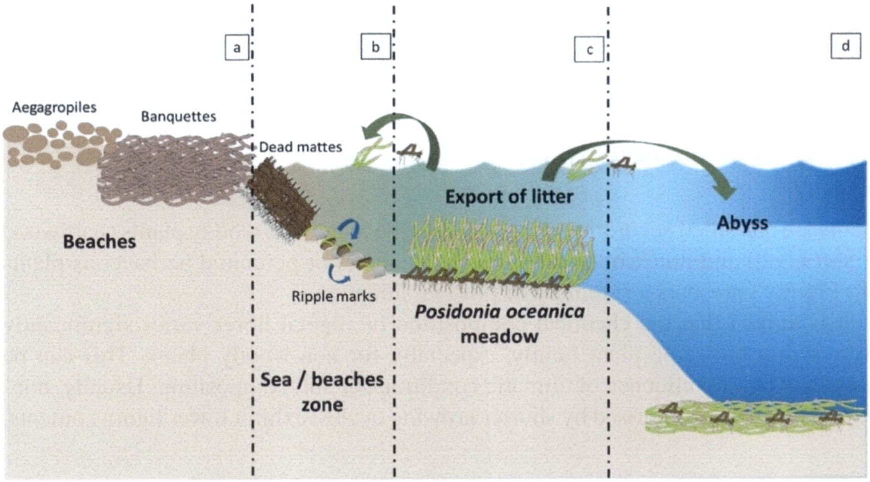

Fig. 2 The export of *P. oceanica* organs from the meadow. **a** beaches with aegagropiles and banquette, **b** the sea/beach zone with dead mattes and meadow organs in ripple marks with the hydrodynamics at shallow depth, **c** *P. oceanica* meadow with its coastline export to the abyss, **d** the abyss with the clusters of the various organs of the meadow [50]

Pergent et al. [66] reported that macro/micro-organisms and abiotic factors can degrade this litter, this latter will form sediments on the seabed from beaches (Fig. 2a) to the abyss (Fig. 2d). Other parts of these leaves arrive on the beaches. It then forms "banquettes" with thickness from few centimetres to several metres (Fig. 2a), that can reduce the beach erosion rate. Otherwise, part of the leaves of *P. oceanica* decompose and turn into plant fibres, under the effect of water, while compacting in a spherical manner around a remnant of the rhizome. Balls, seemingly made of fine hairs, have been formed and carried by the waves to be found on the beaches. These balls are called "aegagropiles" [17, 32].

In antiquity, *Posidonia* leaves were used to reinforce the inside of mattresses [53]. Then, Egyptians used *Posidonia* balls as felting for shoes-manufacturing. Meanwhile, in Venetia, *Posidonia* leaves allowed packing and transporting glassware. At the beginning of the twentieth century, the dry leaves were still used for the construction of roofs or as a food additive for animals. These traditional applications have gradually decreased and therefore the accumulated *Posidonia* is considered to be landfill waste. Recently, some studies have been conducted on the delignification of *Posidonia* to produce cellulose or cellulose derivatives [3, 43], and composites [68]. Researches also suggest the exploitation of *Posidonia* waste as an adsorbent material for dyes [4, 25, 34, 60, 61]. Finally, these materials, after chemical and mechanical treatment, have been studied to produce a renewable, biodegradable nanomaterial called nanocellulose [11–13].

2 *Posidonia Oceanica* Chemical Composition and Fibre Building

2.1 Chemical Composition

Plants are classified according to the origin of the materials: woody plants (hardwood or softwood) and non-woody plants such as annual or perennial herbaceous plants [9]. Figure 3 presents a summary of this classification.

It is known that the chemical composition of vegetal fibres varies significantly even within the same plant family, especially for non-woody plants. This can be explained by the influence of climatic conditions or soil composition. Usually, non-woody plants, characterised by shorter growing cycles, exhibit lower lignin contents,

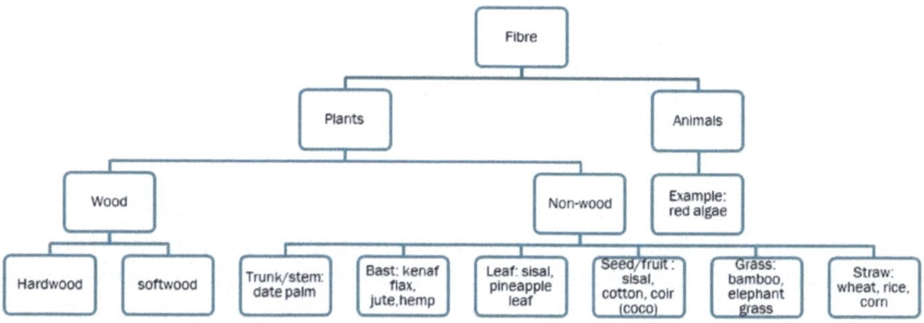

Fig. 3 Classification of fibres from different sources [1, 55]

leading to easier delignification processes for non-woody sources. As an illustration, chemical composition of wood, cane, straw, trunk/stem, bast, seed/fruit and leaf are shown in Table 2.

Table 3 gives indicative values (dry matter content) of the chemical composition of different plants. Cellulose is present with a content of 40 and 45% for wood and a slightly lower content for non-woody plants (around 35%), except for hemp, flux, jute and sisal with content around 70%. Hemicelluloses are also present in significant amounts (up to 30%) in non-woody and hardwood plants, while for softwood, they range between 25 and 35% (Sax, 2004). Finally, for lignin, the percentage varies from 15 to 28% for wood, while for non-wood plants, the lignin content is generally between 10 and 20% (there are some exceptions). It is worth noting that ash content is significantly higher in non-woody plants (it can reach 15% for rice straw, for instance), silica often being the main component. To sum up, the cell wall of vegetal fibres is generally made up of cellulose (from 40 to 60%), hemicelluloses (from 20 to 40%), lignin (from 10 to 25%), together with pectins, waxes and water-soluble substances. The proportion and structure of these components depend mainly on the origin, specie, season, site and part of the biomass [16].

Regarding *P. oceanica*, a more detailed analysis of the literature (see Table 2) shows a cellulose content between around 30 and 50% and hemicelluloses between 20 and 30% for *Posidonia* leaves (POL) and balls (POB), respectively. On the other hand, lignin content varies between 25 and 30%, respectively, for *Posidonia* leaves and balls as reported by different authors [6, 11–13, 19, 43, 60]. The extracted substances' quantity depends on the used solvents (water, sodium hydroxide solution or ethanol-toluene). Ncibi et al. [60] reported 12% of extractives from POL using ethanol/benzene, while Khiari et al. [44] reported 11% of extractives from POB using ethanol/toluene. The study of the chemical composition thus proves that *Posidonia*, such as other non-woody plants, can be considered as a source of cellulose and be valorised and used in the production of lignocellulosic fibres and/or cellulose derivatives.

2.2 Posidonia Oceanica *Structure: Fibre Building*

The cellular structure of natural fibres is quite complex. To the best of our knowledge, in the literature, the hierarchical cell wall structure of *P. oceanica* was not studied. Therefore, the next part is based on general information adapted from other studies on wood or plants. The walls of the natural fibres are formed by a network of cellulose microfibrils, embedded in a matrix composed of lignin and hemicelluloses. Lignin, being the cement between cellulose and hemicelluloses, induces the rigidity of the plant wall (Fig. 4). In 1956, Björkman proposed the existence of covalent bonds between lignin and carbohydrates [14]. Four types of lignin–carbohydrate linkages were suggested: α-ester [27], phenyl glycoside [42] and acetal bonds [84].

In general, wood fibres are characterised by a length of 1–3 mm, a width of 10–50 μm and a wall thickness that varies between 1 and 5 μm [21]. The cellular wall

Table 2 Chemical composition of different lignocellulosic sources (content in %, w/w of raw material)

Wood sources				
Reference	Cellulose	Hemicellulose	Lignin	Ash
Thomas [78]	45	30	20	n.m
Thomas [78]	42	27	28	n.m
Non-wood sources				
Cane				
Phinichka and Kaenthong [67]	42	32	19	n.m
Li et al. [51]	38	28	26	n.m
Rocha et al. [70]	43	25	23	3
Straw				
Liu et al. [52]	45	34	9	6
Khiari et al. [43]	42	29	13	5
Khiari et al. [43]	37–54	16–27	11–20	7–8
Khiari et al. [43]	42	29	13	5
Boufi and Chaker [18]	38	32	19	7
Valdebenito et al. [80]	49	16	7	3
Valdebenito et al. [80]	39	35	10	1
Trunk/stem				
Abraham et al. [2]	70	20	6	n.m
Abraham et al. [2]	68	15	11	n.m
Chaker et al. [20]	46	24	20	7
Luo et al. [54]	30	21	41	n.m
Chaker et al. [20]	38	32	18	8
Seed/fruit				
Housseinpour et al. [38]	49	29	20	7
Barbash et al. [8]	33	24	23	4
Leaf				
Khiari et al. [43]	36	39	20	2
Khiari et al. [43]	73	14	11	n.m
Bast				
Vignon et al. [83]	55	16	4	4
Vignon et al. [83]	44	18	28	2
Serrano et al. [72]	47	33	18	n.m
Khiari et al. [43]	31–9	n.m	15–19	2–5
Khiari et al. [43]	70	22	6	n.m
Khiari et al. [43]	71	21	2	n.m

n.m: not-mentioned.

Table 3 Chemical composition of *P. oceanica* balls (POB) and leaves (POL) reported in different studies

References	Cellulose (%)	Hemicellulose (%)	Lignin (%)	Ash (%)
POB [6]	48	18	23	n.m
POL [60]	38	21	27	2
POB [43]	40	22	30	12
POL [11–13]	31	26	25	5
POB [35]	44	18	29	12

n.m: not-mentioned.

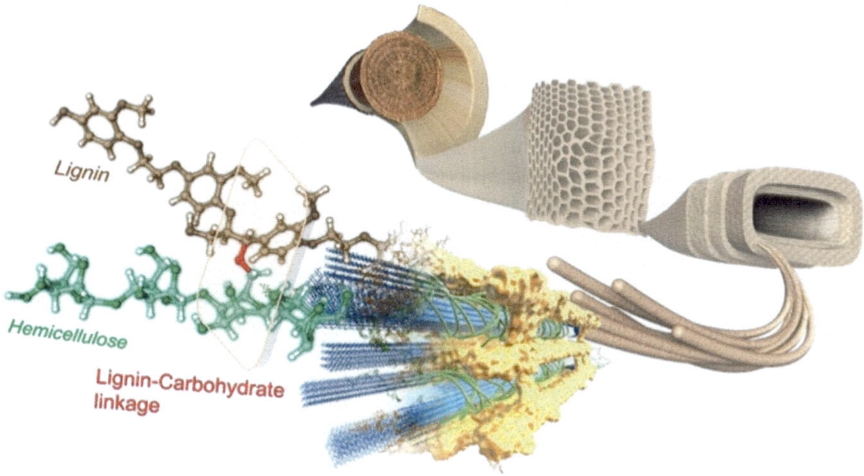

Fig. 4 Structural design of lignin–carbohydrate linkage from [64]

of the fibres is composed of several layers (Fig. 5): the first cell wall constituted by the middle lamella (ML) and the primary wall (P), the secondary wall divided into 3 layers: the outer layer (S1), the middle layer (S2) and the inner layer (S3). Finally, the lumen (W) which is a central cavity around which the wall surrounds, and which ensures water conduction [29]. The middle lamella plays a role of a cement by promoting bonding between the cells thanks to its chemical composition rich in lignin. The primary and secondary walls contain three main components: cellulose, hemicelluloses and a matrix, generally, composed of pectin and lignin in the primary and in the secondary wall, respectively. On the one hand, the primary cell wall (P) has a thickness of about 30–1000 nm and contains cellulose microfibrils transversely located. On the other hand, the secondary cell wall has three layers (S1, S2 and S3), containing most of the cellulose mass in the fibre with a thickness varying from 100 nm (cotton) to 300 nm (spruce wood). Their microfibrils are aligned in parallel

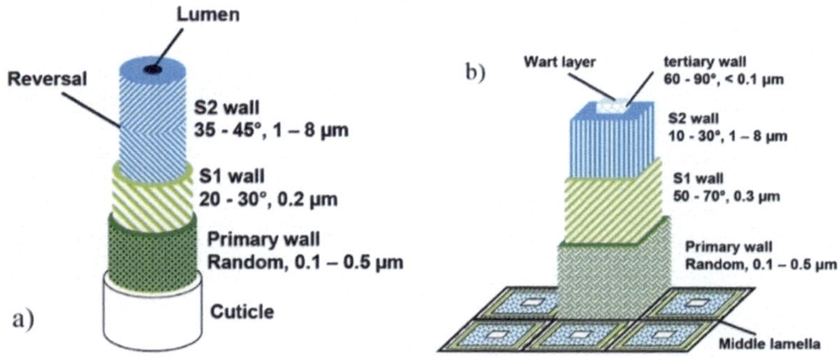

Fig. 5 Structural design of plant cell walls of **a** cotton and **b** wood fibres from [77]

and are packed densely in a flat helix, knowing that the angle relative to the fibre axis varies according to species [44].

Klemm et al. [44] describe cellulose present in the fibres in the form of microfibrillar bands of the order of 100 nm, composed of microfibrils of diameter between 10 and 30 nm. These microfibrils are in turn composed of elementary fibrils, which are considered as the smallest structural units with size between 1.5 and 3.5 nm. These dimensions may vary from one species to another (Fig. 6).

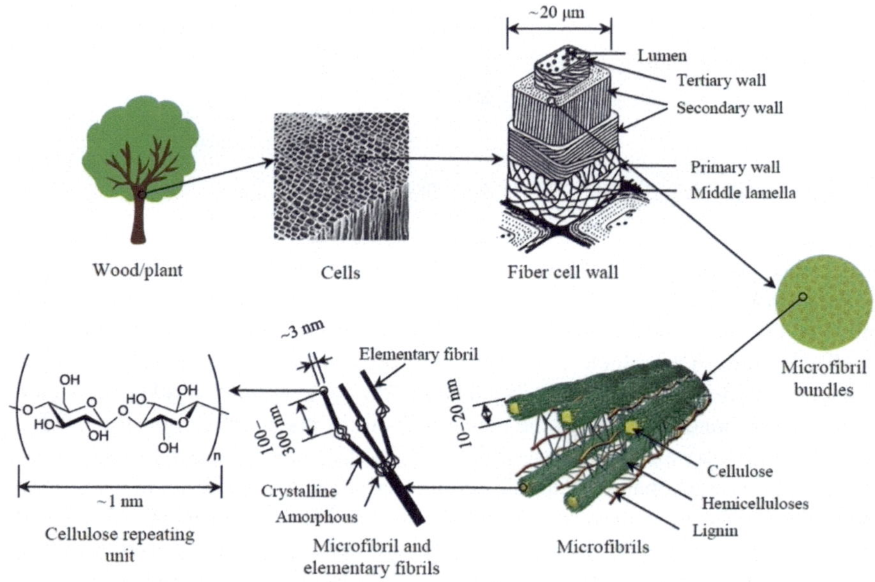

Fig. 6 Hierarchical structures of cell wall of spruce wood from [62]

3 Nanocellulose Fibres

The hierarchical multi-scale structure of the cellulosic polymer in the plant cell wall allowed the production of nanomaterials, namely nanocellulose, combining the interesting properties of cellulose with the advantages of nanomaterials. Thanks to their nanometric size, large specific surface area, high aspect ratio, reactivity, ability to form bonds and create networks, nanocellulose constitutes a bio-based material, with interesting properties, which is progressively used in various applications such as paper, packaging, printed electronics, adhesives, composites, barrier membranes, biomedical engineering, etc. [41, 57]. The nomenclature of nanocellulose has been rather not consistent, and several terms are still used to refer to this material. However, the Technical Association of the Pulp and Paper Industry (TAPPI) has proposed to standardise the terminology for cellulose nanomaterials (Fig. 7).

Figure 8 presents an overview of conventional methods for the isolation of nanocellulose from fibres. Depending on isolation process, nanocellulose is divided into two families, cellulose nanofibrils (CNF), or cellulose microfibrils (CMF), or cellulose micro/nanofibrils (CM/NF) and nanocrystals (CNC) or whisker. These kinds of materials exhibit different morphologies and properties.

The next part will be dedicated to describing the processes of preparation and extraction of nanocellulose.

3.1 Cellulose Nanocrystals

Cellulose nanocrystals, CNC, with rod-like morphology, were first obtained by Ranby in 1949 by hydrolysis with sulphuric acid. In order to produce the cellulose nanocrystals, a treatment of the cellulosic raw material, after a delignification operation, is carried out with a concentrated acid solution which hydrolyses the glycosidic linkage β (1 $\rightarrow$ 4) linking the anhydroglucose units [58]. According to the established protocol and set conditions (time, temperature, acid concentration and acid/

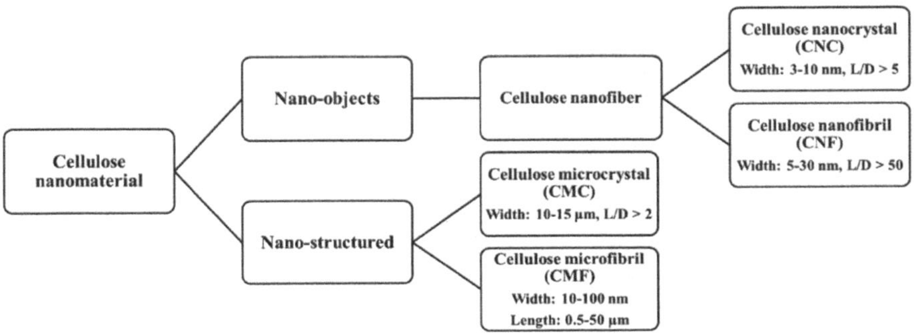

Fig. 7 Standardised terms to classify cellulose nanomaterials according to TAPPI W13021

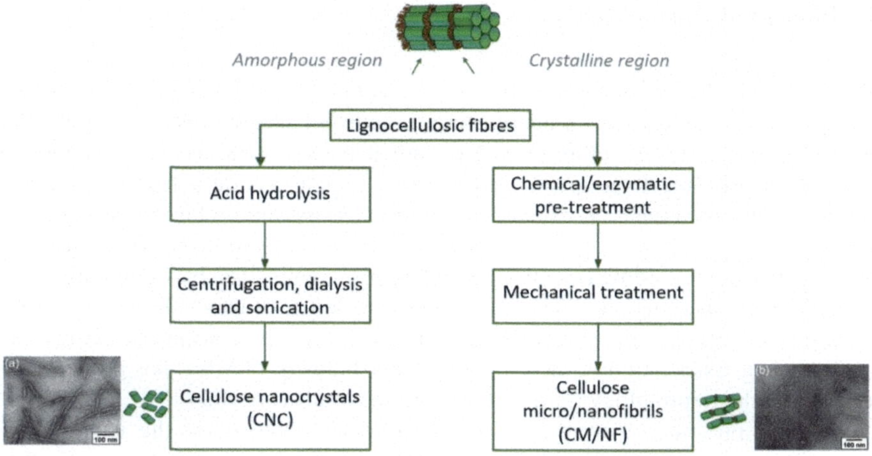

Fig. 8 Schematised method to produce cellulose micro/nanofibrils and nanocrystals, TEM images from Foster et al. [41]

cellulose ratio), sulphuric acid is usually used to keep only the crystalline areas of the cellulose by removing the amorphous regions (Fig. 9). Cellulose nanocrystals CNC can be prepared using several acids. On an industrial scale, the CNC suspension is produced by hydrolysis with hydrochloric acid but the most frequently used acid in the laboratory is sulphuric acid H_2SO_4. Thus, stable aqueous colloidal suspensions of individual nanocrystals can be prepared [69]. The sulphuric acid treatment makes it possible to introduce sulphate groups on the surface of the crystals. These negative charges induce repulsive forces that facilitate the individualisation and the dispersion of CNC in an aqueous medium [69]. These suspensions are stable. In contrast, the treatment of cellulose with hydrochloric acid leads to suspensions that are generally not stable and flocculate. It is also possible to prepare CNC by hydrolysis with hydrochloric acid followed by oxidation with sodium hypochlorite catalysed by the radical TEMPO (2,2,6,6-tetramethylpiperidine-1-oxyl), to introduce carboxylic groups on the surface to facilitate stabilisation in water [56].

Generally, CNC can be produced with different extraction yield, morphological dimensions and degree of crystallinity, depending on the origin of the cellulose and the experimental conditions. The extraction yield of CNC varies between 20 and 45% by mass in relation to the mass of cellulose in the initial material and its crystallinity, except for cotton linters for which the yield can reach 80%. Indeed, the average width of the cellulose nanocrystals obtained is the order of 5–70 nm and their length varies between 100 and 250 nm (for plant cellulose) and can be between 100 nm and several microns in the case of cellulose of tunicates and certain algae [7, 15, 28, 69]. CNC crystallinity (cellulose from wood) is between 70 and 80%, while for CNC crystallinity (cellulose from algae and tunicates) is between 80 and 87% [81].

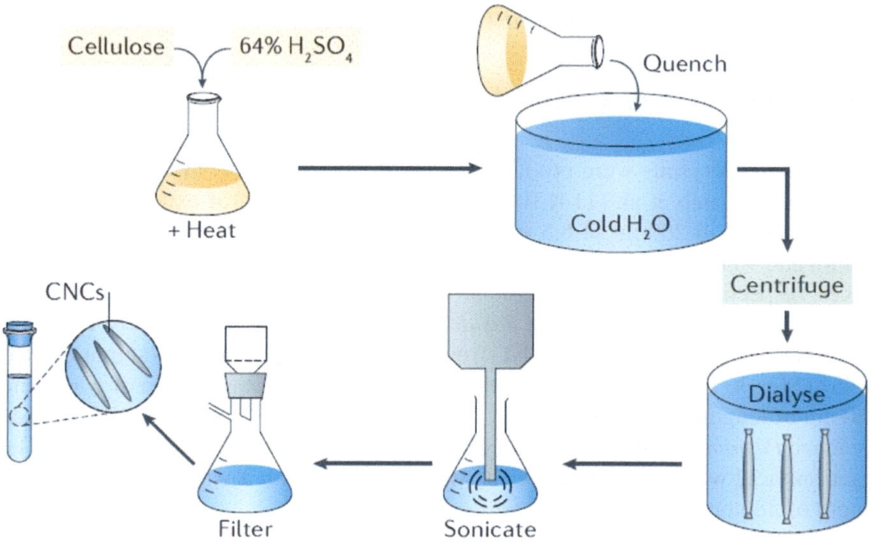

Fig. 9 General laboratory-scale production of cellulose nanocrystals (CNC) from [81]

3.2 Cellulose Micro/nanofibrils

Cellulose micro/nanofibrils (CM/NF) are obtained using mechanical treatments, such as grinding, refining, cryo-crushing, high-pressure homogenisation, etc. These treatments induce fibre delamination and require high energy. To reduce energy consumption, different pretreatments can be used before the mechanical process. Typically, CM/NF have a long web-like structure having a size with width between 5 and 60 nm and a length of few microns [15]. The produced cellulose micro/nanofibrils possess crystalline and amorphous region (parts). Unlike cellulose nanocrystals, the *L*/*d* ratio of cellulose micro/nanofibrils is quite high. This specificity allows it to have a low percolation threshold. It was reported that CM/NF are characterised by gel-like behaviour in water at low concentration while exhibiting pseudoplastic and thixotropic properties [45]. The next part is dedicated to some works that study the valorisation of *Posidonia* into nanocellulose.

4 Valorisation of *Posidonia Oceanica* into Nanocellulose

4.1 Valorisation of Posidonia Oceanica *into CNC*

Cellulose nanocrystals were produced for the first time from *Posidonia* (balls and leaves) by Bettaieb et al. [11–13]. To synthesise them, 10 g of extracted cellulose were dispersed in 200 ml of 6.5 mol/l of sulphuric acid at 55 °C while stirring for 20 min and 40 min. The reaction yield for *Posidonia* was 28% and 30% for balls and leaves, respectively. The width of the produced CNC was 8 nm and 7 nm, while the length was 276 nm and 338 nm for *Posidonia* balls and leaves, respectively. Fortunati et al. [31] succeeded to produce CNC from *Posidonia* balls using 64% sulphuric acid at 45 °C for 30 min. The reaction yield was 15%, and CNC with 5 nm of width and 180 nm of length were obtained (Fig. 10). In 2019, Benito-González et al. [10] used bleached *Posidonia* leaves to prepare CNC using 30% sulphuric acid for 90 min. The reaction yield was 18%, and the CNC were obtained with a width of 10 nm and a length of 586 nm.

Cellulose nanocrystals from *Posidonia* have sizes and properties close to those extracted from various biomass. From sisal, Siqueira et al. [73] obtained whiskers of 7–8 nm in width and 197–347 nm in length. Kunaver et al. [46] reported the production of CNC from eucalyptus wood with width between 14 and 23 nm and a length between 220 and 300 nm. Using rice husk, Islam et al. [39] produced CNC with 50 nm of width and a length of 550 nm, to name but a few examples.

Indeed, cellulose nanocrystals extracted from *Posidonia* balls and leaves may be potentially used as nanofillers in polymer matrices. Fortunati et al. [31] reported the reinforcement of poly(lactic acid) PLA nanocomposite films with functionalised cellulose nanocrystals (prepared by adding a surfactant in CNC suspension from *Posidonia* balls) for food packaging applications. For the different percentages of CNC, the transparency of the CNC composites was maintained with a good dispersion of CNC in the matrix. In the same way, Bettaieb et al. [11–13] prepared

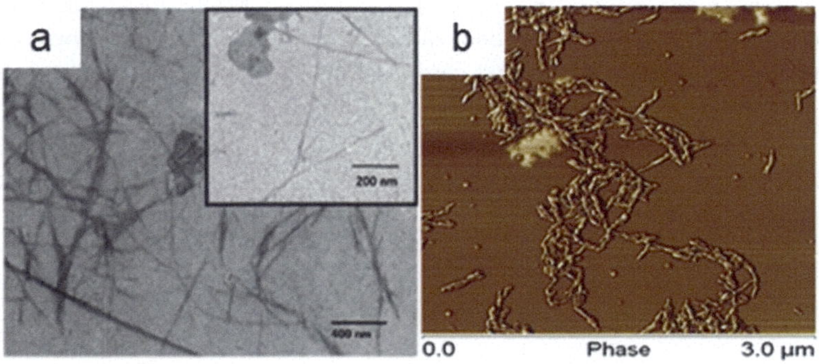

Fig. 10 TEM observation (**a**) and AFM investigation (**b**) of extracted CNC from *Posidonia* from [10, 31] respectively

nanocomposites based on poly(styrene-co-butyl acrylate) and CNC extracted from *Posidonia* balls and leaves. The analyses showed that the thermomechanical properties of the reinforced composites were improved compared to the polymer. The percolation phenomenon explains the strong reinforcing effect of CNC: above the percolation threshold, nanoparticles connect and form a continuous rigid network. In addition, Benito-González et al. [10] used CNC prepared from *Posidonia* leaves to produce films. These films exhibited mechanical properties superior to those of reference biopolymers such as thermoplastic corn starch (TPCS) and PLA and barrier properties comparable to PLA.

4.1.1 Valorisation of *Posidonia Oceanica* into CM/NF

Bettaieb et al. [11–13] reported the isolation of cellulose micro/nanofibrils from *Posidonia* leaves and balls by TEMPO oxidation combined with high pressure homogenisation. The produced CM/NF exhibited quite similar sizes between 5 and 21 nm for *Posidonia* balls and 2 and 15 nm for *Posidonia* leaves (Fig. 11). The obtained cellulose micro/nanofibrils were used to produce nanocomposites (using a commercial latex matrix of poly styrene-co-butyl acrylate), and a high reinforcing effect was observed. The storage modulus increased from 0.133 MPa for the unmodified matrix to about 25 and 60 MPa by adding 5 and 10% of CM/NF, respectively. For the best of our knowledge, this is the first and the only study that present the preparation of CM/NF from *Posidonia.*

5 Conclusion

In this chapter, an overview of the current knowledge on *P. oceanica* was provided. The chapter starts with a description by a detailed explanation of the chemical composition, the hierarchical structure of the fibres and their valorisation into nanomaterials. This literature review identified interesting research questions and challenges that are still unclear: Only one publication has studied *Posidonia* fibres for the production of high-quality modified/pretreated nanocellulose but with high energy consumption.

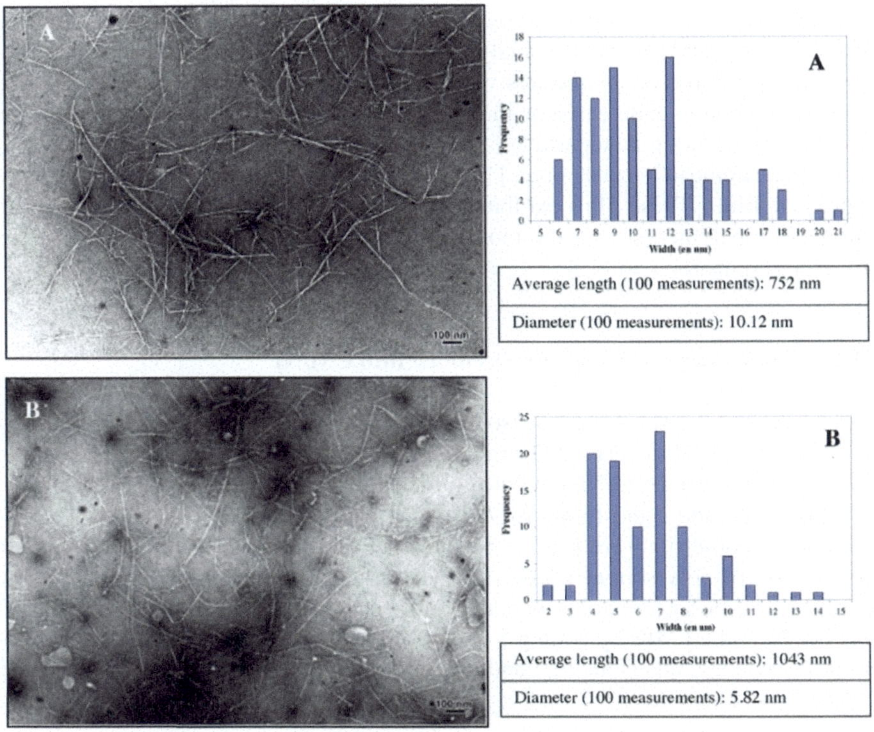

Fig. 11 TEM observations and diameter distribution of the CM/NF-*Posidonia* balls (**a**) and CM/NF-Posidonia leaves (**b**) from [11–13]

References

1. Abdul Khalil HPS, Issam AM, Ahmad Shakri MT, Suriani R, Awang AY (2007) Conventional agro-composites from chemically modified fibres. Ind Crops Prod 26(3):315–323. https://doi.org/10.1016/j.indcrop.2007.03.010
2. Abraham E, Deepa B, Pothan LA, Jacob M, Thomas S, Cvelbar U, Anandjiwala R (2011) Extraction of nanocellulose fibrils from lignocellulosic fibres: a novel approach. Carbohydr Polym 1468–1475
3. Aguir C, Mhenni F (2007) Removal of basic blue 41 from aqueous solution by carboxymethylated Posidonia oceanica. J Appl Polym Sci 103(2):1215–1225. https://doi.org/10.1002/app.23070
4. Aguir C, M'Henni MF (2006) Experimental study on carboxymethylation of cellulose extracted from Posidonia oceanica. J Appl Polym Sci 99(4):1808–1816. https://doi.org/10.1002/app.22713
6. Allègue L, Zidi M, Sghaier S (2015) Mechanical properties of Posidonia oceanica fibers reinforced cement. J Compos Mater 49(5):509–517. https://doi.org/10.1177/0021998314521254
7. Azizi Samir MAS, Alloin F, Dufresne A (2005) Review of recent research into cellulosic whiskers, their properties and their application in nanocomposite field. Biomacromol 6(2):612–626. https://doi.org/10.1021/bm0493685

8. Barbash V, Poyda V, Deykun I (2011) Peracetic acid pulp from annual plants. 6
9. Behzad T, Ahmadi M (2016) Nanofibers. Nanofiber Res-Reach New Height. https://doi.org/10.5772/63704
10. Benito-González I, López-Rubio A, Gavara R, Martínez-Sanz M (2019) Cellulose nanocrystal-based films produced by more sustainable extraction protocols from Posidonia oceanica waste biomass. Cellulose 26(13):8007–8024. https://doi.org/10.1007/s10570-019-02641-4
11. Bettaieb F, Khiari R, Dufresne A, Mhenni MF, Belgacem MN (2015) Mechanical and thermal properties of Posidonia oceanica cellulose nanocrystal reinforced polymer. Carbohyd Polym 123:99–104. https://doi.org/10.1016/j.carbpol.2015.01.026
12. Bettaieb F, Khiari R, Dufresne A, Mhenni MF, Putaux JL, Boufi S (2015) Nanofibrillar cellulose from Posidonia oceanica: properties and morphological features. Ind Crops Prod 72:97–106. https://doi.org/10.1016/j.indcrop.2014.12.060
13. Bettaieb F, Khiari R, Hassan ML, Belgacem MN, Bras J, Dufresne A, Mhenni MF (2015) Preparation and characterization of new cellulose nanocrystals from marine biomass Posidonia oceanica. Ind Crops Prod 72:175–182. https://doi.org/10.1016/j.indcrop.2014.12.038
14. Björkman A (1956) Studies on finely divided wood. Part 1. Extraction of lignin with neutral solvents. Svensk Papperstidning 59(13):477–485. https://www.bcin.ca/bcin/detail.app?id=39370&wbdisable=true
15. Blanco A, Monte MC, Campano C, Balea A, Merayo N, Negro C (2018) Chapter 5—nanocellulose for industrial use: cellulose nanofibers (CNF), cellulose nanocrystals (CNC), and bacterial cellulose (BC). In: Mustansar Hussain C (ed) Handbook of nanomaterials for industrial applications. Elsevier, pp 74–126. https://doi.org/10.1016/B978-0-12-813351-4.00005-5
16. Bledzki A (1999) Composites reinforced with cellulose based fibres. Prog Polym Sci 24(2):221–274. https://doi.org/10.1016/S0079-6700(98)00018-5
17. Boudouresque CF, Bernard G, Bonhomme P, Groupement d'intérêt scientifique Posidonie (Marseille) (2006) Préservation et conservation des herbiers à Posidonia Oceanica. Ramoge pub
18. Boufi S, Chaker A (2016) Easy production of cellulose nanofibrils from corn stalk by a conventional high speed blender. Ind Crops Prod 93:39–47. https://doi.org/10.1016/j.indcrop.2016.05.030
19. Chadlia A, Farouk MM (2010) Chemical modification of Posidonia with cyclic anhydrides: effect on thermal stability. Carbohyd Res 345(2):264–269. https://doi.org/10.1016/j.carres.2009.11.006
20. Chaker A, Alila S, Mutjé P, Vilar MR, Boufi S (2013) Key role of the hemicellulose content and the cell morphology on the nanofibrillation effectiveness of cellulose pulps. Cellulose 20(6):2863–2875. https://doi.org/10.1007/s10570-013-0036-y
21. Chinga-Carrasco G (2011) Cellulose fibres, nanofibrils and microfibrils: the morphological sequence of MFC components from a plant physiology and fibre technology point of view. Nanoscale Res Lett 6(1):417. https://doi.org/10.1186/1556-276X-6-417
22. Den Hartog C (1970) The sea-grasses of the world. North-Holland, Amsterdam. https://ci.nii.ac.jp/naid/10011442938/
23. Duarte CM (1991) Seagrass depth limits. Aquat Bot 40(4):363–377. https://doi.org/10.1016/0304-3770(91)90081-F
25. Dural MU, Cavas L, Papageorgiou SK, Katsaros FK (2011) Methylene blue adsorption on activated carbon prepared from Posidonia oceanica (L.) dead leaves: kinetics and equilibrium studies. Chem Eng J 168(1):77–85. https://doi.org/10.1016/j.cej.2010.12.038
27. Eriksson Ö, Goring DAI, Lindgren BO (1980) Structural studies on the chemical bonds between lignins and carbohydrates in spruce wood. Wood Sci Technol 14(4):267–279. https://doi.org/10.1007/BF00383454
28. Favier V, Chanzy H, Cavaille JY (1995) Polymer nanocomposites reinforced by cellulose whiskers. Macromolecules 28(18):6365–6367. https://doi.org/10.1021/ma00122a053
29. Fengel D, Wegener G (2011) Wood: chemistry, ultrastructure, reactions. Walter de Gruyter
30. Fornes A, Basterretxea G, Orfila A, Jordi A, Alvarez A, Tintore J (2006) Mapping Posidonia oceanica from IKONOS. ISPRS J Photogramm Remote Sens 60(5):315–322. https://doi.org/10.1016/j.isprsjprs.2006.04.002

31. Fortunati E, Luzi F, Puglia D, Petrucci R, Kenny JM, Torre L (2015) Processing of PLA nanocomposites with cellulose nanocrystals extracted from Posidonia oceanica waste: innovative reuse of coastal plant. Ind Crops Prod 67:439–447. https://doi.org/10.1016/j.indcrop.2015.01.075
32. Gobert S, Cambridge MT, Velimirov B, Pergent G, Lepoint G, Bouquegneau JM, Dauby P, Pergent-Martini C, Walker DI (2006) Biology of Posidonia. In: Larkum AWD, Orth RJ, Duarte CM (eds) Seagrasses: biology, ecologyand conservation. Springer Netherlands, pp 387–408. https://doi.org/10.1007/978-1-4020-2983-7_17
33. Green EP, Short FT, UNEP UNE.P, Short FT (2003) World Atlas of seagrasses. University of California Press
34. Guezguez I, Dridi-Dhaouadi S, Mhenni F (2009) Sorption of yellow 59 on Posidonia oceanica, a non-conventional biosorbent: comparison with activated carbons. Ind Crops Prod 29(1):197–204. https://doi.org/10.1016/j.indcrop.2008.05.002
35. Haddar M, Elloumi A, Koubaa A, Bradai C, Migneault S, Elhalouani F (2018) Synergetic effect of Posidonia oceanica fibres and deinking paper sludge on the thermo-mechanical properties of high density polyethylene composites. Ind Crops Prod 121:26–35. https://doi.org/10.1016/j.indcrop.2018.04.075
37. Hooker JD (1844) The botany of the Antarctic voyage of H.M. Discovery ships erebus and terror in the years 1839–1843: under the command of captain Sir James Clark Ross. Reeve
38. Housseinpour R, Latibari AJ, Farnood R, Fatehi P, Sepiddehdam1 SJ (2010) Fiber morphology and chemical composition of rapeseed (Brassica Napus) stems. IAWA J 31(4):457–464. https://doi.org/10.1163/22941932-90000035
39. Islam MdS, Kao N, Bhattacharya SN, Gupta R, Choi HJ (2018) Potential aspect of rice husk biomass in Australia for nanocrystalline cellulose production. Chin J Chem Eng 26(3):465–476. https://doi.org/10.1016/j.cjche.2017.07.004
41. Johan Foster E, J. Moon, R., P. Agarwal, U., J. Bortner, M., Bras, J., Camarero-Espinosa, S., J. Chan, K., D. Clift, M. J., D. Cranston, E., J. Eichhorn, S., M. Fox, D., Y. Hamad, W., Heux, L., Jean, B., Korey, M., Nieh, W., J. Ong, K., S. Reid, M., Renneckar, S., … Youngblood, J. (2018). Current characterization methods for cellulose nanomaterials. Chemical Society Reviews, 47(8), 2609–2679. https://doi.org/10.1039/C6CS00895J
42. Joseleau J-P, Kesraoui R (1986) Glycosidic bonds between lignin and carbohydrates. Holzforschung 40(3):163–168. https://doi.org/10.1515/hfsg.1986.40.3.163
43. Khiari R, Mhenni MF, Belgacem MN, Mauret E (2010) Chemical composition and pulping of date palm rachis and Posidonia oceanica—a comparison with other wood and non-wood fibre sources. Biores Technol 101(2):775–780. https://doi.org/10.1016/j.biortech.2009.08.079
44. Khiari R, Mhenni MF, Belgacem MN, Mauret E (2010) Chemical composition and pulping of date palm rachis and Posidonia oceanica—a comparison with other wood and non-wood fibre sources. Biores Technol 101(2):775–780. https://doi.org/10.1016/j.biortech.2009.08.079
45. Klemm D, Heublein B, Fink H-P, Bohn A (2005) Cellulose: fascinating biopolymer and sustainable raw material. Angew Chem Int Ed 44(22):3358–3393. https://doi.org/10.1002/anie.200460587
46. Klemm D, Kramer F, Moritz S, Lindström T, Ankerfors M, Gray D, Dorris A (2011) Nanocelluloses: a new family of nature-based materials. Angew Chem Int Ed 50(24):5438–5466. https://doi.org/10.1002/anie.201001273
47. Kunaver M, Anžlovar A, Žagar E (2016) The fast and effective isolation of nanocellulose from selected cellulosic feedstocks. Carbohyd Polym 148:251–258. https://doi.org/10.1016/j.carbpol.2016.04.076
48. Kuo J (1978) Morphology, anatomy and histochemistry of the Australian seagrasses of the genus Posidonia könig (posidoniaceae). I. Leaf blade and leaf sheath of Posidonia australis hook F. Aquat Bot 5:171–190. https://doi.org/10.1016/0304-3770(78)90060-8
50. Lahtinen P, Liukkonen S, Pere J, Sneck A, Kangas H (2014) A comparative study of fibrillated fibers from different mechanical and chemical pulps. BioResources 9(2):2115–2127. https://ojs.cnr.ncsu.edu/index.php/BioRes/article/view/BioRes_09_2_2115_Lahtinen_Fibrillated_Fibers_Pulps

51. Lefebvre L, Compère P, Léonard A, Plougonven E, Vandewalle N, Gobert S (2021) Mediterranean aegagropiles from Posidonia oceanica (L.) Delile (1813): A frst complete description from macroscopic to microscopic structure. Mar Biol 168:37. https://doi.org/10.1007/s00227-021-03833-y
52. Li Z, Fei B, Jiang Z (2015) Effect of steam explosion pretreatment on bamboo for enzymatic hydrolysis and ethanol fermentation. BioResources 10(1):1037–1047. https://ojs.cnr.ncsu.edu/index.php/BioRes/article/view/BioRes_10_1_1037_Li_Steam_Explosion_Pretreatment_Bamboo
53. Liu Q, Lu Y, Aguedo M, Jacquet N, Ouyang C, He W, Yan C, Bai W, Guo R, Goffin D, Song J, Richel A (2017) Isolation of high-purity cellulose nanofibers from wheat straw through the combined environmentally friendly methods of steam explosion, microwave-assisted hydrolysis, and microfluidization. ACS Sustain Chem Eng 5(7):6183–6191. https://doi.org/10.1021/acssuschemeng.7b01108
54. de Lumley H, Pillard B, Pillard F (1969) L'habitat et les activités de l'homme du Lazaret. Mémoires de la Société Préhistorique Française 7:183–222. https://hal.archives-ouvertes.fr/hal-00809152
55. Luo H, Zhang H, Yue L, Pizzi A, Lu X (2018) Effects of steam explosion on the characteristics of windmill palm fiber and its application to fiberboard. Eur J Wood Wood Prod 76(2):601–609. https://doi.org/10.1007/s00107-017-1259-7
56. Mohanty AK, Misra M, Drzal LT (2002) Sustainable bio-composites from renewable resources: opportunities and challenges in the green materials world. J Polym Environ 10(1):19–26. https://doi.org/10.1023/A:1021013921916
57. Montanari S, Roumani M, Heux L, Vignon MR (2005) Topochemistry of carboxylated cellulose nanocrystals resulting from TEMPO-mediated oxidation. Macromolecules 38(5):1665–1671. https://doi.org/10.1021/ma048396c
58. Moon RJ, Schueneman GT, Simonsen J (2016) Overview of cellulose nanomaterials, their capabilities and applications. JOM J Miner , Metals Mater Soc (TMS) 68(9):2383–2394. https://doi.org/10.1007/s11837-016-2018-7
60. Ncibi MC, Mahjoub B, Jean-Marius C, Lambert J, Ehrhardt JJ, Bercion Y, Seffen M, Gaspard S (2009) Preparation and characterisation of raw chars and physically activated carbons derived from marine Posidonia oceanica (L.) fibres. J Hazard Mater 165(1–3):240–249. https://doi.org/10.1016/j.jhazmat.2008.09.126
61. Ncibi MC, Mahjoub B, Seffen M (2006) Studies on the biosorption of textile dyes from aqueous solutions using Posidonia oceanica (L.) leaf sheath fibres. Adsorpt Sci Technol 24(6):461–474. https://doi.org/10.1260/026361706780154383
62. Ncibi MC, Mahjoub B, Seffen M (2007) Kinetic and equilibrium studies of methylene blue biosorption by Posidonia oceanica (L.) fibres. J Hazard Mater 139(2):280–285. https://doi.org/10.1016/j.jhazmat.2006.06.029
64. Nechyporchuk O, Pignon F, Belgacem MN (2015) Morphological properties of nanofibrillated cellulose produced using wet grinding as an ultimate fibrillation process. J Mater Sci 50(2):531–541. https://doi.org/10.1007/s10853-014-8609-1
65. Nishimura H, Kamiya A, Nagata T, Katahira M, Watanabe T (2018) Direct evidence for α ether linkage between lignin and carbohydrates in wood cell walls. Sci Rep 8(1):1–11. https://doi.org/10.1038/s41598-018-24328-9
66. Ott JA (1980) Growth and production in Posidonia oceanica (L.) Delile*. Mar Ecol 1(1):47–64. https://doi.org/10.1111/j.1439-0485.1980.tb00221.x
67. Pergent G, Rico-Raimondino V, Pergent-Martini C (1997) Fate of primary production in Posidonia oceanica meadows of the mediterranean. Aquat Bot 59(3):307–321. https://doi.org/10.1016/S0304-3770(97)00052-1
68. Phinichka N, Kaenthong S (2018) Regenerated cellulose from high alpha cellulose pulp of steam-exploded sugarcane bagasse. J Market Res 7(1):55–65. https://doi.org/10.1016/j.jmrt.2017.04.003
69. Rammou E, Mitani A, Ntalos G, Koutsianitis D, Taghiyari HR, Papadopoulos AN (2021) The potential use of seaweed (Posidonia oceanica) as an alternative lignocellulosic raw material for wood composites manufacture. Coatings 11(1):69. https://doi.org/10.3390/coatings11010069

70. Revol J-F, Bradford H, Giasson J, Marchessault RH, Gray DG (1992) Helicoidal self-ordering of cellulose microfibrils in aqueous suspension. Int J Biol Macromol 14(3):170–172. https://doi.org/10.1016/S0141-8130(05)80008-X
71. Rocha GJM, Gonçalves AR, Oliveira BR, Olivares EG, Rossell CEV (2012) Steam explosion pretreatment reproduction and alkaline delignification reactions performed on a pilot scale with sugarcane bagasse for bioethanol production. Ind Crops Prod 35(1):274–279. https://doi.org/10.1016/j.indcrop.2011.07.010
72. Sax DV (2004) Foundations of biogeography: classic papers with commentaries. University of Chicago Press
73. Serrano L, Urruzola I, Nemeth D, Belafi-Bako K, Labidi J (2011) Modified cellulose microfibrils as benzene adsorbent. Desalination 270(1):143–150. https://doi.org/10.1016/j.desal.2010.11.038
77. Sixta, H. (2008). Handbook of Pulp (p. 1352). https://doi.org/10.1002/9783527619887
78. Spence KL, Venditti RA, Rojas OJ, Habibi Y, Pawlak JJ (2010) The effect of chemical composition on microfibrillar cellulose films from wood pulps: water interactions and physical properties for packaging applications. Cellulose 17(4):835–848. https://doi.org/10.1007/s10570-010-9424-8
80. Thomas RJ (1977) Wood: structure and chemical composition. In: Wood technology: chemical aspects, vol 43. American Chemical Society, pp 1–23. https://doi.org/10.1021/bk-1977-0043.ch001
81. Turbak AF, Snyder FW, Sandberg KR (1983) Microfibrillated cellulose, a new cellulose product: properties, uses, and commercial potential. J Appl Polym Sci: Appl Polym Symp; (United States), 37, Article CONF-8205234-Vol.2. https://www.osti.gov/biblio/5062478-microfibrillated-cellulose-new-cellulose-product-properties-uses-commercial-potential
82. Valdebenito F, Pereira M, Ciudad G, Azocar L, Briones R, Chinga-Carrasco G (2017) On the nanofibrillation of corn husks and oat hulls fibres. Ind Crops Prod 95:528–534. https://doi.org/10.1016/j.indcrop.2016.11.006
83. Vanderfleet OM, Cranston ED (2020) Production routes to tailor the performance of cellulose nanocrystals. Nat Rev Mater. https://doi.org/10.1038/s41578-020-00239-y
84. Vasapollo C (2010) Spatio-temporal variability of plant features and motile invertebrates in Posidonia oceanica seagrass meadows

Posidonia Oceanica and Its Derived Activated Carbon as Adsorbents for Phenol Removal from Aqueous Media

Mohamed El Khames Saad, Manel Elakremi, Mongi ben Mosbah, Ramzi Khiari, Elimame Elaloui, Ridha ben Salem, and Younes Moussaoui

Abstract *Posidonia oceanica* dead leaves (DLPO) were used as promise and inexpensive precursor for activated carbon production (CB, CAP). The samples were characterized by using elemental analysis, scanning electron microscopy (SEM), and BET surface analysis. Batch adsorption experiments were applied such as (pH, contact time, adsorbent dose) on the phenol removal efficiency. Results revealed that the phenol adsorption capacity reaches 23.64 mg/g on the CAP which is higher compared to that determined on DLPO (1.21 mg/g) and CB (12.5 mg/g). The equilibrium data follow to the Langmuir model better than Freundlich, Temkin and Dubinin–Radushkevich (D-R) models for both adsorbents. Thermodynamics analysis confirms that the phenol adsorption was spontaneously and exothermic in nature.

Keywords Carbon materials · *Posidonia oceanic* · Activated carbon · Adsorption · Phenol

M. E. K. Saad · M. Mosbah · E. Elaloui
Laboratory for the Application of Materials to the Environment, Water and Energy (LR21ES15), Faculty of Sciences of Gafsa, University of Gafsa, Gafsa, Tunisia

M. Elakremi · R. Salem · Y. Moussaoui (✉)
Organic Chemistry Laboratory (LR17ES08), Faculty of Sciences of Sfax, University of Sfax, Sfax, Tunisia
e-mail: y.moussaoui2@gmail.com

M. Elakremi · M. Mosbah · E. Elaloui · Y. Moussaoui
Faculty of Sciences of Gafsa, University of Gafsa, Gafsa, Tunisia

R. Khiari
Department of Textile, Higher Institute of Technological Studies of Ksar Hellal, 5070 Ksar Hellal, Tunisia

CNRS, Grenoble INP, LGP2, Université Grenoble Alpes, F-3800 Grenoble, France

R. Khiari and M. Jawaid (eds.), *Current Status and Opportunity in Fibre and Composites*, Sustainable Materials and Technology,
https://doi.org/10.1007/978-981-96-4546-6_9

1 Introduction

Phenol and its derivatives are the most widespread forms of organic pollutants in industrial wastewaters produced by petrochemicals, coal mines, polymeric resins manufacturing, pulp and paper, dye synthesis, and pharmaceuticals industries [2, 7, 8, 33, 40, 41]. According to the recommendation of the European and US Environmental Protection Agency (EEA, EPA), these pollutants are acutely toxic to both humankind and the ecological environment even at low concentration [2, 3, 42]. Therefore, the removal of phenol from wastewater is extremely urgent before their injection in the environment. Various methods have been proposed to remove phenol compounds from wastewaters, such as advanced oxidation, membrane processes, solvent extraction, coagulation, and adsorption [3, 4, 9, 12, 20, 29, 30, 38, 39]. Adsorption has been found to be the effective and fundamental method used for wastewater treatment in the industry [9, 16, 27, 39]. Especially in liquid phase, it is used to remove recalcitrant pollutants from effluents. However, due to the high price of commercial activated carbon, many researchers look to find low-cost potential materials in order to reduce the overall cost of adsorbents preparation. Thus, several investigations have reported that the use of abundant and renewable materials has been an attractive approach [1, 4, 8, 10].

2 *Posidonia Oceanica* and Its Derived Carbon Materials

P. oceanica dead leaves (DLPO) are the most abundant and available biomass by-product all over the Mediterranean Sea. Important quantities reaching a thickness of 1–2 m or more are annually accumulated on Tunisian beaches [18, 32]. These wastes have no considerable economic applications and are considered as a harmful solid pollutant to swimmers due to their bad emitted smell. Thus, the valorization of this available and renewable wastes' biomass can be a suitable solution to settle this problem [22, 26, 36]. Various researchers have reported the use of DLPO as such or after treatment for dyes removal [13, 31] and heavy metal [14, 28]. Also, *P. oceanica* is a lignocellulosic material that can be used for particleboards and composites materials [5, 6, 17, 24, 35].

Biochar (CB) was obtained by carbonization of powdered DLPO ($\geq$ 100 μm) at 600 °C at the rate of 10 °C/min under N_2 (100 mL/min) for 1 h. Activated carbon (CAP) was obtained after carbonization of 60% H_3PO_4 impregnated DLPO at a 1/1 weight ratio under the same condition of CB. CB and CAP were then soxhleted with ethanol for 12 h in order to remove tar and impurity, and then they were repeatedly rinsed with distilled water.

The biochemical composition (hollocellulose and Klason lignin), physical, and chemical properties of adsorbents were determined by using common procedures. The Klason lignin was determined after treatment with 72% of sulfuric acid. The acid-insoluble lignin was measured by subjecting the DLPO to acid hydrolysis and filtering

Table 1 Proximate, ultimate, and chemical analyses of DLPO and its derived carbon

Analysis (% wt)	DLPO	CB	CAP
Proximate			
Moisture	3.8	3.3	2.7
Organic content	88.7	89.2	91.1
Ash	6.7	7.3	8.4
Ultimate			
Carbon	48.2	81.9	83.2
Hydrogen	5.5	1.8	1.2
Oxygen	39.3	4.7	2.7
Nitrogen	0.3	0.5	0.5
Sulfur	0.6	0.2	-
H/C	0.11	0.02	0.01
(O + N)/C	0.82	0.06	0.03
Chemical composition			
Klason lignine	26.3	-	-
Holocellulose	59.7	-	-
Surface function			
Acidic groups (mmol/g)	2.84	2.02	1,58
Basic groups (mmol/g)	3.35	1.62	1.38
pH_{PZC}	8.2	5.8	6.5
Structural properties			
BET surface area (m/g)	2	3	147
Cumulative pore volume (cm/g)	0.0058	0.012	0.15

the obtained suspension in order to separate the insoluble lignin. The obtained solid is then dried and weighed. Holocellulose which is comprised of cellulose and hemicellulose was determined after the reaction of DLPO sample with sodium hypochlorite and acetic acid in aqueous solution at 80 °C, as it has been previously described by Wise et al. [45]. The proximate analysis (moisture, ash, organic matter content) were made according to ASTM standard test methods. Elemental analysis was performed with an elemental analyzer (Flash-2000). The acidic and basic surface sites were determined by the acid–base titration method pronounced by the Boehm's method [21, 44]. The pH_{PZC} (point zero charge) was determined by the mass titration method as described by Kaouah et al. [19]. Structural analysis was determined by the N_2 adsorption isotherm (Micromeritics ASAP 2020). The textural morphology was identified by using scanning electron microscopy (SEM) technique.

Table 1 summarizes the biochemical composition, physical, and chemical properties of DLPO, CB, and CAP. The high organic contains and the low ash content, as well as the important hollocellulose and lignin contents (59.7 and 26.3%, respectively), make the DLPO a suitable precursor for preparing activated carbon.

Notable differences of chemical component were observed after pyrolysis. The carbon contains of CB and CAP (81.9 and 83.2%, respectively) increased compared to DLPO (48.2%). Hydrogen and oxygen contents decreased accordingly. The H/C ratio decrease from 0.11 in the DLPO to 0.02 and 0.01 in CB and CAP, respectively. This is attributed mostly to the increase of the aromaticity [19]. Besides, the polarity index ((O + N)/C) decreases from 0.82 in DLPO to 0.06 and 0.03 in CB and CAP, respectively.

This is likely due to the loss of volatiles during the pyrolysis which eliminate the non-carbon species (Oxygen and Hydrogen) and enriched carbon. The amounts of acidity function were measured as 2.84, 2.02, and 1.58 mmol/g, while the basic functions were 3.25, 1.62, and 1.38 mmol/g, respectively, for the DLPO, CB, and CAP. The pH_{PZN} values were taken as 8.2, 5.8, and 6.5 for the DLPO, CB, and CAP, respectively. Based on the results of the BET analysis and the pore size distribution, CAP has a remarkable development of textural proprieties of compared to the LDPO and CB. CAP reached the highest specific surface area (147 m^2/g) and a cumulative pore volume of 0.15 cm^3/g. This can be explained by micropore development through reaction between H_3PO_4 and biochemical composition during activation process [34]. Based on the results of SEM analysis, there were significant differences in the surface morphology between both adsorbents. It can be seen that both LDPO and CB (Fig. 1a and Fig. 1b) are low porous materials. However, CAP (Fig. 1c) appears to have a highly porous surface, and there is a suitable probability for the adsorption of small organic molecules typically for phenol from liquid phase.

3 Phenol Adsorption

To study the feasibility of using DLPO as bio-sorbent and as promising raw material for the activated carbon, phenol was chosen as model of organic pollutants to evaluate the performance of the adsorption of DLPO and its derived carbons (CB and CAP).

Batch adsorption studies were conducted at 298–313 K temperature by agitating 5 mL of desired phenol concentration solution at desired pH (2–12), contact time (10–120 min), and variable adsorbent doses (0.025–0.3 g) in flat bottom flask. The pH was adjusted by dilute HCl or NaOH solutions. The equilibrium concentration of phenol was measured at the maximum wavelength ($\lambda_{max} = 270$ nm) using a UV/Vis spectrometer (BECKMAN DU 800). The equilibrium adsorption capacity, Q_e (mg/g), and the phenol removal percentage, (% R), were determined as follows [2, 8, 37, 46].

$$Q_e = \frac{(C_i - C_e) \times V}{m} \tag{1}$$

$$\% R = \frac{(C_i - C_e)}{C_i} \times 100 \tag{2}$$

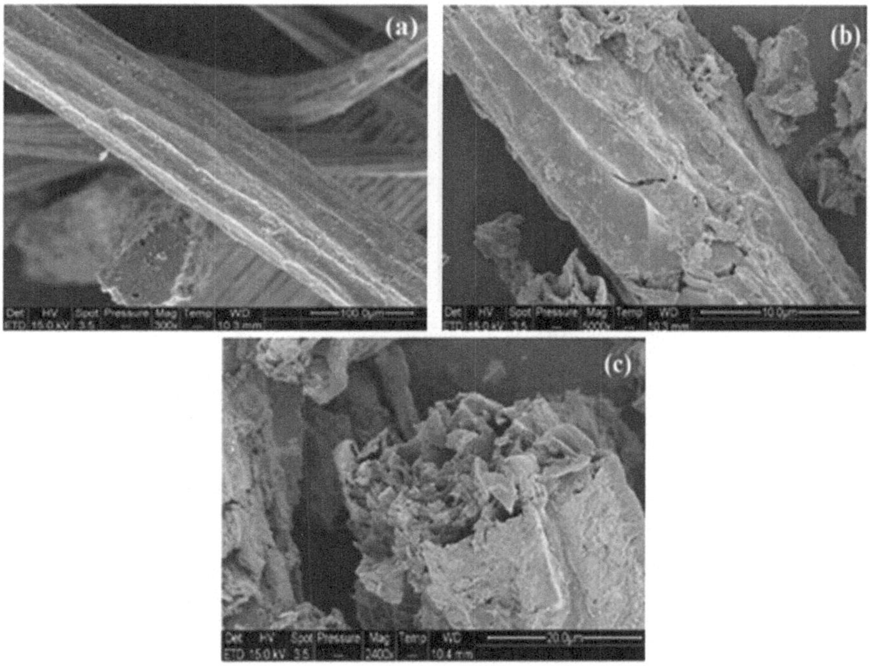

Fig. 1 Scanning electron micrographs of **a** LDPO, **b** CB, and **c** CAP

where C_i and C_e (mg/L) are the concentrations at initial and equilibrium, respectively. V (L) is the solution volume; m (g) is CAP dose.

3.1 Optimization of Experimental Parameters Affecting Phenol Adsorption

From Fig. 2a, the pH of solution has an important effect on the phenol removal adsorption. For the LDPO, CB, or CAP samples, in acidic medium, the phenol uptake increased to attain an optimum at pH = 6 and proportionally decreases to record its lowest value in alkaline environments. This can be explained by the electrostatic repulsion between the negative surface charge and phenol ions (anion-phenolate) in alkaline medium [46]. Therefore, pH solution was chosen as an optimum value to investigate the influence of other parameters.

The influence of contact time was examined and shown in Fig. 2c. Results reveal that the phenol removal onto both adsorbents increases with time until reaching a plateau phase. The phenol uptake was very fast in the first 20–30 min, and then was gradually slowly to reach equilibrium in about 40 min. This is attributed to the availability of more sites on the exterior surface adsorbent at beginning time. Then,

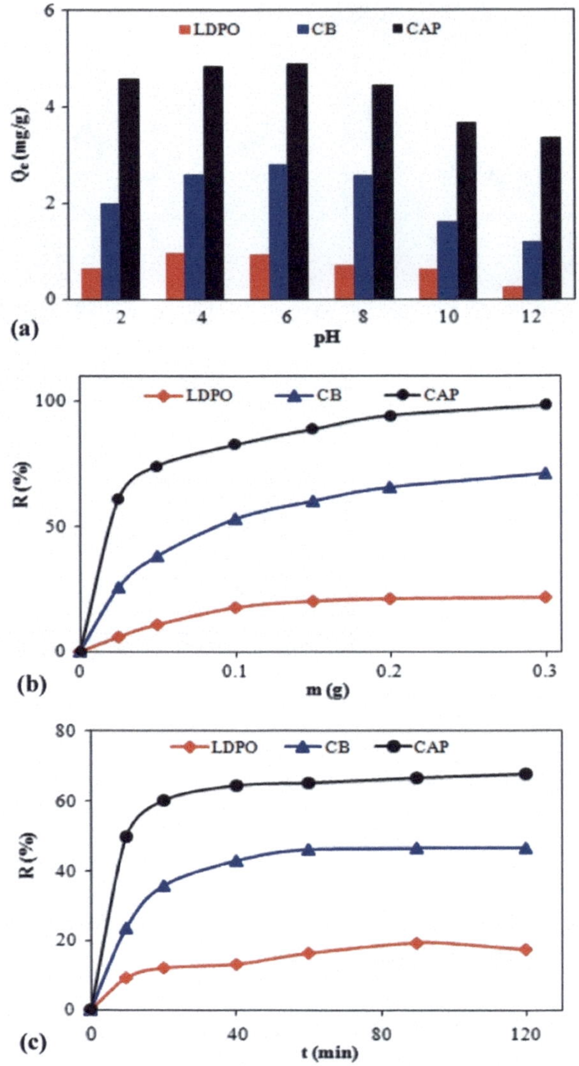

Fig. 2 Effect of the solution pH **a**, adsorbent dose **b**, and contact time **c** on the removal of phenol

phenol entered onto the pores and was absorbed by the interior surface. This step takes a relatively long contact time. The equilibrium time considered for the further work has been taken as 60 min to ensure equilibrium state.

Figure 2b proves that adsorbent dose and nature showed a remarkable influence on the uptake of phenol. The phenol removal onto the studied adsorbents increased with increasing adsorbent dose up to 0.15 g significantly and after that remained slowly. This result was expected because the increase in the dose of adsorbent solution does increase the adsorption sites on the surface and to a state of the solid; the adsorbent which generally has heterogeneous and large surface area has the maximum retention of organic molecules [21, 23]. Therefore, the maximum phenol removal efficiencies were about 20, 60, and 87%, respectively, for 0.15 g of LDPO, CB, and CAP. The highest phenol removal rate onto CAP could be there due to the large surface area and the competition of the CAP active site with phenol. Based on this ascertainment, 0.6 g of adsorbent was considered as a sufficient dose for phenol removal in this study.

3.2 Adsorption Isotherms

Obtained experimental data were fitted by nonlinear Dubinin–Radusckevich, Freundlich Langmuir and Temkin equations (Table 2).

$B = \mathrm{RT}/b_T$ is the constant related to heat of sorption (J/mol); $\xi = \mathrm{RT}\ \ln(1 + (1/C_e))$ is the D–R isotherm constant; R: universal gas constant (8.314 J/mol K); T: absolute temperature (K).

According to the results summarized in Table 3, the values of the correlation coefficients (R^2) are significantly higher for the Langmuir and D-R models compared to those of the Freundlich, Temkin models with all adsorbents, indicating that Langmuir and D-R models give the better fitting in the adsorption of phenol on DLPO, CB, and CAP, respectively.

According to the Fig. 3, it can be seen that Langmuir model describes the experimental data very well. Therefore, the maximum adsorption capacities (Q_m) determined from the Langmuir model for phenol were 1.21, 12.5, and 25 mg/g onto LDPO, CB, and CAP, respectively, consisting of a monolayer adsorption and the adsorbents sites are of identical affinities. The essential characteristics of Langmuir adsorption isotherm can be expressed by a dimensionless equilibrium parameter called separation factor (or equilibrium parameter), R_L, as reported in the literature [12, 21, 37].

$$R_L = \frac{1}{1 + K_L.C_0} \tag{3}$$

Table 2 Equation and models

Models	Linear equation	Plot
Langmuir [25]	$\frac{C_e}{Q_e} = \frac{C_e}{Q_m} + \frac{1}{Q_m K_L}$	$\frac{C_e}{Q_e}$ versus C_e
Freundlich [15]	$\log Q_e = \log K_F + \frac{1}{n} \log C_e$	$\log Q_e$ versus $\log C_e$
Temkin [43]	$Q_e = B \ln A_T + B \ln C_e$	Q_e versus $\ln C_e$
Dubinin–Radushkevich [11]	$\ln Q_e = \ln Q_s - (K\xi^2)$	$\ln Q_e$ versus ξ^2

Table 3 Calculated constants and correlation coefficients of the four isotherm models for the adsorption of phenol onto LDPO, CB, and CAP

Models		LDP O	CB	CAP
Langmuir	Q_m (mg/g)	1.21	12.5	25
	K_L (L/mg)	0.11	0.07	0.03
	R^2	0.99	0.99	0.99
Freundlich	K_F ((mg/g)(L/mg)$^{1/n}$)	0.20	1.34	2.19
	n	0.43	0.52	0.44
	R^2	0.81	0.82	0.83
Temkin	B (J/mol)	0.26	2.74	4.89
	A_T (L/g)	1.16	0.76	0.38
	b_T [(L/mg)$^\beta$]	9529.12	904.22	506.86
	R^2	0.93	0.96	0.96
D-R	q_s (mg/g)	0.99	9.02	18.17
	K (mol^2 kJ^{-2})	3.63	3.83	19.21
	E (kj/mol)	2.69	2.76	6.19
	R^2	0.98	0.96	0.98

The calculated values of the separation factor R_L for LDPO ($0.74 > R_L > 0.22$), CB ($0.42 > R_L > 0.06$), and CAP ($0.22 > R_L > 0.03$) were found to be between 0 and 1, revealing the favorable aspect of the adsorption.

In the D-R adsorption isotherm, the energy of adsorption ($E = (2\ \mathrm{K})^{-0.5}$) is corresponding to the necessary energy for one mole of adsorbate to transferred from the solution interface to the surface of adsorbent. The E value is less than 8 kJ/mol for physical adsorption and more than 8 kJ/mol for chemical adsorption. Moreover, from the E value in D-R model (< 8 kJ/mol for all adsorbents) (Table 3), the adsorption is deduced to be reinforced by physical process.

3.3 Thermodynamic Analysis

The thermodynamic parameters for the phenol adsorption process such as standard Gibbs free energy $\Delta G°$ (kJ/mol), standard enthalpy $\Delta H°$ (kJ/mol), and standard entropy $\Delta S°$ (kJ/mol K) were evaluated using the following equations and calculated from a graph of ln K_L versus $1/T$ [8, 12, 21].

$$\Delta G° = -\mathrm{RT}\ \ln K_D \tag{4}$$

where K_D is the equilibrium constant for adsorption, which can be calculated as follows [8]:

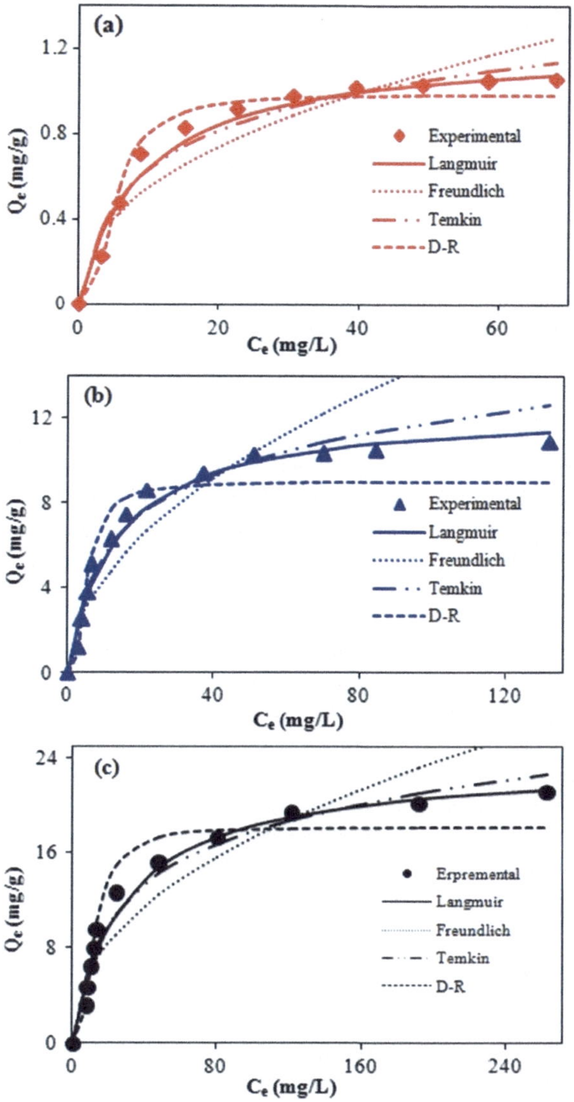

Fig. 3 Modeling results of the adsorption of phenol onto **a** LDPO, **b** CB, and **c** CAP

$$K_D = \frac{q_e}{C_e} \tag{5}$$

The relationship among the entropy change (ΔS°), the enthalpy change (ΔH°), and ΔG° is given by:

$$\Delta G^\circ = \Delta H^\circ - T\Delta S^\circ \tag{6}$$

Table 4 Thermodynamic parameters for the adsorption of phenol onto LDPO, CB, and CAP

Adsorbent	ΔH° (kJ/mol)	ΔS° (kJ/mol K)		ΔG° (kJ/mol)		
			T (K)	298	303	313
LDPO	−26.77	−0.84		−1.82	−1.40	−0.56
CB	−29.88	−0.83		−5.06	−4.82	−4.34
CAP	19.48	−0.48		−6.26	−6.67	−7.47

$$\ln K_D = \frac{\Delta S^\circ}{R} - \frac{\Delta \mathrm{H}^\circ}{\mathrm{RT}} \tag{7}$$

, where R is the ideal gas constant (8.314 J/(mol K)) and T is the absolute temperature (K).

The calculated values of ΔG°, $\Delta \mathrm{H}^\circ$, and ΔS° were given in Table 4. Generally, the ΔG° value between 0 and − 20 kJ/mol indicates a weak electrostatic interaction between adsorption sites and their adsorbate (physical adsorption), while a set ranging from − 80 to − 400 kJ/mol indicates that the adsorption involves charge sharing or transferring from the adsorbent surface to the adsorbate to form a coordinate bond (chemisorption) [12, 21, 23, 37, 46].

In this study, phenol adsorption was conducted at three temperatures of 298, 303, and 313 K. The results, listed in Table 4, prove that the value of ΔG° range from − 0.56 to − 1.82 kJ/mol, − 3.78 to − 5.03 kJ/mol, and − 4.34 to − 5.06 kJ/mol for LDPO, CB, and CAP respectively, when the temperature increases from 298 to 313 K. From the thermodynamic parameters, it is apparent that the phenol adsorption process was spontaneous and was dominated by physical sorption. The negative value of ΔH° indicated that the adsorption process was exothermic in nature. The positive value of ΔS° showed the decrease in the degree of freedom at the adsorbent/solution interface during the adsorption process.

4 Conclusion

According to the overall results, activated carbon prepared from *P. oceanica* seems to be a promising, cost-effective, eco-friendly adsorbent for the phenol removal from aqueous medium. Langmuir isotherm model perfectly describes experimental data compared to other theoretical models. The mechanism of the adsorption of phenol on the LDPO, CB, and CAP is favorable and mainly with physical in nature.

References

1. Abu-Nada A, Abdala A, Gordon McKay (2021) Removal of phenols and dyes from aqueous solutions using graphene and graphene composite adsorption: a review. J Environ Chem Eng 9:105858. https://doi.org/10.1016/j.jece.2021.105858
2. Abussaud B, Asmaly HA, Ihsanullah STA, Gupta VK, Laoui T, Atieh MA (2016) Sorption of phenol from waters on activated carbon impregnated with iron oxide, aluminum oxide and titanium oxide. J Mol Liq 213:351–359. https://doi.org/10.1016/j.molliq.2015.08.044
3. Al Bsoul A, Hailat M, Abdelhay A, Tawalbeh M, Al-Othman A, Al-kharabsheh IN, Al-Taani AA (2021) Efficient removal of phenol compounds from water environment using Ziziphus leaves adsorbent. Sci Total Environ 761:143229. https://doi.org/10.1016/j.scitotenv.2020.143229
4. Ben Mosbah M, Mechi L, Khiari R, Moussaoui Y (2020) Current state of porous carbon for wastewater treatment. Processes 8:1651. https://doi.org/10.3390/pr8121651
5. Benito-Gonzalez I, Lopez-Rubio A, Galarza-Jimenez P, Martínez-Sanz M (2021) Multifunctional cellulosic aerogels from *Posidonia oceanica* waste biomass with antioxidant properties for meat preservation. Int J Biol Macromol 185:654–663. https://doi.org/10.1016/j.ijbiomac.2021.06.192
6. Bettaieb F, Khiari R, Dufrence A, Mhenni MF, Putaux JL, Boufi S (2015) Nanofibrillar cellulose from Posidonia oceanica: properties and morphological features. Ind Crop Prod 72:97–106. https://doi.org/10.1016/j.indcrop.2014.12.060
7. Bin-Dahman OA, Saleh TA (2020) Synthesis of carbon nanotubes grafted with PEG and its efficiency for the removal of phenol from industrial wastewater. Environ Nanotechnol Monit Manage 13:100286. https://doi.org/10.1016/j.enmm.2020.100286
8. Cheng WP, Gao W, Cui X, Ma JH, Li RF (2016) Phenol adsorption equilibrium and kinetics on zeolite X/activated carbon composite. J Taiwan Ins Chem Eng 62:192–198. https://doi.org/10.1016/j.jtice.2016.02.004
9. Dengfeng Z, Peili H, Wei L (2016) Behavior of phenol adsorption on thermal modified activated carbon. Chin J Chem Eng 24:446–452. https://doi.org/10.1016/j.cjche.2015.11.022
10. Dhahri R, Guizani M, Yilmaz M, Mechi L, Alsukaibi AKD, Alimi F, Ben Salem R, Moussaoui Y (2022) Optimization of the preparation of activated carbon from prickly pear seed cake for the removal of Lead and Cadmium ions from aqueous solution. Sustainability 14:3245. https://doi.org/10.3390/su14063245
11. Dubinin MM, Radushkevich LV (1947) Equation of the characteristic curve of activated charcoal. Proc USSR Phys Chem 55:331–333
12. Elhleli H, Mannai F, ben Mosbah M, Khiari R, Moussaoui Y (2020) Biocarbon derived from Opuntia ficus indica for p-nitrophenol retention. Processes 8:1242. https://doi.org/10.3390/pr8101242
13. Elmorsi RR, Abou-El-Sherbini KS, El-Dein WAS, Lotfy HR (2022) Activated eco-waste of *Posidonia oceanica* rhizome as a potential adsorbent of methylene blue from saline water. Biomass Convers Biorefin. https://doi.org/10.1007/s13399-022-02709-5
14. Elmorsi RR, El-Wakeel ST, Shehab El-Dein WA, Lotfy HR, Rashwan WE, Nagah M, Abou-El-Sherbini KS (2019) Adsorption of methylene blue and Pb^{2+} by using acid-activated Posidonia oceanica waste. Sci Rep 9:1–12. https://doi.org/10.1038/s41598-019-39945-1
15. Freundlich HMF (1906) Uber die adsorption in losungen. Z Phys Chem 57:385–470
16. Galdino AL, Oliveira JCA, Magalhaes ML, Lucena SMP (2021) Prediction of the phenol removal capacity from water by adsorption on activated carbon. Water Sci Technol 84:135–143. https://doi.org/10.2166/wst.2021.196
17. Garcia DG, Carrillo LQ, Montanes N, Fombuena V, Balart R (2017) Manufacturing and characterization of composite fibreboards with Posidonia oceanica wastes with an environmentally-friendly binder from epoxy resin. Materials 11:35. https://doi.org/10.3390/ma11010035
18. Jebara A, Hassen W, Ousleti A, Mabrouk L, Jaziri A, Di Bella G,·Ben Mansour H (2022) Multidrug-resistant epi-endophytic bacterial community in *Posidonia oceanica* of Mahdia coast

as biomonitoring factor for antibiotic contamination. Arch Microbiol 204:229. https://doi.org/10.1007/s00203-022-02842-9
19. Kaouah F, Boumaza S, Berrama T, Trari M, Bendjam Z (2013) Preparation and characterization of activated carbon from wild olive cores (*oleaster*) by H_3PO_4 for the removal of basic red 46. J Clean Prod 54:296–306. https://doi.org/10.1016/j.jclepro.2013.04.038
20. Khadhri N, Ben Mosbah M, Rabaaoui N, Khadimallah MA, Moussaoui Y (2021) Advancements of Electrochemical removal of o-methylphenol from aqueous using BDD anode compared to Pt one: kinetics and mechanism determination. Iran J Chem Chem Eng 40:1446–1456. https://doi.org/10.30492/ijcce.2020.39880
21. Khadhri N, Saad MK, Ben Mosbah M, Moussaoui Y (2019) Batch and continuous column adsorption of indigo carmine onto activated carbon derived from date palm petiole. J Environ Chem Eng 7:102775. https://doi.org/10.1016/j.jece.2018.11.020
22. Khiari R, Belgacem MN (2017) Potential for using multiscale Posidonia oceanica waste: current status and prospects in material science. In: Jawaid M, Tahir PMD, Saba N Woodhead publishing series in composites science and engineering, lignocellulosic fibre and biomass-based composite materials. Woodhead Publishing, p 447–471. https://doi.org/10.1016/B978-0-08-100959-8.00021-4
23. Kumar A, Jena HM (2016) Removal of methylene blue and phenol onto prepared activated carbon from Fox nutshell by chemical activation in batch and fixed-bed column. J Clean Prod 137:1246–1259. https://doi.org/10.1016/j.jclepro.2016.07.177
24. Kuqo A, Korpa A, Dhamo N (2019) Posidonia oceanica leaves for processing of PMDI composite boards. J Compos Mater 53:1697–1703. https://doi.org/10.1177/0021998318808024
25. Langmuir I (1918) The adsorption of gases on plan surfaces of glass, mica and platinum. J Am Chem Soc 40:1361–1403. https://doi.org/10.1021/ja02242a004
26. Levent C, Zeynelabidin K, Hakan A, Hatice D, Demir GK (2011) Thomas and artificial neural network models for the fixed-bed adsorption of methylene blue by a beach waste *Posidonia oceanica* (L.) dead leaves. Chem Eng J 171:557–562. https://doi.org/10.1016/j.cej.2011.04.030
27. Mandal A, Bar N, Das SK (2020) Phenol removal from wastewater using low-cost natural bioadsorbent neem (Azadirachta indica) leaves: adsorption study and MLR modeling. Sustainable Chem Pharm 17:100308. https://doi.org/10.1016/j.scp.2020.100308
28. Marta I, Carmen G, Paula M, Francisco JH (2010) Modeling of copper fixed-bed biosorption from wastewater by *Posidonia oceanica*. Bioresource Technol 101:510–517. https://doi.org/10.1016/j.biortech.2009.08.018
29. Moreira FC, Boaventura RA, Brillas E, Vilar VJ (2017) Electrochemical advanced oxidation processes: a review on their application to synthetic and real wastewaters. Appl Catal B Environ 202:217–261. https://doi.org/10.1016/j.apcatb.2016.08.037
30. Naguib DM, Badawy NM (2020) Phenol Removal from Wastewater using Waste Products. J Environ Chem Eng 8:103592. https://doi.org/10.1016/j.jece.2019.103592
31. Najoua BD, Latifa B, Samar M, Ramzi K, Mohamed FM (2013) Macroscopic and microscopic studies of methylene blue sorption onto extracted celluloses from *Posidonia oceanica*. Ind Crops Prod 45:106–113. https://doi.org/10.1016/j.indcrop.2012.12.007
32. Ontoria Y, Gonzalez-Guedes E, Sanmarti N, Bernardeau-Esteller J, Ruizet JM, Romero J, Pérez M (2019) Interactive effects of global warming and eutrophication on a fastgrowing Mediterranean seagrass. Mar Environ Res 145:27–38. https://doi.org/10.1016/j.marenvres.2019.02.002
33. Priya DS, Sureshkumar MV (2020) Synthesis of Borassus flabellifer fruit husk activated carbon filter for phenol removal from wastewater. Int J Environ Sci Technol 17:829–842. https://doi.org/10.1007/s13762-019-02325-3
34. Qin C, Chen Y, Gao J (2014) Manufacture and characterization of activated carbon from marigold straw (*Tagetes erecta* L) by H_3PO_4 chemical activation. Mater Lett 135:123–126. https://doi.org/10.1016/j.matlet.2014.07.151
35. Rammou E, Mitani A, Ntalos G, Koutsianitis D, Taghiyari HR, Papadopoulos AN (2021) The potential use of seaweed (*Posidonia oceanica*) as an alternative lignocellulosic raw material for wood composites manufacture. Coatings 11:69. https://doi.org/10.3390/coatings11010069

36. Ruiz JM, Marin-Guirao L, Garcia-Munoz R, Ramos-Seguraet A, Bernardeau-Esteller J, Pérez M, Sanmarti N, Ontoria Y, Romero J, Arthur R, Alcoverro T, Procaccini G (2017) Experimental evidence of warming-induced flowering in the Mediterranean seagrass Posidonia oceanica. Mar Pollut Bull 134:49–54. https://doi.org/10.1016/j.marpolbul.2017.10.037
37. Saad MK, Mnasri N, Mhamdi M, Chafik T, Elaloui E, Moussaoui Y (2015) Removal of methylene blue onto mineral matrices. Desalin Water Treat 56:2773–2780. https://doi.org/10.1080/19443994.2015.1012338
38. Saad MK, Rabaaoui N, Elaloui E, Moussaoui Y (2016) Mineralization of p-methylphenol in aqueous medium by anodic oxidation with a boron-doped diamond electrode. Sep Purif Technol 171:157–163. https://doi.org/10.1016/j.seppur.2016.07.018
39. Said KAM, Ismail AF, Karim ZA, Abdullah MS, Hafeez A (2021) A review of technologies for the phenolic compounds recovery and phenol removal from wastewater. Process Saf Environ Prot 151:257–289. https://doi.org/10.1016/j.psep.2021.05.015
40. Saleh TA, Adio SO, Asif M, Dafalla H (2018) Statistical analysis of phenols adsorption on diethylenetriamine-modified activated carbon. J Clean Prod 182:960–968. https://doi.org/10.1016/j.jclepro.2018.01.242
41. Singh R, Dutta RK, Naik DV, Ray A, Kanaujia PK (2021) High surface area Eucalyptus wood biochar for the removal of phenol from petroleum refinery wastewater. Environ Chall 5:100353. https://doi.org/10.1016/j.envc.2021.100353
42. Stoycheva I, Tsyntsarski B, Petrova B, Georgiev G, Budinova T, Petrov N, Trzebicka B, Pusz S, Kumanek B, Szeluga U (2022) Investigation of the possibilities for removal of phenolic toxic compounds from water by nanoporous carbon from polymer by-products. Appl Sci 12:2243. https://doi.org/10.3390/app12042243
43. Temkin MJ, Pyzhev V (1940) Recent Modifications to Langmuir isotherms. Acta Physiochimca URSS 12:217–225
44. Tsechansky L, Graber ER (2014) Methodological limitations to determining acidic groups at biochar surfaces via the Boehm titration. Carbon 66:730–733. https://doi.org/10.1016/j.carbon.2013.09.044
45. Wise LE, Murphy M, D'Addieco AA (1946) Chlorite holocellulose: its fractionation and bearing on summative wood analysis and on studies on the hemicellulose. Paper Trade J 122:35–43
46. Zhongxin L, Manglai G, Senfeng Y, Qiang Y (2015) Adsorption of phenols on reduced-charge montmorillonites modified by bispyridinium dibromides: mechanism, kinetics and thermodynamics studies. Colloids Surf A Physicochem Eng Asp 482:222–230. https://doi.org/10.1016/j.colsurfa.2015.05.014

Opportunities and Challenges of Using *Posidonia oceanica* for Bioenergy and Agronomic Applications

Mohamed Ali Wahab, Yasmin Cherni, Zeineb Louati, and Naceur Jedidi

Abstract *Posidonia oceanica* is a native Mediterranean seagrass forming big grasslands throughout the coastline and playing a primary role in the Mediterranean ecosystems, oxygen regeneration, water filtration, and increasing the total organic carbon in seabed. Nevertheless, large volumes of Posidonia wrack (leaves and rhizomes) deposit onto beaches Posidonia, during the winter, poses a range of challenges for local authorities, such as odor and visual appearance which reduce beach use and imply substantial economic costs for its removal. *P. oceanica* wrack are often removed as part of beach cleaning activities in many localities and are disposed into landfills. Such conventional management strategy is responsible for the loss of valuable organic carbon waste that could be converted into biofuel and soil organic amendment. Through the last decades, *Posidonia* was used as feedstock for both biochemical and thermochemical conversion processes to produce biofuel and materials. Furthermore, attempts were also made to produce a compost-based *Posidonia* with good agronomic quality. Based on the findings of previous studies, *P. oceanica* could be considered as a potential biomass resource to develop biofuel and organic agriculture sectors in the Mediterranean region by adopting a circular economy approach. Nevertheless, modification methods are required to enhance the biological conversion of Posidonia into biogas and bioethanol. Further investigations are required to determine the salinity risk associated with the use of soil organic amendment produced from *P. oceanica*.

Keywords *Posidonia oceanica* · Biofuel · Biochemical conversion · Thermochemical conversion · Composting

---

M. A. Wahab (✉) · Y. Cherni · Z. Louati · N. Jedidi
Laboratory of Wastewater Treatment and Valorization, Water Research and Technologies Center, CERTE, Technopark Tourist Route of Soliman Nabeul, Soliman, Tunisia
e-mail: mohamedali.wahab@certe.rnrt.tn

Carthage University, Carthage, Tunisia

Z. Louati
Faculty of Sciences Bizerte, Carthage University, Carthage, Tunisia

R. Khiari and M. Jawaid (eds.), *Current Status and Opportunity in Fibre and Composites*, Sustainable Materials and Technology,
https://doi.org/10.1007/978-981-96-4546-6_10

1 Introduction

During the last century, human society has developed an economic system based on fossil fuels, chemical fertilizers, and pesticides to respond to the high need for food and energy due to rapid industrialization and the growth of the global population. Such unsustainable economic system led to many environmental issues such as the depletion of nonrenewable fuel sources, climate change, and environmental pollution. In fact, about eighty percent of global energy supply is covered by fossil fuels [6]. This global energy demands are increasing with the increase of population and development of the industrial sector, resulting in overexploitation of nonrenewable reserves of fossil fuels [83, 81]. The use of these fuels is the primary contributor to global warming through the increase of greenhouse gas (GHG) emissions. As a consequence, it is mandatory to develop the biomass-based biofuel sector as an alternative to fossil-based fuels in order to meet the increasing demand for energy in a sustainable manner [34]. Because first-generation biofuel can compete with food, producing energy from waste biomass is a sustainable alternative.

Recently, biorefinery was proposed as an effective tool to promote circular economy and overcome environmental pollution by recycling organic waste into biofuel and biocommodities [95]. Furthermore, biorefineries are considered as a waste management strategy since the feedstocks used are mainly organic waste generated from agriculture, forest, aquatic, and agro industrial. Biorefinery is an integrated process combining different biochemical and thermochemical processes generating biofuels and products that could be used for industrial and agriculture applications as alternatives to synthetic chemicals with high carbon footprint.

Waste biorefineries are gaining popularity as long-term waste management solutions around the world.

Anaerobic fermentation is the main biochemical process generating biomethane, biohydrogen, and bioethanol. The main thermochemical processes are pyrolysis, hydrothermal carbonization, and gasification, producing mainly biochar in addition to liquid fuel and gas. By-products yield and proprieties vary depending on the composition of the substrate and operating conditions. Biochemical conversion processes involve the use of microorganisms which require a lot of time going from hours to weeks depending on the rate of digestibility of the substrate. On the other hand, thermochemical processes generate many by-products in less time under high temperatures [2]. Nevertheless, these processes are a high consumer of energy and involve the use of a solvent or a catalyst. Thereby, combining thermochemical and biochemical routes in biofuel processing can be beneficial by incorporating the advantages of both methods [94].

Biomass-based fuels are mainly derived from three biofuel generation. The first-generation biofuels are generated from edible energy crops such as sugarcane, sorghum corn, sunflower, and canola [55]. Second-generation biofuels can be produced from crop waste which contributes to reducing the pressure on edible energy crops and competition for land and food. Nevertheless, lignocellulosic biomass requires physicochemical and enzymatic pretreatments to improve their degradability and to increase the energy yield recovered [39]. The third-generation biofuels are produced from algae representing a sustainable solution to overcome the drawbacks of the first- and second-generation biofuels [8].

Posidonia oceanica (PO) dead leaves are a marine organic waste that could be considered as a futuristic biomass for biofuel generation using biochemical and thermochemical conversion processes. PO dead leaves are a lignocellulosic biomass that has a similar advantage to algae as compared to terrestrial biomass.

PO biomass has a good heating value making it a perfect feedstock for biofuel production through thermochemical processes, in addition to its high carbon content allowing its use as substrate for the production of composts [24].

This chapter presents the physicochemical properties of *P. oceanica* and its potential to be used as a substrate for biofuel through biochemical and thermochemical routes and its application as fertilizer and soil amendment.

2 *P. oceanica* A Potential Marine Biomass for Biofuel Production

Marine ecosystems generate a huge quantity of organic waste which makes their management and treatment an economic and environmental problem for the coastal community. Marine organic waste accumulated on the beaches is mainly composed of seaweeds and seagrass wrack. Contrary to seaweed which are highly biodegradable, seagrass wrack has an important percentage of lignin and cellulose against a low nutrient content which reduces their decomposition and increases their persistence on coastal beaches. Seagrass wrack results from the accumulation of dead seagrass leaves reaching the coast through the wave movements before their breaking down [49].

In the Mediterranean Sea, seagrass wrack originated from *P. oceanica* which accumulates between autumn and winter when plants renew their leaves and completely cover beaches for long period creating beds of two meters in height [9]. *P. oceanica* banquette plays the role of natural dock, acting as physical barrier to the action of wave which reduces erosion [85].

PO is a phanerogamic plant that plays an important role in preserving marine ecosystems in the Mediterranean Sea, reducing coastal erosion, and contributing to CO_2 sequestration [47]. PO grasslands are critical to marine ecological balance since they provide habitat and food for a variety of fish [82, 63]. PO seagrass also plays a major function in carbon sequestration in Mediterranean Sea through photosynthesis process and organic litter accumulation in seabed [47]. In this frame, PO loses a high part of leaves and rhizomes, which reach the coastline and have a negative impact on beach quality, particularly in touristic areas. The biodegradation of PO residues generates odor mainly H_2S, and contributes to the growth of beach flies [57], which affects the touristic activities on coasts and beaches. Furthermore, PO decomposition process contributes to the generation of GHG which depends on the season and the biochemical properties of the PO residues [40].

For instance, seagrass wrack can generate up to 176 mg of methane daily per square meter depending on the temperature. In addition, some studies revealed that the decomposition of these residues has contributed to increasing GHG emissions and global temperatures [11]. Over one year, the beach-cast seagrass emissions of methane can reach 0.14 mmol CH_4/m^2g [53]. Accordingly, seagrass can be considered a suitable feedstock to generate biomethane under controlled conditions. Deciphering the effect of biotic and abiotic factors on the fluctuations of emissions can help to better control the anaerobic fermentation of PO residues to enhance biomethane production.

The removal and management of excessive accumulations of wrack represent an economic issue for the coastal city municipalities due to the high investment costs required to ensure the collection and disposal of huge quantities of PO wrack in landfills or through conversion waste facilities [41]. Falco et al. [29] have estimated the total removed amount of *P. oceanica* banquette applied on 44 beaches (out of 116) of the island of Sardinia at 106,180 m^3 indicating that more than eighty percent of these waste was discarded in unofficial areas.

PO residues are used for a variety of applications, such as material in buildings, mulching in plant production, amendment for soil, fibers for textile industry, packaging protector, feed for livestock, and in pulp and paper production [15]. However, due to the development of tourism activities on the most Mediterranean coast, *P. oceanica* leaves are more and more assimilated as waste and dumped in landfills which resulted in the loss of massive quantities of organic matter [12].

Recently, the recycling of PO as marine bioresource for biofuel is more considered in several Mediterranean countries (Pardilhó et al. 2022). Due to its high availability and renewal character, using PO as substrate for biofuel production is attracting more interest from several researcher [37]. Furthermore, many authors have investigated the biochemical composition of PO and its potential application in energy and industrial sectors.

According to Table 1, the proximate composition of *P. oceanica* analyzed by different authors indicates that this marine biomass could be considered as a good feedstock for biofuel. In fact, PO has a low moisture content (lower than 13%) which makes it a good feedstock for thermochemical conversion processes especially pyrolysis and gasification. Furthermore, PO has a high volatile matter content which could be converted into biogas thorough anaerobic fermentation processes and bio-oil through pyrolysis [94].

Besides its important volatile matter content which exceeded 50% of the total weight, *P. oceanica* has a high carbon content ranging between 30 and 60% as indicated in Table 2. Such high carbon content makes *P. oceanica* more interesting in terms of biofuel production as compared to seaweed which has low carbon content and high ash content [95]. Moreover, *P. oceanica* has a low oxygen, sulfur, and nitrogen content which could produce a good quality bio-oil in comparison to seaweed and might avoid nitrogen and sulfur oxide emissions during the combustion of the pyrolytic oil [94].

Table 1 Proximate composition of *P. oceanica* from literature

Proximate analysis	Moisture (wt%)	Volatile matter (wt%)	Fixed carbon (wt%)	Ash (wt%)	HHV (MJ/kg)	References
Raw Posidonia	8.54	51.61	12.07	44.46	–	Moltó et al. [54]
	12.7	56.9	16.4	14.0	–	Pedicini et al. [64]
	5.5	54.4	14.0	26.1	12.816	Plis et al. [70]
	10.7	53.1	19.2	17.0	14.88	Chiodo et al. [22]
	12.67	60.95	12.36	14.02	17.2	Maisano et al. [43]

Table 2 Elemental composition of *P. oceanica* from literature

Ultimate analysis (wt%)	C	H	N	S	O	References
Raw Posidonia	29.28	3.62	0.77	0.87	–	Moltó et al. [54]
	40.3	6.0	1.1	0.3	25.7	Pedicini et al. [64]
	35.5	3.6	0.27	0.4	28.85	Plis et al. [70]
	31.31	3.33	1.08	0.14	36.44	Chiodo et al. [22]
	58.9	8.3	1.3	0.2	31.3	Jeguirim et al. [32]
	46.1	6.82	1.28	0.3	31.41	Maisano et al. [43]
	42.1	6.44	1.52	3.78	34.66	Ncibi et al. [59]

PO heating value varies from 12 to 18 MJ/kg (Table 1) due to its important carbon content. The high heating value of *P. oceanica* is similar to some woody feedstocks and straw [23].

The high heating value of *P. oceanica* allowed its application as feedstock for hydrothermal gasification in Turkey [27]. Moreover, *P. oceanica* has been identified as an excellent additive during the combustion of solid fuels such as coal [88].

3 Biochemical Conversion of *P. oceanica* Residue

3.1 *Pretreatment and Fermentation of* P. oceanica *Residue*

The production of bioethanol from marine organic waste is gaining popularity as clean and sustainable energy source. Bioethanol has less environmental issues than petroleum products emitting less atmospheric pollutants [5]. Thus, biotechnology advances towards improving the sustainability of bioethanol production are critical to ensure green energy transition. Bioethanol is conventionally made from first-generation feedstocks such as wheat, corn, and sugarcane [1]. However, because these biomasses are also edible crops, their use for biofuel could compromise food security. Marine biomass could replace these feedstocks for bioethanol production. To date, seaweeds which have low concentrations of lignin and high carbohydrate content, are considered the most promising marine substrate for bioethanol production [84]. Despite its high lignin content, *P. oceanica* residue was also proposed as a substrate for fermentative microorganisms (bacteria and yeast) to produce bioethanol after some pretreatment. Indeed, the composition of lignocellulosic waste is made of three main components; cellulose, hemicellulose, and lignin. Carbohydrates in *P. oceanica* represent more than eighty percent display of its total dry weight [48]. Furthermore, *P. oceanica* is a highly available seagrass with a high amount of holocellulose (65–75%) which makes it a unique candidate for both utilization of sugars and subsequently bioethanol production [35]. In light of the information gathered from the literature, several studies have determined PO composition presented in Table 3. The results demonstrate that PO balls contain a high concentration of lignin and cellulose compared to raw leaves.

The breakdown of carbohydrates in PO through pretreatment and hydrolysis can generate sugars that could be fermented to produce ethanol. However, the high lignin content (a complex heteropolymer) reduces the biodegradation rate of PO through fermentation which requires the application of chemical and enzymatic pretreatments. Pretreatments increase the accessibility of PO lignocellulosic matrix to enzymatic hydrolysis through the removal of lignin and the increase its porosity. Pretreatments reduce cellulose crystallinity which increases the effectiveness of hydrolytic microorganisms to convert complex carbohydrate polymers into monomeric sugars for fermentation [45]. Previous studies indicated that PO delignification and bleaching could produce 60% of cellulose with alkali solution and 45% of pure cellulose [10]. Pure microcrystalline cellulose was also extracted from *P. oceanica* when delignified and treated with alkali and acid treatments [82]. Physicochemical pretreatment methods such as alkali and acid treatments, steam treatment, and ammonia fiber explosion are the most pretreatment methods used to produce reducing sugar for fermentation processes. Nevertheless, chemical methods generate some inhibitors during the degradation of lignin. In fact, the breakdown of lignin through chemical pretreatment generates several aliphatic and aromatic by-products

Table 3 *P. oceanica* biomass composition

	Lignin (%)	Hemicellulose (%)	Cellulose (%)	References
Raw balls	23.8	13.6	42.9	Pilavtepe et al. [67]
Raw balls	28.2	23.3	32.5	Tarchoun et al. [82]
Raw leaves	29.3	25.7	31.4	Bettaieb et al. [10]
Raw balls	29.8	21.8	40.0	Bettaieb et al. [10]
Raw leaves	27	21	38	Ncibi et al. [59]

in addition to the further degradation of hexose and pentose sugars to weak organic acids and various furan products.

Enzymatic hydrolysis is proposed as an alternative to reduce inhibitory product formation. Among the potent enzymes produced by several different species of fungi, cellulases were applied to improve bioethanol production from *P. oceanica.* Bacteria isolated from Tunisian desert with high cellulase activity were selected to produce ethanol using *P. oceanica* leaves as a low-cost cellulosic substrate [80].

TL13 strain was used to produce an enzyme able to degrade polysaccharide from raw PO into fermentable sugars for bioethanol production reaching about 34% of the theoretical maximum. Such promising results open the doors in front of the development of seagrass-based biorefinery using Microbacterium biocatalysts [80].

In 2013, Pilavtepe and co-workers investigated the pretreatment of *P. oceanica* waste using acid and enzymatic pretreatments to produce bioethanol. In this study, reducing sugar concentrations have reached a maximum of 27.6 and 28.4 g/100 g when using 2% H_2SO_4 and 16 FPU/g of enzyme concentration, respectively. When enzymatic and acid hydrolysis were applied consecutively reducing sugar reached 52.3 g/100 g. At 400 °C and a solid/liquid ratio of 0.08 (w/v) the concentration of reducing sugar in liquid fraction was about 14 g/L generating 0.46 $kg/m^3/h$ and 0.73 $kg/m^3/h$ of bioethanol in flasks and 2 L CSTR, respectively. The potential of PO biomass for bioethanol production was also studied using consecutive supercritical CO_2 and supercritical water hydrolysis [27].

Rajkumar et al. [72] have investigated bioethanol production through the pretreatment of seagrass by acid and cellulase isolated from marine actinobacterial strain.

Masri et al. [48] have tested the use of seagrass hydrolysate as substrate for fermentation to produce biodiesel using oleaginous yeast. *P. oceanica* hydrolysate was obtained at 1.5% w/dw biomass, 15 mg/g biomass, and thermophilic conditions for three days. Oleaginous yeast *Trichosporon oleaginosus* grown faster in *P. oceanica* hydrolysate than cultivation medium made of six other seagrass hydrolysates. Glucose in *P. oceanica* hydrolysate represented 56%, making it a perfect substrate for fermentation processes. *P. oceanica* hydrolysate contained high glucose concentration (28.5 g/L), various nutrients, trace elements, and low nitrogen content needed for induction of lipogenesis in oleaginous microorganisms. *Trichosporon oleaginosus* grown in PO hydrolysate accumulated 45% of lipids which is close to the lipids content (55%) accumulated using standard medium. *P. oceanica* hydrolysate produces lipids mainly in triglycerides form generating biodiesel of high quality. The price of microbial oil produced from *P. oceanica* is around $7.3 kg/L.

Using *Streptomyces roseochromogenes* ATCC 13,400, *P. oceanica* was able to produce extracellular melanin at 0.098 g/L/h [73].

3.2 *Anaerobic Fermentation of* P. oceanica

Anaerobic fermentation is another biochemical conversion process of biomass that has become more widely implemented and used, mostly as a result of the demand for more sustainable methods of waste reduction and increased generation of renewable energy. Climate change and the growing demand for non-fossil fuel sources to fulfill industry activities have been the main drivers of the emergency of anaerobic processes for biogas production [60].

Without oxygen, anaerobic bacteria can turn organic substrate into biogas (between 50 and 75% CH_4 and 25 to 50% CO_2), which can help to reduce GHG emissions and mitigate climate change [33].

Temperature, pH, C/N ratio, alkalinity, organic loading rate, hydraulic retention duration, and volatile fatty acid concentration are only a few of the operational factors that impact anaerobic digestion. Methane and carbon dioxide make up the majority of biogas, with smaller amounts of hydrogen sulfide, nitrogen, hydrogen, ammonia, and water vapors (Abbasi et al. 2012).

Anaerobic fermentation process can be divided into four stages: hydrolysis, acidogenesis, acetogenesis, and methanogenesis involving different group of microorganisms. Hydrolytic bacteria transform big molecules into small components like sugar during the initial stage of anaerobic digestion known as hydrolysis. Fermentative bacteria convert sugar and other streamlined molecules into propionic, butyric, and other acids by the following process, known as acidogenesis. The process of acetogenesis is when homoacetogens transform an acidogenesis result into acetate. Acetoclastic and hydrogen-utilizing methanogens perform methanogenesis, which is the stage by which CH_4 is created from acetate and carbon dioxide, as the process's last step (carbon dioxide reducing methanogens). While hydrogen-using methanogens make methane using carbon dioxide and hydrogen as the electron acceptor and donor, respectively, acetoclastic methanogens transform acetate into methane and carbon dioxide [79].

Different lignocellulosic biomasses were used as substrates to produce biomethane through AD process. However, high lignin content reduces the efficiency of hydrolytic bacteria breakdown hemicellulose and cellulose which requires the application of chemical and biological pretreatments [90].

Posidonia's suitability as a substrate for biomethane production may be impacted primarily by lignin content and the salinity of the unwashed residues [91]. In fact, excessive salinity and concentrations of certain ions might hinder the anaerobic microbes activity, resulting in the reduction of the process efficiency [21].

Biochemical methane potential (BMP) tests were used to evaluate methane production from seagrass waste at mesophilic temperatures and high salinity [53]. The results of BMP indicated that the methane production increased when temperature rises from 35 °C to 30 °C reaching 0.90–1.37 mL/g VS per day. The metagenomics results indicated that at low salinity levels, Ruminococcaceae known for their ability to degrade carbohydrates were dominant, while at high salinity levels, Petrotogaceae known for their ability to ferment carbohydrates were more abundant. The maximum methane production obtained from PO residue was about 263 mL CH_4/gVS, depending on moisture content and inoculum/substrate ratio [52].

According to Khiari et al. [36], lignin can represent roughly 30% of the PO dry weight which requires pretreatments to improve its anaerobic digestion. Similarly, De Sanctis and Di Iaconi [76] have indicated that the PO digestibility under anaerobic conditions was poor due to the high content of lignin. Thermal pretreatment combined with acid treatment (0.4% HCl) of *P. oceanica* was efficient in increasing CH_4 production by five folds as compared to thermal pretreatment only [75].

In another study, washing salt combined with thermal treatment have not a significant effect on the methane produced. However, the combination of thermal treatment at 132 °C and hydrochloric acid (0.4% w/w) allowed the enhancement of PO digestibility and as consequence the reduction of VS by 50% [76].

Based on a quantitative study conducted by Balata and Tola [7], it indicated that for industrial scale, it is important to investigate the energy recovered from PO co-digested with other local abundant organic waste. In this frame, seagrass was used as co-substrate with fish wastes and macroalgae to produce biogas as a sustainable circular economy concept in Small Island with high industrial activity in aquaculture and seafood production [58].

A simplified energy analysis conducted by Misson et al. [52] indicated that for 8.5 tons of seagrass produced per year, it is possible to ensure a yearly income of 190,000 € from methane-generated wastewater treatment plants with anaerobic digestion unit process and reduce the cost for landfill disposal estimated at more than 600,000 €/y.

4 Thermochemical Conversion *P. oceanica*

4.1 Thermal Properties of P. oceanica

The thermal proprieties of marine biomass are determined by the thermogravimetric analysis (TGA) [94]. TGA technique ensures to evaluate the behavior of PO during the pyrolysis process [70]. This analysis is applied to monitor the weight change of a sample (heated at a constant rate) during a period of against time or a temperature gradient. TGA is a destructive analysis using a tiny sample for a quantitative or qualitative analysis. With TGA it is possible to determine the proximate composition of a sample in addition to its stability [95].

The derivative thermogravimetric (DTG) curve indicates if the sample decomposition over a gradient of temperature occurs easily at a low temperature and allows to determine the volatile matter released through the pyrolysis reactions [22].

The determination of the thermal degradation stages allows to better understand the transformation of *P. oceanica* component during the pyrolysis. PO decomposition was performed by heating the sample from ambient temperature to 800 °C with at a slow heating rate of 10 °C/min in the presence of N_2 to determine its proximate composition [64]. Three main degradation steps were determined from the TGA profiles of *P. oceanica.* The first stage for T < 120 °C was attributed to water and carbohydrates release while volatile matter release occurred up to 600 °C, representing the main mass loss of the *P. oceanica.* During the last step between 600 and 800 °C, the carbonaceous matter formed in the second stage can decompose. Chiodo et al. [22] revealed that at 320 °C, *P. oceanica* weight loss is more important than that of an algal biomass.

Typically, the thermal stability of marine biomass is lower than terrestrial waste, which may limit the energy required for pyrolysis and, as a consequence, reduce the operating costs of the thermal conversion of *P. oceanica.* Plis et al. [70] have investigated the thermal stability of *P. oceanica* using TGA. Between 313 and 450 K, the cellulose decomposition represented only 2.2% of the mass loss while in the second (450 K and 750 K) and third stage (750 K and 1273) the mass loss was more important with 27.5% and 19.7%, respectively, due to hemicellulose and lignin decomposition. For temperatures ranging between 613 and 973 K, the thermal degradation was reduced due to the slow carbonization of lignin. After the pyrolysis process, only 50% of the feedstock was decomposed.

Plis et al. [70] have investigated the thermal decomposition of PO which was divided into four steps. During the first step was the moisture removed resulting in a reduction of 1.6% of the feedstock mass between 313 and 450 K. In the second stage, the mass loss reached 28.6% between 450 and 685 K resulting from hemicellulose and cellulose decomposition. For temperature range between 685 and 850 K, occurred the decomposition of the formed char into volatiles and solid residues. In the fourth step, the leftover char and the mineral inorganic residues decomposed resulting in a residual mass of about 46%.

Plis et al. [70] indicated that PO volatile metal and carbonate decomposed between 1023 and 1073 K. In the 500 K to 700 K range, CH_4, NH_3, H_2O, CO, and CO_2 were produced while CH_4 and CO_2 were generated in a distinct range of temperature between 900 and 1050 K.

Jeguirim et al. [32] have compared the thermal degradation proprieties of *P. oceanica* under N2 atmosphere using TGA analysis. Results showed that under inert and oxidative atmosphere, moisture was released during the first step followed by the release of volatiles. Under inert atmosphere, the char was formed through the mechanism of nucleation, while under oxidative atmosphere, the char decomposed through area phase boundary-controlled reaction.

The thermal characterization of *P. oceanica* was also investigated under different oxygenation conditions [54]. During the combustion, the volatiles generated were oxidized in short time. Over 500 °C, the residual mass left was the result of the decomposition of the char formed. However, in inert atmosphere (nitrogen only), from 250 °C to 350 °C, the low molecular weight compounds began to volatilize and the residual fraction represent the ash and solid carbonaceous residues that were not oxidized in the presence of N_2.

The thermal analysis of PO gasification was performed in nitrogen-steam atmosphere [44]. DTG analysis highlighted that char gasification reaction occurred between 840 and 1135 K. The gasification of the formed char was detected between

836 and 920 K due to the important ash content (14 wt%) in PO which increased thermal conductivity and intra-particle heating transfer [89]. Accordingly, the gasification of *P. oceanica* occurred at low temperatures as compared to woody biomass [44].

4.2 Pyrolysis and Carbonization of P. oceanica

Pyrolysis was widely used for the thermal degradation of lignocellulosic waste. Endothermic reactions result in the production of three by-products; char, oil, and gas under inert atmosphere [31]. Factors such as feedstock nature and operating conditions influence pyrolytic product yield. During the pyrolysis steps, the biomass can undergo different thermochemical reactions such as isomerization, aromatization, depolymerisation, and charring. Typically, biomass pyrolysis occurs in five steps. First, the heat is transferred from the heating source to the feedstock then the reaction starts with increasing the temperature leading to the generation of vapors, volatiles, and carbonaceous residue formation. During the third step, a heat transfer occurred between unpyrolysed fuel and hot volatiles which condensed during the fourth step to form tar through secondary reactions. During the last step, primary reactions occur in conjugation with autocatalytic secondary pyrolysis reactions [26].

Among the pyrolysis by-products, biochar, a porous carbonaceous product, was recently used as a material in different research areas including wastewater treatment, soil amendment, and bioremediation [56]. Thanks to its physicochemical properties, such as high porosity and specific surface area [3], biochar has received significant attention due to its role in mitigating global warming, increase carbon sequestration, combat soil degradation, enhance crops production, and reduce environmentally [96].

Producing Posidonia-based biochar was investigated by Macreadie et al. [41], who revealed that the pyrolysis of PO to produce biochar could be considered ecofriendly alternative to landfilling. In this frame, Zaafouri et al. [97] showed that the highest biochar yield produced from *P. oceanica* was about 44% at 450 °C and decreased with increasing the heating temperature to 32.6% at 600 °C.

High yield of biochar could be generated from *P. oceanica* reaching 75% at temperatures lower than 500 °C [54]. Increasing temperature have tendency to decrease H/C ratio resulting from the drop of organic carbon content from 38 to 22% against an increase of the pH from 8.02 to 10.38 due to the high accumulation of ash (57%) [54].

On the other hand, increasing temperature to 600 °C increased the surface area from 41 to 2810 m^2/g and the carbon content from 58 to 93 wt%, after chemical activation with KOH [64]. At 4:1 KOH/biochar ratio, the biochar micropores increased and allowed the condensation of larger aromatic networks. However, at 2:1 ratio, the mesoporosity of the biochar increased in addition to the increase of organic radical content by almost four orders of magnitude resulting from the high ash content of the starting biochar [68].

Biochar-based *P. oceanica* could also be used for electrochemical applications. Indeed, activated carbon from *P. oceanica* prepared by pyrolysis was used as electrodes for supercapacitor devices with an excellent electrochemical reversibility. The characterization of the Posidonia-activated carbon revealed a polycrystalline graphitic structure and a layered structure with micropores in addition to a high carbon percentage (60%) [13].

The biochar produced from PO was also proposed as promising sorbent material for the elimination of different kinds of pollutants from wastewater. Biochar-based Posidonia demonstrated a high adsorption capacity of trace metals from wastewater. The pyrolysis of PO followed by a chemical activation produced an activated biochar with a specific surface area and a cumulative pore volume of 1563 m^2/g and 0.74 cm^3/g, respectively allowing a rapid and high adsorption capacity for Cr(VI) of 120 mg/g [4]. Cataldo et al. [19] have also activated PO-based biochar to remove Cd^{2+}, Pb^{2+}, and Cu^{2+} ions. In addition to their high adsorption capacity towards heavy metals removal, PO-based biochar was also efficient to adsorb hydrocarbons from synthetic solution [20].

The production of bio-oil from *P. oceanica* was also investigated by catalytic pyrolysis process resulting in 81% and79% conversion rate using Ni/HZSM-5and CeO_2, respectively, allowing the generation of high-quality bio-oil with lower content of acids and oxygen and a higher hydrocarbon content, combined with higher heating value [44].

4.3 Gasification and Combustion of **P. oceanica**

In addition to pyrolysis, PO residue was also used as a feedstock in gasification and combustion [70]. However, the high ash content (16–36%) of *P. oceanica* limits its application in direct combustion [88].

Plis et al. [70] have investigated PO combustion using a thermogravimetric analysis and a fluidized-bed reactor. The results indicated a good combustion performance of *P. oceanica* in the fluidized-bed reactor occurring in two stages. First, a short stage of 30 s during which devolatilisation reaction occurred followed by a combustion stage of char, characterized by a relatively low SO_2 emission.

Accordingly, the low sulfur and nitrogen content of *P. oceanica* allows the use of this marine biomass as a promising feedstock for combustion or co-combustion [88].

Furthermore, *P. oceanica* gasification into a syngas composed of a mixture of carbon monoxide, hydrogen, carbon dioxide, methane was also proposed as a route for the energetic valorization of this marine biowaste. PO gasification in the presence of oxygen and steam was investigated by the addition of different amounts of dolomite. The reforming of syngas produced was characterized by a H_2/CO ratio ranging from 0.3 to ca. 3 [25].

PO hydrothermal gasification was also promising in terms of hydrogen and/or methane production. Temperatures between 300 and 600 °C and biomass concentrations of 0.04 to 0.12 g/mL were tested in a batch reactor to determine the yield and composition of gaseous products without adding any catalysts. The highest yields of hydrogen and methane was obtained at the highest tested temperature with a biomass concentration of 0.08 g/mL [27].

The gasification of PO, Citrus peels, and woody biomass performed using a Bench-scale fluidized-bed gasifier revealed that PO decomposition occurred at low temperatures as compared to woody biomass due to the high ash content in PO biochar. Furthermore, PO gasification generates higher yields of syngas (2.6 Nm^3/kg) and hydrogen (0.6 Nm^3/kg) as compared to woody biomass [43].

5 Agronomic Application of *P. oceanica*

Seagrass contributes to the biogeochemcial processes in marine ecosystem and coast environment as litter degradation releases soluble organic carbon, nitrogen, and phosphorus to shallow coastal waters [71]. Furthermore, half of the carbon in seabed comes from the coastal vegetated ecosystems [69]. Up to 83% of total organic matter in seabed have for origin seagrass, while the rest comes from algae [78].

Seagrass litter is characterized with valuable macro- and microelements which allow its use as a fertilizer in addition to its ability to enhance soil water holding capacity and to improve crop growth [28]. Accordingly, *P. oceanica* fiber which has a high organic matter content could be valorized as a low-cost substrate for soilless culture. In this frame, fibers of *P. oceanica* were used as culture substrate to germinate Lens seeds and compared to those seeded on cotton. The results indicated that PO fibers can be used as alternative substrate to cotton to grow crops [46]. The effect of *P. oceanica* wrack on seed germination and seedling survival, plant growth, nutrient acquisition, and soil moisture indicated the contribution of such substrate in the enhancement of fitness and productivity of *Cakile maritima* and *Elymus farctus* [87]. Posidonia leaves were also useful in agriculture as a fertilizer, particularly in soil with high sand content and poor soils [30].

Using seagrass wrack as organic amendment was also proposed to ameliorate the physical and biological properties of soil [28]. The value of seagrasses for agricultural purposes has been stressed further on the basis of evidence describing their elemental composition [50]. The use of PO as mulching material for crops is also an old practice in the Mediterranean region.

Composting Posidonia residues might also help to improve the local soil's physical and nutritional characteristics [17]. Through enhancing organic matter's biostability and maturity, composting strives to advance the humification process, enhancing its application as an alternative fertilizer [14]. Composts often increase the total organic carbon of the soil in addition to essential nutrients while preventing harmful environmental impacts such as nitrogen immobilization, phytotoxicity, and soil compaction. Using PO-based compost as soil amendment for tomato culture revealed a positive effect on the availability of nutrients and enhancement of water retention in soil in addition increasing the production yield and nutritional quality of tomato fruits [30]. Furthermore, PO-based compost has multiple benefits such as increasing the availability of nutrients to plants, reduction of phytosanitary product application, and reduction of water evaporation in hot season. Some research has been done to assess the usage of PO-based compost as a partial peat alternative in soilless cultivation [51] and for biological pest control [38]. While Mininni et al. [51] assessed its application in the growth of potted basil. Posidonia residue-based compost, which has been utilized in horticulture for greenhouse tomato and lettuce cultivation as well as for nursery production, has been shown in previous research to exhibit favorable chemical characteristics without phytotoxicity occurrences [23]. In addition, compost made from Posidonia is a viable nursery growth for food production, including tomato, lettuce, basil, and melon (Montesano et al. 2018).

PO co-composting with other organic biomasses was investigated. Castaldi et al. [18] investigated the impact of a co-compost of PO and sewage sludge on soil properties and plant growth (Ocimum basilicum L.). Peruzzi et al. [65] have also prepared a PO-based substrate through the composting of PO and decontaminated sediments.

Furthermore, the recycling of PO residues as compost has contributed to the reduction of landfilling cost of marine biomass by 65% [28]. Nevertheless, using PO-based compost for soil amendment might present certain challenges due to its high salt content, inappropriate physical characteristics, and fluctuating quality and composition [74]. The only expenses associated with using this material as a compost are those related to transportation and washing [77].

In addition to Posidonia-based compost, predictive models, estimated a low heavy metal contamination of PO-based digestate which could represent another sustainable route for PO reuse in agriculture [52].

Moreover, the organic carbon content of Posidonia-based biochar was more than 10% in addition, to its low toxicity. Using PO-based biochar for contaminated environment remediation can also reduce the amount of waste dumped in landfills [54].

The use of *P. oceanica* fibers as a sorbent material for the removal of nutrients from secondary wastewater and the reuse of the loaded fiber with nutrients as fertilizer and compost was also proposed [92, 93]. Phosphorus-loaded *P. oceanica* based biochar has also shown a beneficial effect on germination and plant growth [66].

6 Conclusions and Challenges of *P. oceanica* Uses in Biofuel Production and Agriculture

The management of the huge quantities of PO residue represents an economic issue for Mediterranean countries. To date, landfill is the common practice applied for the management of this challenging marine organic waste. However, this practice led to the loss of high and renewable organic matter that could have an added value in agriculture and biofuel sectors. Alternative management strategies were proposed for *P. oceanica* valorization into biofuel and soil amendment and fertilizer, including anaerobic digestion composting, and thermochemical conversion processes. However, these alternatives to landfill have to be evaluated from ecological and economical point of view. Mainardis et al. [42] have conducted an economic assessment using life cycle costing (LCC) approach taking into account the direct and indirect impacts. ReCiPe 2016H methods were used to evaluate the environmental impacts of the PO management scenarios. According to the LCA findings, anaerobic digestion was among the best routes to achieve the best ecological performances. On the other hand, composting was picked as the best recycling method in terms of economic benefits and fertilizing value.

A system dynamics model was used to determine the effects of seagrass meadow changes on biomass production and the value of restoration [86]. Overall, the model revealed complexities in stakeholder perspective within the waste-to-resource framework and resulted in a total net present cost of 3.161,462.40 € for the baseline scenario, 1,488,277.28 € for the scenario of wrack valorization, and 1,231,325.12 € for the scenario of wrack valorization and meadow restoration. The obtained information can be of help for designing efficient systems for biofuel production from seagrass wrack, as well as for selecting the most appropriate managing route among the currently available technologies (such as on-site environmental preservation, composting, and anaerobic digestion). Nevertheless, the preparation process of *Posidonia* should be also considered when elaborating these models as high quantity of fresh water is required to remove salts which have an inhibitory effect on biological processes [30]. As a solution, Parente et al. [62] proposed to store PO residues in open sky to allow the rain to wash the salt excess and to boost the composting process. Moreover, low-cost enzymes from microorganisms tolerating high salt concentration could also be an alternative to solve this problem as well as increase rates and improve yields of biofuel produced from this important marine waste.

Finally, Posidonia-based biorefinery integrating different biomass conversion processes especially anaerobic digestion and pyrolysis could be considered as a promising management strategy allowing to maximize energy recovery from this marine organic waste while producing digestate and biochar that could be used as fertilizer and soil amendment. Such biorefinery could contribute to mitigating climate change, increasing carbon sequestration, combating soil degradation, and reducing the environmental and socioeconomic impacts associated with this waste in the Mediterranean region.

References

1. Akbarian A, Andooz A, Kowsari E, Ramakrishna S, Asgari S, Cheshmeh ZA (2022) Challenges and opportunities of lignocellulosic biomass gasification in the path of circular bioeconomy. Bioresour Technol 362:127774.

2. Aliyu M, Iwabuchi K, Itoh T (2021) Upgrading the fuel properties of hydrochar by co-hydrothermal carbonisation of dairy manure and Japanese larch (Larix kaempferi): product characterisation, thermal behaviour, kinetics and thermodynamic properties. Biomass Convers Biorefinery 1–6
3. Amalina F, Abd Razak AS, Krishnan S, Sulaiman H, Zularisam AW, Nasrullah M (2022) Biochar production techniques utilizing biomass waste-derived materials and environmental applications–a review. J Hazard Mater Adv 7:100134
4. Asimakopoulos G, Baikousi M, Salmas C, Bourlinos AB, Zboril R, Karakassides MA (2021) Advanced Cr(VI) sorption properties of activated carbon produced via pyrolysis of the "Posidonia oceanica" seagrass. J Hazard Mater 405:124274. https://doi.org/10.1016/j.jhazmat.2020.124274
5. Assad H, Kaya S, Kumar PS, Vo DV, Sharma A, Kumar A (2022) Insights into the role of nanotechnology on the performance of biofuel cells and the production of viable biofuels: a review. Fuel 323:124277
6. Balachandar P, Ganesan S, Jayakumar N, Subramanian S (2017) Economic/environmental dispatch of an interconnected power system considering Multiple Fuel Sources. In: 2017 International Conference on Circuit, Power and Computing Technologies (ICCPCT) 2017 Apr 20 (pp 1–7). IEEE
7. Balata G, Tola A (2018) Cost-opportunity analysis of the use of Posidonia oceanica as a source of bio-energy in tourism-oriented territories. the case of Alghero. J Clean Prod 172:4085–4098. https://doi.org/10.1016/j.jclepro.2017.02.072
8. Behera S, Singh R, Arora R, Sharma NK, Shukla M, Kumar S (2015) Scope of algae as third generation biofuels. Front Bioeng Biotechnol 2:90
9. Beltrán-Arredondo LI, Hernández-Leyva SR, Maldonado-Mendoza IE, Reyes-Moreno C, Contreras-Andrade I, Castro-Martínez C (2020) Valorisation of agroindustrial residues acid hydrolyzates as carbon sources for ethanol production by native yeast strains with different fermentative capabilities. Biotecnia 22(2):78–87
10. Bettaieb F, Khiari R, Hassan ML, Belgacem MN, Bras J, Dufresne A, Mhenni MF (2015) Preparation and characterization of new cellulose nanocrystals from marine biomass Posidonia oceanica. Ind. Crops Prod. 72:175–182. Special issue derived from International conference on bio-based materials and composites. https://doi.org/10.1016/j.indcrop.2014.12.038
11. Björk M, Rosenqvist G, Gröndahl F, Bonaglia S (2023) Methane emissions from macrophyte beach wrack on Baltic seashores. Ambio 52:171–181. https://doi.org/10.1007/s13280-022-01774-4
12. Blanco-Murillo F, Fernández-Torquemada Y, Garrote-Moreno A, Sáez CA, Sánchez-Lizaso JL (2022) Posidonia oceanica L.(Delile) meadows regression: long-term affection may be induced by multiple impacts. Mar Environ Res 174:105557
13. Boukmouche N, Azzouz N, Bouchama L, Chopart JP, Bouznit Y (2014) Activated carbon derived from marine Posidonia Oceanica for electric energy storage. Arab J Chem 7:347–354. https://doi.org/10.1016/j.arabjc.2012.12.010
14. Brunetti F (2024) Dai margini alla profondità: la rappresentazione del rapporto fra potere e spazialità oceanica. Iconocrazia 2(26):41–57
15. Bussotti S, Guidetti P, Rossi F (2022) Posidonia oceanica wrack beds as a fish habitat in the surf zone. Estuar Coast Shelf Sci 272:107882
16. Calvo R (2018) Thermal insulation role and possible exploitation of Posidonia oceanica detritus in the Mediterranean area. Flora Mediterr 28:279-285
17. Castaldi P, Melis P (2002) Composting of Posidonia oceanica and its use in agriculture. In: Microbiology of composting (pp 425–434). Berlin, Heidelberg: Springer Berlin Heidelberg
18. Castaldi P, Garau G, Melis P (2004) Influence of compost from sea weeds on heavy metal dynamics in the soil-plant system. Fresen Environl Bull 13(11): 1322–1328
19. Cataldo S, Chiodo V, Crea F, Maisano S, Milea D, Pettignano A (2018) Biochar from byproduct to high value added material—a new adsorbent for toxic metal ions removal from aqueous solutions. J Mol Liq 271:481–489. https://doi.org/10.1016/j.molliq.2018.09.009

20. Cataldo S, Muratore N, Giannici F, Bongiorno D, Chiodo V, Maisano S, Pettignano A (2022) Hydrocarbons removal from synthetic bilge water by adsorption onto biochars of dead Posidonia oceanica. Environ Sci Pollut Res 29:90231–90247. https://doi.org/10.1007/s11356-022-21998-x
21. Chen CY, Yang MH, Yeh KL, Liu CH, Chang JS (2008) Biohydrogen production using sequential two-stage dark and photo fermentation processes. Int J Hydrogen Energy 33(18):4755–4762
22. Chiodo V, Zafarana G, Maisano S, Freni S, Urbani F (2016) Pyrolysis of different biomass: direct comparison among Posidonia oceanica, Lacustrine alga and White-Pine. Fuel 164:220–227. https://doi.org/10.1016/j.fuel.2015.09.093
23. Cocozza C, Parente A, Zaccone C, Mininni C, Santamaria P, Miano T (2011) Chemical, physical and spectroscopic characterization of Posidonia oceanica (L.) Del. residues and their possible recycle. Biomass Bioenerg 35:799–807. https://doi.org/10.1016/j.biombioe.2010.10.033
24. Cocozza C, Parente A, Zaccone C, Mininni C, Santamaria P, Miano T (2011) Comparative management of offshore posidonia residues: composting vs. energy recovery. Waste Manage 31:78–84. https://doi.org/10.1016/j.wasman.2010.08.016
25. Conesa JA, Domene A (2015) Gasification and pyrolysis of Posidonia oceanica in the presence of dolomite. J Anal Appl Pyrol 113:680–689. https://doi.org/10.1016/j.jaap.2015.04.019
26. Dabros TMH, Stummann MZ, Høj M, Jensen PA, Grunwaldt J-D, Gabrielsen J, Mortensen PM, Jensen AD (2018) Transportation fuels from biomass fast pyrolysis, catalytic hydrodeoxygenation, and catalytic fast hydropyrolysis. Prog Energy Combust Sci 68:268–309. https://doi.org/10.1016/j.pecs.2018.05.002
27. Deniz İ (2018) Hydrothermal liquefaction of marine biomass: an integrated process. Acad Platform-J Eng Sci 6:36–39. https://doi.org/10.21541/apjes.320484
28. Emadodin I, Reinsch T, Rotter A, Orlando-Bonaca M, Taube F, Javidpour J (2020) A perspective on the potential of using marine organic fertilizers for the sustainable management of coastal ecosystem services. Environ Sustain 3:105–115. https://doi.org/10.1007/s42398-020-00097-y
29. Falco GD, Simeone S, Baroli M (2008) Management of beach-cast Posidonia oceanica seagrass on the island of Sardinia (Italy, Western Mediterranean). J Coast Res 24:69–75. https://doi.org/10.2112/06-0800.1
30. Grassi A, Buonerba A, Sanchez SO (2015) Bio-sourced polyolefins. Biorefineries Introduction 31:165–96
31. Hu X, Gholizadeh M (2019) Biomass pyrolysis: a review of the process development and challenges from initial researches up to the commercialisation stage. J Energy Chem 39:109–143. https://doi.org/10.1016/j.jechem.2019.01.024
32. Jeguirim M, Elmay Y, Limousy L (2016) Thermal degradation kinetics and mechanisms of Posidonia oceanica under inert and oxidative atmospheres. Int J Green Energy 13:665–671. https://doi.org/10.1080/15435075.2014.892003
33. Kainthola J, Kalamdhad AS, Goud VV (2019) A review on enhanced biogas production from anaerobic digestion of lignocellulosic biomass by different enhancement techniques. Process Biochem 84:81–90
34. Karkal SS, Kudre TG (2020) Valorization of fish discards for the sustainable production of renewable fuels. J cleaner Prod 275:122985
35. Khiari R, Marrakchi Z, Belgacem MN, Mauret E, Mhenni F (2011) New lignocellulosic fibres-reinforced composite materials: a stepforward in the valorisation of the Posidonia oceanica balls. Compos Sci Technol 71(16):1867–1872
36. Khiari R, Mhenni MF, Belgacem MN, Mauret E (2010) Chemical composition and pulping of date palm rachis and Posidonia oceanica–A comparison with other wood and non-wood fibre sources. Bioresour Technol 101(2):775–80
37. Khokhlova A, Gudnitz MN, Ferriol P, Tejada S, Sureda A, Pinya S, Mateu-Vicens G (2022) Epiphytic foraminifers as indicators of heavy-metal pollution in Posidonia oceanica seagrass meadows. Ecol Ind 140:109006
38. Kouki S, Saidi N, Ben Rajeb A, Brahmi M, Bellila A, Fumio M, Hefiène A, Jedidi N, Downer J, Ouzari H (2012) Control of Fusarium Wilt of Tomato Caused by Fusarium oxysporum F. Sp. Radicis-Lycopersici Using Mixture of Vegetable and Posidonia oceanica Compost. Appl Environ Soil Science. 2012(1):239639

39. Kumari D, Singh R (2022) Rice straw structure changes following green pretreatment with petha wastewater for economically viable bioethanol production. Sci Rep 12(1):10443
40. Liu S, Trevathan-Tackett SM, Jiang Z, Cui L, Wu Y, Zhang X, Li J, Luo H, Huang X (2022) Nutrient loading decreases blue carbon by mediating fungi activities within seagrass meadows. Environ Res 212:113280
41. Macreadie PI, Trevathan-Tackett SM, Baldock JA, Kelleway JJ (2017) Converting beach-cast seagrass wrack into biochar: a climate-friendly solution to a coastal problem. Sci Total Environ 574:90–94. https://doi.org/10.1016/j.scitotenv.2016.09.021
42. Mainardis M, Magnolo F, Ferrara C, Vance C, Misson G, De Feo G, Speelman S, Murphy F, Goi D (2021) Alternative seagrass wrack management practices in the circular bioeconomy framework: a life cycle assessment approach. Sci Total Environ 798:149283. https://doi.org/10.1016/j.scitotenv.2021.149283
43. Maisano S, Urbani F, Cipitì F, Freni F, Chiodo V (2019) Syngas production by BFB gasification: experimental comparison of different biomasses. Int J Hydrogen Energy 44:4412–4422. European fuel cell conference & exhibition 2017. https://doi.org/10.1016/j.ijhydene.2018.11.148
44. Maisano S, Urbani F, Mondello N, Chiodo V (2017) Catalytic pyrolysis of Mediterranean sea plant for bio-oil production. Int J Hydrogen Energy 42:28082–28092. Advanced materials for fuel cells and electrolyzers: the E-MRS 2016 fall meeting symposium Q, 19–22 September 2016, Warsaw, Poland. https://doi.org/10.1016/j.ijhydene.2017.07.124
45. Malik K, Sharma P, Yang Y, Zhang P, Zhang L, Xing X, Yue J, Song Z, Nan L, Yujun S, El-Dalatony MM, Salama E-S, Li X (2022) Lignocellulosic biomass for bioethanol: Insight into the advanced pretreatment and fermentation approaches. Ind Crops Prod 188:115569. https://doi.org/10.1016/j.indcrop.2022.115569
46. Marilés B-A, Jaime G-P, Eva B-B, Ignacio M, Pablo D-G (2019) Fibers of the seagrass Posidonia oceanica as substrate for germination of lentil seeds. SN Appl Sci 1:1414. https://doi.org/10.1007/s42452-019-1420-5
47. Martínez-Abraín A, Castejón-Silvo I, Roiloa S (2022) Foreseeing the future of Posidonia oceanica meadows by accounting for the past evolution of the Mediterranean Sea. ICES J Mar Sci 79(10):2597–2599
48. Masri MA, Younes S, Haack M, Qoura F, Mehlmer N, Brück T (2018) A seagrass-based biorefinery for generation of single-cell oils for biofuel and oleochemical production. Energy Technol 6:1026–1038. https://doi.org/10.1002/ente.201700604
49. Mateo MA (2010) Beach-cast cymodocea nodosa along the shore of a semienclosed bay: sampling and elements to assess its ecological implications. J Coast Res 26:283–291. https://doi.org/10.2112/08-1100.1
50. Milchakova NA, Böer B, Boyko LI, Mikulich DV (2014) The chemical composition and technological properties of seagrasses a basis for their use (a review). Sabkha Ecosystems: Volume IV: Cash Crop Halophyte and Biodiversity Conservation, 313–323
51. Mininni C, Bustamante MA, Medina E, Montesano F, Paredes C, Pérez-Espinosa A, Moral R, Santamaria P (2013) Evaluation of posidonia seaweed-based compost as a substrate for melon and tomato seedling production. J Hortic Sci Biotechnol 88(3):345–351
52. Misson G, Mainardis M, Incerti G, Goi D, Peressotti A (2020) Preliminary evaluation of potential methane production from anaerobic digestion of beach-cast seagrass wrack: the case study of high-adriatic coast. J Clean Prod 254:120131. https://doi.org/10.1016/j.jclepro.2020.120131
53. Misson G, Mainardis M, Marroni F, Peressotti A, Goi D (2021) Environmental methane emissions from seagrass wrack and evaluation of salinity effect on microbial community composition. J Clean Prod 285:125426. https://doi.org/10.1016/j.jclepro.2020.125426
54. Moltó J, Montalbán MG, Núñez SS, Jordá JD (2022) Revalorization of Posidonia oceanica waste for the thermochemical production of biochar. Appl Sci 12:7422. https://doi.org/10.3390/app12157422
55. Moodley P (2021) Sustainable biofuels: opportunities and challenges. Sustain Biofuels 1–20

56. Mukherjee A, Gómez-Sala B, O'Connor EM, Kenny JG, Cotter PD (2022) Global regulatory frameworks for fermented foods: a review. Front Nutr 9:902642
57. Mutlu E, Olguner C, Gökoğlu M, Özvarol Y (2022) Seasonal growth dynamics of Posidonia oceanica in a pristine Mediterranean gulf. Ocean Sci J 57(3):381–397
58. Nazurally N (2018) Anaerobic digestion of fish waste and seagrass/macroalgae: potential sustainable waste management for tropical small island developing states. J Mater Cycles Waste Manag 20:1724–1735. https://doi.org/10.1007/s10163-018-0738-1
59. Ncibi MC, Jeanne-Rose V, Mahjoub B, Jean-Marius C, Lambert J, Ehrhardt JJ, Bercion Y, Seffen M, Gaspard S (2009) Preparation and characterisation of raw chars and physically activated carbons derived from marine Posidonia oceanica (L.) fibres. J Hazard Mater 165:240–249. https://doi.org/10.1016/j.jhazmat.2008.09.126
60. O'Connor J, Mickan BS, Rinklebe J, Song H, Siddique KH, Wang H, Kirkham MB, Bolan NS (2022) Environmental implications, potential value, and future of food-waste anaerobic digestate management: a review. J Environ Manag 318:115519
61. Pardilhó S, Boaventura R, Almeida M, Dias JM (2022) Marine macroalgae waste: a potential feedstock for biogas production. J Environ Manage 304:114309
62. Parente A, Serio F, Montesano FF, Mininni C, Santamaria P (2013) The compost of Posidonia residues: a short review on a new component for soilless growing media. In: International Symposium on Growing Media and Soilless Cultivation 1034 , pp 291–298
63. Pazzaglia J, Badalamenti F, Bernardeau-Esteller J, Ruiz JM, Giacalone VM, Procaccini G, Marín-Guirao L (2022) Thermo-priming increases heat-stress tolerance in seedlings of the Mediterranean seagrass P. oceanica. Mar Pollut Bull 174:113164
64. Pedicini R, Maisano S, Chiodo V, Conte G, Policicchio A, Agostino RG (2020) Posidonia oceanica and wood chips activated carbon as interesting materials for hydrogen storage. Int J Hydrogen Energy 45:14038–14047. https://doi.org/10.1016/j.ijhydene.2020.03.130
65. Peruzzi E, Macci C, Doni S, Zelari L, Masciandaro G (2020) Co-composting as a management strategy for Posidonia oceanica residues and dredged sediments. Waste Biomass Valor 11:4907–4919. https://doi.org/10.1007/s12649-019-00822-7
66. Photiou P, Vyrides I (2021) Recovery of phosphate from dewatered anaerobic sludge and wastewater by thermally treated P. oceanica residues and its potential application as a fertilizer. J Environ Manag 298:113441. https://doi.org/10.1016/j.jenvman.2021.113441
67. Pilavtepe M, Celiktas MS, Sargin S, Yesil-Celiktas O (2013) Transformation of Posidonia oceanica residues to bioethanol. Ind Crops Prod 51:348–354. https://doi.org/10.1016/j.indcrop.2013.09.020
68. Pizzanelli S, Maisano S, Pinzino C, Manariti A, Chiodo V, Pitzalis E, Forte C (2022) The effect of activation on the structure of biochars prepared from wood and from Posidonia Oceanica: a spectroscopic study. Physchem 2:286–304. https://doi.org/10.3390/physchem2030021
69. Piñeiro-Juncal N, Kaal J, Moreira JCF, Martínez Cortizas A, Lambais MR, Otero XL, Mateo MA (2021) Cover loss in a seagrass Posidonia oceanica meadow accelerates soil organic matter turnover and alters soil prokaryotic communities. Org Geochem 151:104140. https://doi.org/10.1016/j.orggeochem.2020.104140
70. Plis A, Lasek JA, Zuwała J, Yu C-C, Iluk A (2016) Combustion performance evaluation of Posidonia oceanica using TGA and bubbling fluidized-bed combustor (batch reactor). J Sustain Min 15:181–190. https://doi.org/10.1016/j.jsm.2017.03.006
71. Prasad MHK, Ganguly D, Paneerselvam A, Ramesh R, Purvaja R (2018) Seagrass litter decomposition: an additional nutrient source to shallow coastal waters. Environ Monit Assess 191:5. https://doi.org/10.1007/s10661-018-7127-z
72. Rajkumar J, Dilipan E, Ramachandran M, Panneerselvam A, Thajuddin N (2021) Bioethanol production from seagrass waste, through fermentation process using cellulase enzyme isolated from marine actinobacteria. Vegetos 34:581–591. https://doi.org/10.1007/s42535-021-00239-5
73. Restaino OF, Scognamiglio M, Mirpoor SF, Cammarota M, Ventriglia R, Giosafatto CVL, Fiorentino A, Porta R, Schiraldi C (2022) Enhanced Streptomyces roseochromogenes melanin production by using the marine renewable source Posidonia oceanica egagropili. Appl Microbiol Biotechnol 106:7265–7283. https://doi.org/10.1007/s00253-022-12191-8

74. Ribeiro J, Bentes L, Coelho R, Gonçalves JM, Lino PG, Monteiro P, Erzini K (2006) Seasonal, tidal and diurnal changes in fish assemblages in the Ria Formosa lagoon (Portugal). Estuar Coast Shelf Sci 67(3):461–74
75. De Sanctis M, Chimienti S, Pastore C, Piergrossi V, Di Iaconi C (2019) Energy efficiency improvement of thermal hydrolysis and anaerobic digestion of Posidonia oceanica residues. Appl Energy 252:113457. https://doi.org/10.1016/j.apenergy.2019.113457
76. De Sanctis M, Di Iaconi C (2019) Evaluation of Posidonia oceanica residues as feedstock for anaerobic digestion. Bioresour Technol Rep 8:100317. https://doi.org/10.1016/j.biteb.2019.100317
77. Serio F, Gara LD, Caretto S, Leo L, Santamaria P (2004) Influence of an increased NaCl concentration on yield and quality of cherry tomato grown in posidonia (Posidonia oceanica (L) Delile). J Sci Food Agric 84(14):1885–1890
78. Serrano O, Rozaimi M, Lavery PS, Smernik RJ (2020) Organic chemistry insights for the exceptional soil carbon storage of the seagrass Posidonia australis. Estuar Coast Shelf Sci 237:106662. https://doi.org/10.1016/j.ecss.2020.106662
79. Shah FA, Mahmood Q, Rashid N, Pervez A, Raja IA, Shah MM (2015) Co-digestion, pretreatment and digester design for enhanced methanogenesis. Renew Sustain Energy Rev 2:627–642
80. Souii A, Guesmi A, Ouertani R, Cherif H, Chouchane H, Cherif A, Neifar M (2020) Carboxymethyl cellulase production by extremotolerant bacteria in low-cost media and application in enzymatic saccharification of stevia biomass. Waste Biomass Valor 11:2111–2122. https://doi.org/10.1007/s12649-018-0496-2
81. Sui Y, Liu C, Masse RC, Neale ZG, Atif M, AlSalhi M, Cao G (2020) Dual-ion batteries: the emerging alternative rechargeable batteries. Energy Storage Mater 25:1–32
82. Tarchoun AF, Trache D, Klapötke TM (2019) Microcrystalline cellulose from Posidonia oceanica brown algae: extraction and characterization. Int J Biol Macromol 138:837–845. https://doi.org/10.1016/j.ijbiomac.2019.07.176
83. Tohjima Y, Patra PK, Niwa Y, Mukai H, Sasakawa M, Machida T (2020) Detection of fossil-fuel CO2 plummet in China due to COVID-19 by observation at Hateruma. Sci Rep 10(1):18688
84. Toor M, Kumar SS, Malyan SK, Bishnoi NR, Mathimani T, Rajendran K, Pugazhendhi A (2020) An overview on bioethanol production from lignocellulosic feedstocks. Chemosphere 242:125080
85. Tursi A, Mastrototaro F, Montesanto F, De Giosa F, Lisco A, Bottalico A, Chimienti G (2022) The status of posidonia oceanica at Tremiti Islands marine protected area (Adriatic Sea). Biology 11(6):923
86. Vance C, Mainardis M, Magnolo F, Sweeney J, Murphy F (2022) Modeling the effects of ecosystem changes on seagrass wrack valorization: merging system dynamics with life cycle assessment. J Clean Prod 370:133454. https://doi.org/10.1016/j.jclepro.2022.133454
87. Vecchio SD, Marbà N, Acosta A, Vignolo C, Traveset A (2013) Effects of Posidonia oceanica beach-cast on germination, growth and nutrient uptake of coastal dune plants. PLoS ONE 8:e70607. https://doi.org/10.1371/journal.pone.0070607
88. Voca N, Grubor M, Peter A, Kricka T (2019) Evaluation of Posidonia oceanica waste as a biomass source for energy generation. Bioenerg Res 12:1104–1112. https://doi.org/10.1007/s12155-019-10032-7
89. Volpe MG, Fabbrocini A, Siano F, Coccia E, Scordella G, Licchelli C, De Sio F, Paolucci M (2018) Gonad quality of sea urchin Paracentrotus lividus cultured in an offshore pilot-scale trial on the south-east Italian coast. Aquac Nutr 24(5):1444–1455
90. Wahab MA, Habouzit F, Bernet N, Jedidi N, Escudié R (2016) Evaluation of a hybrid anaerobic biofilm reactor treating winery effluents and using grape stalks as biofilm carrier. Environ Technol 37:1676–1682. https://doi.org/10.1080/09593330.2015.1127291
91. Wahab MA, Habouzit F, Bernet N, Steyer J-P, Jedidi N, Escudié R (2014) Sequential operation of a hybrid anaerobic reactor using a lignocellulosic biomass as biofilm support. Biores Technol 172:150–155. https://doi.org/10.1016/j.biortech.2014.08.127

92. Wahab MA, Hassine RB, Jellali S (2011) Posidonia oceanica (L.) fibers as a potential low-cost adsorbent for the removal and recovery of orthophosphate. J Hazard Mater 191:333–341. https://doi.org/10.1016/j.jhazmat.2011.04.085
93. Wahab MA, Jellali S, Jedidi N (2010) Effect of temperature and pH on the biosorption of ammonium onto Posidonia oceanica fibers: equilibrium, and kinetic modeling studies. Biores Technol 101:8606–8615. https://doi.org/10.1016/j.biortech.2010.06.099
94. Wahab MA, Kebelmann K, Schartel B, Griffiths G (2021) Improving bio-oil chemical profile of seaweeds through anaerobic fermentation pre-treatment. Energy Convers Manage 245:114632. https://doi.org/10.1016/j.enconman.2021.114632
95. Wahab MA, Kebelmann K, Schartel B, Griffiths G (2022) Valorization of macroalgae digestate into aromatic rich bio-oil and lipid rich microalgal biomass for enhanced algal biorefinery performance. J Cleaner Prod 341:130925
96. Yu S, Park J, Kim M, Ryu C, Park J (2019) Characterization of biochar and byproducts from slow pyrolysis of hinoki cypress. Bioresour Technol Rep 6:217–22
97. Zaafouri K, Trabelsi AB, Krichah S, Ouerghi A, Aydi A, Claumann CA, Wüst ZA, Naoui S, Bergaoui L, Hamdi M (2016) Enhancement of biofuels production by means of co-pyrolysis of Posidonia oceanica (L.) and frying oil wastes: experimental study and process modeling. Bioresour Technol 207:387–98

Posidonia oceanica as a Potential Bioindicator and an Adsorbent Material for the Removal of Inorganic Pollutants from Wastewater

Mohamed Ali Wahab, Yasmin Cherni, Amani Haddouk, and Naceur Jedidi

Abstract Over the past 50 years, Mediterranean marine ecosystems are facing more threats than ever before, associated with human activity, climate change, and pollution. Among these environmental issues, pollution by inorganic contaminants such as nutrients and trace elements has devastating effects on coastal marine ecosystems by causing algal blooms and contaminating sediment and marine food web. These inorganic pollutants enter the Mediterranean Sea through sewage discharge and agriculture runoff and affect *Posidonia oceanica* seagrass growth and distribution. In order to overcome the negative impacts of these contaminants on marine life, *P. oceanica* fiber was recently proposed as a low-cost adsorbent material for the removal of inorganic pollutants from wastewater before discharge into the Mediterranean Sea. Such a promising strategy could reduce the concentration of pollutants entering the sea while ensuring the better management of huge quantities of *Posidonia* fibers depositing onto Mediterranean beaches. During the last decade, natural and modified *Posidonia* fibers were tested for their capacities to remove inorganic contaminants from both synthetic and real wastewater. Studies on Posidonia fibers' adsorption capabilities have primarily focused on testing with synthetic solutions, where researchers have carefully examined key parameters such as pH, contact time, pollutant concentration, and the amount of adsorbent material used. However, research investigating how these fibers perform in removing inorganic pollutants from actual wastewater remains limited. In addition, further investigations should be performed under dynamic conditions using fixed bed columns to determine the economic viability and the effectiveness of this treatment approach at a larger scale.

M. A. Wahab (✉) · Y. Cherni · A. Haddouk · N. Jedidi
Laboratory of Wastewater Treatment and Valorization, Water Research and Technologies Center, CERTE, Technopark Tourist Route of Soliman Nabeul, Soliman, Tunisia
e-mail: mohamedali.wahab@certe.rnrt.tn

Carthage University, Carthage, Tunisia

A. Haddouk
Faculty of Sciences Bizerte, Carthage University, Carthage, Tunisia

R. Khiari and M. Jawaid (eds.), *Current Status and Opportunity in Fibre and Composites*, Sustainable Materials and Technology,
https://doi.org/10.1007/978-981-96-4546-6_11

Keywords *Posidonia oceanica* · Wastewater · Marine pollution · Bioindicator · Biosorption · Inorganic contaminants

1 Introduction

The endemic seagrass *Posidonia oceanica* is widely distributed throughout the Mediterranean coastal waters, forming vast meadows that cover an area of 12,247 km^2 [4]. These meadows offer a range of environmental and socioeconomic benefits, including high primary productivity, serving as a source of food, shelter, and breeding grounds for marine species, carbon sequestration, oxygen generation, and coastal protection through sediment entrapment [21]. *P. oceanica* meadows serve a dual role in coastal protection against erosion. This protection occurs both through their living underwater root and stem structures (matte), as well as through the natural beach deposits (banquettes) formed by accumulated dead leaves that wash up on the shoreline [93]. Due to their significant ecological value in marine biodiversity protection and carbon capture, *P. oceanica* meadows constitute a vital ecosystem essential for maintaining marine health. Recognizing their crucial role, these meadows have been designated as a priority habitat for conservation under the Habitats Directive (Dir. 92/43/CEE) [4].

Beyond its already established roles in supporting marine ecosystem productivity and mitigating climate change, *P. oceanica* also serves as an exceptional bioindicator for seawater quality assessment, due to its high sensitivity to pollutants [18]. *P. oceanica* demonstrates rapid responses to environmental stressors, including pollutant accumulation in its tissues, changes in temperature and salinity, and exhibits decreased survival rates when water quality deteriorates [15]. The decline of *P. oceanica* has been directly linked to human activities that degrade water quality, demonstrating this seagrass's vulnerability to pollution [45]. Throughout the last decades, seagrass *P. oceanica* was facing increasing pressure due to the intensive economic activities characterizing the Mediterranean coasts. As a consequence, *Posidonia* meadows are declining in many parts of the Mediterranean because of pollution, tourism, and fishing activities. In the past 50 years, the overall decline in distribution area was estimated at 34% which makes this important aquatic plant in grave danger of disappearing in the next few years [101]. At least 1.5% of seagrass beds are lost every year which will cause an increase in ecological and economic losses on Mediterranean coasts [108]. Given *P. oceanica*'s extremely slow growth rate, once meadows are damaged, their complete restoration is essentially impossible within human timescales, making their degradation effectively permanent [28, 45].

The effective management of seagrass meadows requires a comprehensive strategy that addresses multiple factors. This includes reducing inputs of nutrients and organic matter from three main sources: agricultural runoff, aquaculture operations, and urban waste. Additionally, it's critical to prevent sediment accumulation, as this reduces underwater light penetration essential for seagrass growth and survival

[108]. A significant but unquantified volume of both inorganic and organic pollutants enters aquatic ecosystems through discharges from wastewater treatment plants [32]. Improper wastewater discharge in Mediterranean Sea may introduce persistent organic pollutants and inorganic contaminants which strongly affect marine life including *P. oceanica* seagrass [90]. It is therefore necessary to develop proper and sustainable treatment methods for the removal of pollutants from wastewater before their discharge in Mediterranean Sea. However, the conventional biological and physiochemical treatment processes present many drawbacks such as the use of huge amounts of chemicals, high operational costs associated with aeration, and generate high amount of sludge in addition to their low efficiencies to remove soluble inorganic and emerging organic pollutants [95].

Research priorities have evolved toward developing sustainable and eco-friendly systems for eliminating both organic and inorganic pollutants from wastewater effluents. This shift serves two key purposes: reducing treatment costs and protecting marine ecosystems from pollution. The adsorption process has gained increasing attention as a promising wastewater treatment method due to its cost-effectiveness and high efficiency in removing dissolved pollutants [74, 78, 100]. Adsorption processes present many advantages as compared to conventional wastewater treatment methods such as their easy operation, simplicity of design, and high efficiency [26, 30].

The most effective materials for contaminant removal through adsorption are commercial activated carbon, activated alumina, and zeolites. These adsorbents successfully eliminate various pollutants, including ammonium, heavy metals, dyes, pharmaceutical compounds, and persistent organic pollutants [43, 54, 58, 75]. Despite their superior performance in pollutant removal, these adsorbents' high costs make their application in wastewater treatment economically unfeasible as a practical solution [26]. As consequence, organic wastes, such as plant fibers, agro-food wastes, crops residues, forest by-products and chitosan, have emerged as alternative non-conventional adsorbent materials for the removal of soluble pollutants from wastewater [2, 5, 88, 103, 107, 111]. These biological materials are cheap and abundant and often have high affinity and selectivity.

In Mediterranean coasts, the dead leaves of *P. oceanica* known also as Posidonia fiber are one of the most abundant plant wastes. Posidonia fiber is a lignocellulosic material with physicochemical characteristics allowing its use as a natural low-cost adsorbent material [105]. Research has demonstrated that Posidonia fibers possess exceptional biosorption capabilities, effectively removing various types of pollutants from wastewater, including nutrients, heavy metals, dyes, and organic contaminants [8, 18, 24, 25, 33, 34, 104–106]. This chapter investigates the dual role of *P. oceanica* in marine environments and wastewater treatment. It examines how marine pollution impacts this seagrass species and explores its utility as a marine environment bioindicator. Additionally, the chapter evaluates its potential as a biosorbent material, specifically focusing on removing inorganic contaminants from both synthetic solutions and actual wastewater. The fundamental principles and various mechanisms of biosorption onto *P. oceanica* are also analyzed and discussed.

2 Marine Pollution: Sources and Impacts

Marine ecosystems have experienced intensifying human impacts over recent decades, driven by population growth and industrial expansion. These pressures have resulted in increasing pollutant loads entering oceans and seas through inadequately treated or completely untreated urban and industrial wastewater discharges [17]. The contamination of marine ecosystems by organic and inorganic pollutants has evolved into a worldwide crisis, with far-reaching consequences for marine biodiversity, human health, and environmental stability. A wide range of contaminants—including nutrients, dyes, heavy metals, surfactants, pesticides, and pharmaceutical compounds—are released into coastal waters. These substances pose dual threats: they are toxic to living organisms and persist in the environment, leading to both immediate harm and long-term disruption of food web dynamics and ecosystem balance [86]. When nutrients and trace elements accumulate through different trophic levels in marine food chains, they can cause significant harm to human health and result in negative economic impacts.

2.1 Nutrient Pollutants and Eutrophication

Nitrogen and phosphorus represent essential major nutrients that play fundamental roles in all living organisms, as they are critical for energy transfer processes and the synthesis of vital cellular components, including proteins and nucleic acids [52, 60, 65, 107]. Despite being essential nutrients, excessive levels of nitrogen and phosphorus disrupt natural nutrient cycles that exist between living organisms, soil, water, and the atmosphere. This disruption triggers multiple serious consequences for both ecosystems and human society, such as the proliferation of harmful algal blooms, contamination of surface and groundwater resources, reduction in biodiversity, degradation of soil quality, and increased emission of greenhouse gases [37, 91, 112].

Marine pollution with nutrients is mainly due to the discharge of untreated sewage in water bodies. Nutrients in wastewater come mainly from household sources (urine, faeces, detergents, and greywater) and agriculture sources (fertilizer, herbicides, manure, and food waste). The concentration of pollutants varies significantly depending on the wastewater source, with total ammonia–nitrogen levels ranging from 190 to 5190 mg/L, and total phosphorus concentrations varying between 32 and 2765 mg/L [73]. The discharge of nutrient-rich wastewater into water bodies can trigger severe eutrophication problems [36]. The discharge of wastewater to the environment should be strictly controlled and the concentration of nitrogen and phosphorus in sewage discharged into the sea should be lower than 1–3 mg N/L and 0.1 mg P/L, respectively [55, 62, 109]. The Mediterranean Sea faces significant environmental challenges due to nutrient pollution from land-based sources, particularly in regions with high population density and intense economic activity. A substantial portion of the total nitrogen entering the Mediterranean Sea comes directly from wastewater discharges [102]. During the period 2003–2007, the Mediterranean Sea received significant annual nutrient loads, with discharges measuring 1.87 Tg of total nitrogen, 1.22 Tg of nitrates, 0.11 Tg of total phosphorus, and 0.03 Tg of orthophosphate per year [67]. Excessive nutrient enrichment, particularly of nitrogen and phosphorus, has led to widespread degradation of coastal waters and has been identified as a primary factor in the decline of seagrass populations throughout the Mediterranean Sea [20].

Nutrient over-enrichment primarily causes seagrass decline through two main mechanisms: it stimulates excessive algal growth that creates high biomass overgrowth, and it increases water turbidity [31]. Seagrass decline is further complicated by additional human-induced factors that interact with nutrient enrichment: the introduction of invasive species, rising temperatures from global warming, and disruptions in food web balance. These stressors work in combination to accelerate seagrass habitat degradation [79]. Excessive nutrient supply contributes to seagrass decline through three key mechanisms: direct inhibition by nitrates, reduction in available light, and disruption of nitrification processes [47]. Nitrification serves as a vital process in coastal zones, playing a fundamental role in both marine primary production and the nitrogen cycle of coastal ecosystems [71]. When wastewater discharges disrupt nitrification processes, they alter the seasonal patterns of nitrogen cycling. This disruption affects both the abundance and diversity of phytoplankton, creating a ripple effect that impacts other marine organisms throughout the food web [9].

High nutrient levels can make ecosystems more susceptible to invasion by opportunistic plant species. These invasive species are typically fast-growing and excel at rapid nutrient uptake, channeling their energy into quick expansion, thereby outcompeting native species [96]. Recently, the application of slow-release fertilizer (struvite) to the seagrass's rhizosphere was found to improve seagrass growth in restoration (with 16–114% more aboveground biomass) and reduce excessive nutrient discharge in the surrounding ecosystem which may reduce the proliferation of invasive species [68].

2.2 Marine Pollution by Trace Elements

Trace heavy metal contamination has become a global issue, degrading the aquatic environment and posing a serious threat to the aquatic faunal communities and human health through food chain contamination. Metals can be classified into two categories based on their biological effects: essential and non-essential metals. Essential metals serve crucial roles as protein cofactors across various biological processes and contribute positively to species growth and feed efficiency [56]. Essential metals remain harmless when present at trace levels within organisms, but can become toxic once their concentrations exceed specific threshold limits [3]. Non-essential metals are inherently toxic to organisms even in minute quantities, as they serve no biological function in living systems [12, 40, 41].

The accumulation of trace elements in aquatic environments stems from both human-driven and natural sources. While volcanic eruptions represent a natural source, most trace elements originate from human activities including agricultural operations, industrial and metallurgical processes, fossil fuel burning, eroding landfills, port activities, and wastewater discharge from both industrial and domestic sources [3, 56].

Once discharged in aquatic environment, trace elements are accumulated in the sediments. To determine trace element concentrations and to assess the contamination and ecological risks associated with metal enrichment of marine sediments, several geochemical approaches can be used including the Contamination Factor (Cf), Geoaccumulation Index (Igeo), Potential Ecological Risk Index (PERI), Pollution Load Index (PLI), and many other indices [10, 57].

Heavy metals in aquatic systems pose a serious environmental and health risk through a process of bioaccumulation. Once these metals reach certain concentration levels, they are absorbed and stored in the tissues of aquatic organisms. As they move up through the food chain, these metals become increasingly concentrated. Ultimately, this creates significant human health risks when contaminated aquatic organisms are consumed as food [39].

Heavy metals severely impact aquatic organisms, particularly affecting fish through disrupted growth, impaired reproduction, and damaged embryonic and larval development. These effects cause both permanent losses to the aquaculture industry and create significant health risks for people who consume affected fish [99].

3 Potential Use of *P. oceanica* as a Bioindicator of Marine Environment Contamination

P. oceanica plays multiple vital roles in Mediterranean marine systems: it supports coastal primary production, occupies a crucial position in the marine food web's trophic structure, and facilitates nutrient recycling throughout the ecosystem [16, 87]. *P. oceanica* is especially vulnerable to human-induced pressures originating from coastal areas and shows particular sensitivity to any alterations in its environmental conditions [17, 29]. As a consequence, they can incorporate large amounts of trace elements and pollutants from the environment. Indeed, *P. oceanica* has a high trace element bioaccumulation capacity through their leaves and roots which are sites of ionic uptake from water column and sediment [42]. Due to its sensitivity to environmental changes, *P. oceanica* serves as an invaluable bioindicator, becoming an essential tool for ecological research focused on coastal ecosystem conservation.

3.1 Impact of Excess Nutrient on P. oceanica

Elevated nutrient concentrations in seawater impact submerged aquatic vegetation communities through multiple mechanisms [49]. Elevated nutrient levels in marine waters can influence seagrass through key ecological interactions, specifically by altering competitive relationships between species and affecting herbivory patterns [96]. Excessive nutrients trigger seagrass mortality through shading, as they stimulate the rapid growth of competing organisms like phytoplankton, epiphytes, and opportunistic macroalgae in both water columns and sediments. These fast-growing species, thriving on increased nutrient availability, block vital light from reaching the seagrass [97].

Research shows a dual effect of nutrient enrichment on seagrass: while increased nutrients in sediments can actually enhance seagrass growth without harmful effects, excessive nutrients in the water column produce negative impacts. This water column enrichment stimulates epiphytic growth and phytoplankton blooms that reduce light penetration, ultimately leading to significant seagrass decline [22, 81]. The survival, growth, and depth distribution of *P. oceanica* directly depend on underwater light availability for photosynthesis. Global reports indicate that reduced underwater light levels have led to widespread seagrass loss [64]. Higher nutrient levels encourage increased herbivore consumption of seagrass, likely because the elevated nutrients enhance the nutritional value of the seagrass as a food source [72]. This increased herbivory results in a significant decrease in photosynthetic biomass, ultimately leading to seagrass death.

Nitrogen availability variation in marine ecosystems results in a high inter- and intra-specific variability [11]. While moderate increases in nitrogen promote seagrass growth, however, large increases in nitrogen availability are deleterious to seagrass [81]. Alterations in the ecosystem are caused by the high availability of nitrogen which has a direct effect on *P. oceanica* through ammonia toxicity. Due to the lack of inhibitory feedback mechanisms, the down-regulation of nitrogen uptake is out of the ability of the seagrass [11]. High levels of ammonium, when it becomes the dominant nitrogen form in marine environments, negatively impact plant growth and survival through two mechanisms: the direct toxic effects of ammonium itself and the substantial energy costs required by plants to assimilate this form of nitrogen [81]. Plants meet the high energy requirements and carbon skeleton needs for ammonium processing by redirecting resources away from growth and basic metabolic maintenance. Seagrasses exhibit complex physiological responses to increased nutrient levels resulting from eutrophication. These responses slow their growth and follow clear annual patterns, which are primarily controlled by the interaction between physical factors like light and temperature. The process involves efficient resource management, where the plants store and remobilize nutrients and carbon during alternating periods of abundance and scarcity [20]. Studies examining the transcriptional responses of different *P. oceanica* organs to varying levels of nutrient enrichment have shown that different parts of the plant respond differently to nutrient stress.

Specifically, leaf tissues demonstrated greater sensitivity to nutrient stress compared to shoot apical meristem [81].

3.2 *Impact of Trace Elements Accumulation on* P. oceanica

P. oceanica has proven to be an effective metal bioindicator, leading to its widespread use in environmental monitoring programs that track metal pollution [83]. Historical trace element contamination can be detected and studied using *P. oceanica*, which serves as a biological indicator of past pollution events. Research conducted by Malea and colleagues in 2019 studied the concentrations of multiple metals (cadmium, cobalt, chromium, copper, nickel, and lead) in dead sheaths of *P. oceanica*. Using lepidochronology dating techniques, they examined these concentrations across different years at eight locations along the Thracian coast of the Aegean Sea, an area considered uncontaminated [69].

P. oceanica shoots were employed to trace historical mercury contamination by comparing two contrasting locations: Rosignano in Tuscany, Italy (an impacted site) and Tonnara (a pristine site) (Corsica, France) [63]. The lepidochronological analysis of mercury concentrations revealed higher levels in dead sheaths compared to living ones, with concentrations stabilizing after the third lepidochronological year. Throughout Mediterranean history, *P. oceanica* has played a crucial role in trace element cycling due to its significant capacity to accumulate these elements [92].

P. oceanica effectively mirrors trace element levels found in sediments, with its leaves serving particularly well as bioindicators for short-term monitoring studies. Luy and colleagues [66] documented how trace elements vary spatially and are distributed differently across various *P. oceanica* tissues. Luy et al. [66] research showed that trace elements primarily accumulated in the photosynthetic tissues of *P. oceanica*, indicating their uptake occurs from the surrounding water column. The accumulation process appears to be gradual, as evidenced by higher concentrations in adult leaves compared to intermediate leaves.

Richir and Gobert [87] conducted an in-situ study examining how *P. oceanica* responds to trace element exposure. They contaminated selected portions of seagrass beds with a mixture of 15 trace elements (aluminum, vanadium, chromium, manganese, iron, cobalt, nickel, copper, zinc, arsenic, molybdenum, silver, cadmium, lead, and bismuth) to analyze both the uptake and release patterns of these elements in the plant. The results revealed a fast accumulation kinetics in shoots occurred immediately from the beginning of the contamination. Furthermore, the leaf age seemed to have a direct effect on the accumulation kinetic of trace elements which was more rapid in the case of younger growing leaves forming new tissues as compared to the older senescent leaves. The study revealed that trace elements accumulated in *P. oceanica* leave primarily through the water column, as no contamination was detected in the surface sediments. The presence of certain trace elements in rhizomes was attributed to translocation processes from the leaves to the rhizomes.

The elevated accumulation of trace elements in *P. oceanica* leaves is attributed to their high regeneration rate, which enables the plant to mitigate the adverse effects of toxic concentrations on photosynthetic processes [17]. Thus, the accumulation of trace elements in the temporary organs of *P. oceanica* (leaves) can be viewed as a bioremediation strategy, aiding in the removal of metals from the water column through the high turnover rate of its leaves.

Such a strategy may reduce the trace elements contamination of column water and sediment against an increase of these contaminants in the banquettes through the transportation of dead leaves litter loaded with trace elements to the beach. Shabaka et al. [93] reported that trace element concentrations in seawater and sediments were lower than those reported for on-shore samples collected from west Mediterranean Coast of Egypt. Similarly, Castillo et al. [23] observed that the Metal Pollution Index (MPI) of *P. oceanica* litter was higher than that of *P. oceanica* tissue collected from unpolluted regions in Spain, France, and Italy, yet significantly lower than the MPI recorded in heavily polluted areas of Libya.

4 Potential Use of *P. oceanica* Fiber as an Adsorbent Material for the Removal of Inorganic Pollutants from Wastewater

4.1 Wastewater Types

According to the literature, wastewater can be categorized into four types: rainwater, agricultural wastewater, domestic wastewater, and industrial wastewater [80]. Typically, the chemical composition and flow conditions of wastewater are analyzed to determine its characteristics, which are then used to guide the design of wastewater treatment plants [94]. The flow conditions of wastewater are influenced by seasonal variations, with the wet season being the most likely to result in an influx of storm runoffs. The chemical quality of wastewater is assessed based on its organic and inorganic constituents. Key parameters typically measured to evaluate its chemical characteristics include pH, alkalinity, Biochemical Oxygen Demand (BOD), Chemical Oxygen Demand (COD), Total Solids (TS), Volatile Solids (VS), Total Nitrogen (TN), and Total Phosphorus (TP). Consequently, wastewater represents a complex and significant challenge for both human health and environmental sustainability [76]. As a result, there are several kinds of wastewater to treat, and each has distinctive qualities that involve innovative and environmentally friendly alternatives. Nutrient compounds present in domestic and agricultural wastewater can cause eutrophication, which promotes the growth of algae at the expense of other aquatic species [48]. In a similar line, wastewater with high levels of trace metals is a significant environmental toxin that also affects human health. On the other hand, industrial sectors such as agro-food, textile, and pulp industries are considered among the largest consumers of water and are also identified as major contributors to chemical pollution [46]. For instance, colored effluent from the textile or paper industries, which has a significant visual impact due to its color, is considered by many researchers as an indicator of the presence of hazardous pollution and poisonous substances [80]. Wastewater treatment performance is highly challenging because the effluent, depending on its source, comprises a variety of contaminants. Because there are so many different types of wastewater to treat, each has distinctive qualities that necessitate specific treatments available that are also environmentally friendly.

4.2 Wastewater Treatment

When water becomes polluted and decontamination is necessary, the most appropriate purification method should be selected to meet the decontamination goals as outlined by legislation. A typical purification process consists of four stages: pretreatment, primary treatment, secondary treatment (chemical and biological), and tertiary or final treatment (physical and chemical). Depending on the context, the first two stages are often grouped together under the concept of pretreatment or the preliminary step [89]. Pretreatment involves the removal of suspended substances and floating solid particles from the effluent. This stage, which can be performed mechanically or physically, is necessary before proceeding to secondary treatment, as particulate pollution (such as colloids, suspended solids, fats, etc.) can interfere with subsequent treatment processes, reduce their efficiency, or damage decontamination equipment. In some cases, primary chemical treatment may also be needed, such as pre-reducing a high organic load, oxidation, and pH adjustment [70].

Industrial effluents often have highly variable pH levels, with many being alkaline and containing a high organic load. As a result, pretreatment of these effluents is essential before proceeding to secondary treatment to ensure the effectiveness of the subsequent processes and to protect the treatment systems from potential damage [89]. However, these treatments alone are often insufficient to meet legislative requirements. The pre-treated effluent must undergo secondary purification before being discharged into the environment or reused. Chemical pollutants can be removed through biological, physical, or chemical methods. In some cases, a final or tertiary treatment is necessary to eliminate any remaining pollutants that may have been produced during the secondary purification process [70]. To date, the primary tertiary treatments applied at some industrial plants include adsorption, ion exchange, membrane filtration (such as ultrafiltration and reverse osmosis), and advanced oxidation. These methods are used to further purify effluent by removing remaining pollutants that may not have been fully addressed during earlier treatment stages. These alternatives may be effective in the treatment of various effluents, but they are expensive. Furthermore, these alternatives generate a lot of sludge, which causes obvious disposal issues. Thus, biosorption has been widely used to treat a wide range of organic and inorganic pollutants because it is cheap, simple, and efficient [70].

4.3 Biosorption

Recently, several studies have been investigated for the development of low-cost and effective alternatives for wastewater treatment. In this frame, biosorption has been proven to be an efficient technique for removing pollutants from wastewater [13]. Biomass is the key component in a biosorption process. It refers to both intact living cells and biologically derived materials, such as waste or charcoal, which can undergo varying degrees of transformation. In the case of dead biomass, pollutants bind to it passively (independently of metabolism) through ionic, chemical, or physical mechanisms. These interactions allow dead biomass to effectively capture and remove pollutants from the environment [27]. However, the process becomes more complex with living biomass because its metabolic activity is integrated with the passive mechanisms of biosorption. This metabolic activity enables pollutants to be actively transported across the cell membrane and accumulated inside the cell through bioaccumulation. The use of living biomass as a biosorbent has the potential to remove a greater quantity of pollutants, offering a significant advantage over using dead biomass for biosorption [85]. However, other benefits and drawbacks of using one type of biomass over another must be considered. Since most applications rely on dead biomass, several advantages are realized: toxicological issues are avoided, no maintenance is required, and the biomass can be stored for extended periods without losing effectiveness. Additionally, regeneration is more feasible, and dead biomass can function across a wider range of environmental variables, making it a practical and efficient option for many biosorption processes [1]. Furthermore, dead biomass can be cut and ground to achieve the desired particle size, enhancing its versatility. Despite the many advantages of dead biomass, the use of living biomass offers a significant benefit. As previously mentioned, the cells are metabolically active, allowing pollutants to be actively transported into the cell interior. This increases the efficiency of the process, as bioaccumulation enhances the initial biosorption, leading to a more effective removal of pollutants [6]. In this case, the process involves two steps. The first step, which is independent of metabolism, involves the pollutant binding to the cell surface (biosorption in the strict sense). The second step, which is metabolism-dependent, involves the active transport of the pollutant through the cell membrane into the cell interior. It is important to note that some pollutants may also pass through the membrane via passive diffusion, bypassing the active transport mechanism.

Continuing with the benefits and drawbacks of living biomass, an additional advantage of active systems is the potential for biotransformation or biodegradation, which can enhance the ability to eliminate a greater amount of pollutants in certain cases. However, there are some drawbacks associated with using living biomass. To utilize living biomass, culture systems, nutrient supply, and methods for cell harvesting are required, which can increase the overall cost of the process [110]. In addition to biomass from microorganisms, a wide range of materials have been evaluated as biosorbents for removing various pollutants, including agro-industrial waste materials, sludge, polysaccharides, plant-derived materials, and biopolymers. It is important to note that over the years of biosorption development, numerous biological materials of diverse origins have been used to remove different contaminants from wastewater. Many of these materials are considered waste; therefore, using them as biosorbents provides a dual benefit. Not only does it reduce the cost of wastewater treatment by utilizing low-cost adsorbent materials, but it also promotes better waste management, diverting these materials from incineration or landfill disposal.

Biosorption involves an interaction between the biomass and the pollutant, often involving multiple types of interactions simultaneously. As a result, the factors that influence these interactions—such as pH, temperature, biomass type, and pollutant concentration—will directly affect the efficiency of the process. Understanding and controlling these factors is essential for optimizing biosorption performance [110]. In a biosorption process, various interactions can occur, including ion exchange, complexation/coordination, electrostatic interactions, chemisorption, physisorption, microprecipitation, and reduction (Fig. 1). Considering these interactions, factors such as pH, temperature (thermodynamic studies), contact time, shaking speed, initial pollutant concentration, and the amount of biosorbent are well-known and extensively evaluated to optimize the efficiency of the biosorption process [85].

Other, less studied factors that influence the biosorption process include the type and number of functional groups in the biomass, such as carboxyl, amino, phosphoryl, or sulfonate groups, which affect the biosorption of certain pollutants, particularly

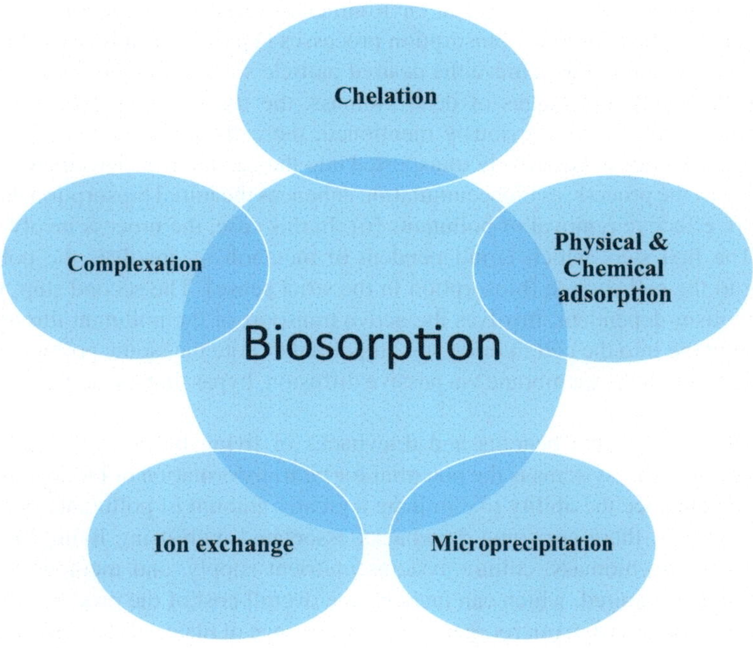

Fig. 1 Various mechanisms occurred through biosorption process

metals. Additionally, factors such as ionic strength, the presence of dissolved organic matter (which can alter metal adsorption), and competition with other pollutants can also significantly impact the efficiency of the biosorption process. These factors can modify the binding sites and interactions between the biomass and pollutants, thereby influencing the overall performance [85]. When employing a living system, it is crucial to also evaluate potential toxic effects and the organism's response to the pollutant.

In the field of biosorption process research, mathematical modeling plays a crucial role in understanding and optimizing different adsorption mechanisms. Researchers employ both kinetic and isotherm models to analyze the complex interactions between biosorbents and target substances. Traditional kinetic models, such as pseudo-first order, pseudo-second order, and intraparticle models, provide valuable insights into the process dynamics, helping scientists characterize the speed of adsorption, determine equilibrium times, and identify potential rate-limiting stages. Additional models like the Brouers-Sotolongo family kinetic model further enhance researchers' ability to comprehend these intricate processes.

While isotherm models continue to rely primarily on established approaches like Langmuir and Freundlich models, the current research focus is centered on developing innovative biosorbents. These new materials must not only demonstrate improved efficiency and cost-effectiveness but also show promise in real-world applications. For biosorbents utilizing dead biomass, the emphasis remains on maintaining low production costs rather than achieving multiple-use capabilities. The ultimate goal is to create materials that can effectively address environmental and industrial challenges, proving their practical value through rigorous testing and performance evaluation [110]. In fact, *P. oceanica* fibers are a promising biosorbent due to their distinctive characteristics including: a cost-efficient biomaterial, good chemical reagent resistance and stability, the appropriate size and strength to resist the pressures of the biosorption process, powerful adsorption capacity, being both effective and stable and a great potential for regeneration [38, 107].

4.4 Physicochemical Characteristics and Preparation of **P. oceanica** *fibers as Biosorbent Material*

According to Bettaieb et al. [14], the general morphology of *P. oceanica* leaves and longitudinal cross-sectional views demonstrated a particularly porous structure. Wahab et al. [106] have also investigated the microscopic structure of the *P. oceanica* fibers using scanning electron microscopy (SEM). The microscopic examination of the fibers revealed a distinctive cylindrical morphology with a notably irregular external surface. These surfaces were marked by distinctive furrows, which were attributed to the effects of dehydration and cell wall plasmolysis in the dead leaf material. Upon closer investigation of these grooves, researchers discovered a precise structural feature: aligned and evenly spaced pores positioned centrally within the furrow formations [106].

P. oceanica is also characterized by relatively high concentrations of extractives in balls and leaves with 10.7% and 19.7%, respectively. The ash content, which is higher in leaves (12%) than balls (10%) was attributed to the sand-contaminated balls or the chemical composition of the marine environment. Balls have significantly more holocellulose and cellulose than leaves [59].

Elemental analysis was used to identify the chemical composition indicating that silicon (2.1% w/w in oven-dry raw material) is the main inorganic component of *P. oceanica* [14]. Using flame atomic absorption spectroscopy, researchers analyzed the cation exchange characteristics of *P. oceanica* fibers. The analysis revealed a specific profile of exchangeable cations, with calcium predominating at 2.3 cmol+/kg. Magnesium was the second most abundant cation at 1.6 cmol+/kg, followed by sodium at 1.4 cmol+/kg, and potassium at the lowest concentration of 1.25 cmol+/kg [106].

Fourier transform infrared (FTIR) spectroscopic analysis of the *P. oceanica* surface identified several key functional groups, including phenolic, carboxylic, hydroxyl, and carbonyl groups. The spectroscopic examination also revealed a distinctive peak associated with pectin, which corroborates the earlier observation of the high calcium cation content in the fibers [106].

According to the literature, the preparation of *P. oceanica* for biosorption process required a mechanically or manually milled to separate the fibers from the ball. Furthermore, to separate fibers from sand and marine seaweed and to reduce the salinity of the biosorbent involved usually a washing step with distilled water before usage for biosorption assays [98].

To enhance the sorption capacity of biomass, researchers can employ various chemical modification techniques. These methods fundamentally transform the biomass by altering its functional groups and surface topography, which ultimately improves its ability to bind pollutants. Recently published research has explored several modification approaches, including esterification, graft polymerization, surface coating, and treatments using chemical agents like nitric acid (HNO_3), phosphoric acid (H_3PO_4), hydrogen peroxide (H_2O_2), and sodium hypochlorite (NaOCl). Each of these techniques aims to optimize the biomaterial's surface properties, creating more effective binding sites and increasing the overall sorption efficiency for targeted pollutants. The biosorption potential of *P. oceanica* leaf sheaths was enhanced using chemical pretreatment with H_3PO_4 and HNO_3 solutions allowing an increase of the adsorption efficiency up to 80% [77]. Recently, Krika et al. [61] investigated the potential use of NaOH pretreatment to enhance the biosorption capacity of *P. oceanica* to remove organic emergent pollutants. The biosorption capacities were improved by 50%.

4.5 *Biosorption Capacity of* P. oceanica *Fibers to Remove Inorganic Pollutants from Synthetic and Real Wastewater*

In addition to its high capacity to accumulate nutrients and trace elements from marine environments as living organisms, *P. oceanica* have also demonstrated high biosorption capacities to remove inorganic contaminants from wastewater as dead leaves. Through the last two decades, many studies have investigated the potential of using *P. oceanica* as a low-cost biosorption material to treat organic and inorganics contaminant from synthetic solutions and real wastewater. Table 1 summarizes some selected treatment processes using *P. oceanica* as bioadsorbent.

The biosorption of nutrients by *P. oceanica* was mainly applied to remove ammonium and phosphorus from synthetic and urban effluents. Wahab et al. [106] have investigated the effect of pH and temperature on the removal of ammonium from synthetic solution under batch conditions. The outcomes demonstrated that ammonium biosorption occurred for a wide pH range from 6 to 10. The authors of this study indicated that the abundance of carboxyl and hydroxyl groups in the biopolymers of *P. oceanica* may function as proton donors allowing both the binding of ammonium ions to the deprotonated hydroxyl and carboxyl groups and avoiding the volatilization of ammonia under basic pH. Furthermore, FTIR analysis revealed the presence of pectin confirming the high quantity of calcium in *P. oceanica* which allows the biosorption of ammonium through ion exchange. Contrary to ammonium, the highest biosorption capacity of orthophosphate was recorded at pH 3 and decreased with increasing the pH due to the competition of hydroxide ions with orthophosphate ions for adsorption sites [104].

The study examined the influence of temperature on the biosorption capacity of *P. oceanica* for nutrient removal from synthetic solutions, focusing on ammonium and orthophosphate. The findings demonstrated a consistent response across both nutrients: as the temperature increased up to 60 °C, the biosorption capacity improved significantly. This enhancement can be attributed to two primary mechanisms: first, the rising temperature increased the kinetic energy of the adsorbent particles, promoting more active interactions, and second, the higher temperature facilitated the rupture of bonds in functional groups on the adsorbent surface, particularly affecting pectin molecules. These molecular-level changes ultimately resulted in improved nutrient removal efficiency [104]. As the temperature increased, the pore dimensions of the fibers expanded, which significantly reduced intraparticle diffusion resistance. This geometrical transformation means that larger pore sizes create easier pathways for ion movement. Moreover, the elevated temperature induced a swelling effect within the internal structure of *P. oceanica* fibers, facilitating deeper penetration of orthophosphate ions. By creating more accessible and enlarged internal spaces, the temperature-induced structural changes enhanced the potential for more efficient ion adsorption and transport through the fiber matrix [104, 106].

The modeling studies revealed that the pseudo-second-order model accurately described both ammonium and orthophosphate biosorption, predicting chemisorption interactions at earlier stages and intraparticle diffusion at later stages [104,

Table 1 Sorption performance of nutrients and trace elements by *P. oceanica (PO)* according to the literature

PO Type	Wastewater type (Real/ Synthetic)	Treatment conditions	Parameter of interest	Sorption performance	Reference
Raw dried biomass	Synthetic	Adsorption batch experiments: [NH_4^+]: 15–100 mg/L; PH:7; [Adsorbent]: 2 g/L; Temperature:20 ± 2 °C	Ammonium	1.97 mg/g	Wahab et al. [106]
Raw dried biomass	Synthetic and secondary treated wastewater	CSTR system: [PO fiber]:10 g L^{-1}	Orthophosphates	1–3.03 mg/g	Wahab et al. [105]
Raw dried biomass	Synthetic	Adsorption batch experiments: [PO_4^{3-}]: 15–100 mg/L; PH:7; [Adsorbent]: 2 g/L; Temperature:20 ± 2 °C	Orthophosphates	7.45 mg/g	Wahab et al. [104]
Thermally treated PO (500 °C, 1 h)	Anaerobic effluent wastewater and leached solution from dewatered anaerobic sludge	Adsorption batch experiments: 20 g/L of thermally treated PO; contact time 24 h; shaking incubator operating at 100 rpm at 33 ◦C	Orthophosphates	179.1 mg/g	Photiou and Vyrides [84]
Raw dried biomass	Synthetic	PH: 3; [biosorbent]: 10 g/L; room temperature (batch experiment)	Uranium (VI) Chromium (VI) Lead (II)	5.67 mg g^{-1} 14.48 mg g^{-1} 140 mg g^{-1}	Pennesi et al. [82]
Raw dried biomass	Synthetic	Contact time:80 min; PH:6–8	Pb (II) Cu (II) Ni(II) Zn (II)	48.33 mg g^{-1} 43.9 mg g^{-1} 41.02 mg g^{-1} 37.90 mg g^{-1}	Boulaiche et al. [19]
Acid-activated PO	Synthetic	[biosorbent]: 3.0 and 0.5 g L^{-1}; PH:2–12, contact time: 30 min	Pb^{2+}	631.13 mg g^{-1}	Elmorsi et al. [35]
HCL-pretreated PO	Synthetic	[biosorbent]:100 mg; Temperature: 20 ± 1 °C; Reaction time:24 h; Stirring speed:150 rpm; PH: 4 and 5	Pb^{2+}	109–140 mg/g	Allouche et al. [7]

(continued)

Table 1 (continued)

PO Type	Wastewater type (Real/ Synthetic)	Treatment conditions	Parameter of interest	Sorption performance	Reference
Grafted PO	Synthetic	[biosorbent]:200 mg; PH: 2–6; Temperature:30 °C	Pb^{2+}	218.4 mg/g	Chadlia et al. [25]
Activated PO	Synthetic	PH:3; Temperature: 21 °C; Reaction time:48 h	Cr^{6+}	120 mg/g	Asimakopoulos et al. [8]

106]. At various temperatures, the ammonium biosorption was governed by the film diffusion process.

The Langmuir isotherm accurately represented equilibrium data, confirming the mono-layer coverage. According to the Langmuir isotherm model, *P. oceanica* fibers exhibit maximal adsorption capacities for ammonium and orthophosphate of 1.97 mg/g and 7.45 mg/g, respectively [104, 106]. High orthophosphate adsorption capacity of 179.1 mg/g was obtained after a thermally pre-treated *P. oceanica* (at 500 °C for one hour) treating anaerobic effluent wastewater and leached solution from dewatered anaerobic sludge [84].

The combined analyses using scanning electron microscopy (SEM), energy dispersive spectroscopy (EDS), and Fourier transform infrared spectroscopy (FTIR) conducted before and after nutrient biosorption unveiled the primary mechanisms of orthophosphate adsorption onto *P. oceanica*. These mechanisms primarily involve two key processes: ligand exchange, where orthophosphate ions replace chloride (Cl^-), sulphate (SO_4^{2-}), and hydroxyl (OH^-) ions, and precipitation reactions with calcium. These molecular interactions explain the efficient removal of orthophosphate by the *P. oceanica* fibers, highlighting the complex surface chemistry that enables nutrient capture [104]. For ammonium removal, the primary adsorption mechanisms involve two distinct processes: surface ion exchange with magnesium and calcium cations, and surface complexation through interactions between ammonium ions and specific functional groups. These functional groups include oxygen–hydrogen (O–H), nitrogen–hydrogen (N–H), carbonyl (C=O), carbon–hydrogen (C–H) stretching, and nitrogen–hydrogen (N–H) bending. These molecular-level interactions demonstrate the complex surface chemistry that enables *P. oceanica* fibers to effectively capture and remove ammonium ions from solution [106].

In the practical operation of full-scale biosorption processes, continuous-flow systems such as fixed bed columns and continuous stirred tank reactor (CSTR) are often used. Dynamic flow tests using CSTR were carried out to assess the biosorption capacity of *P. oceanica* to remove orthophosphate from synthetic and secondary treated wastewater [105]. The results showed that the removal efficiency of phosphorus from synthetic solutions is about 80% for 10 g/L of *P. oceanica.* The biosorption capacity using CSTR was about 3.03 mg/g for orthophosphate concentration of 50 mg/L which is close to the capacity obtained using batch system (3.5 mg/g). However, when using secondary treated wastewater, the removal efficiency of orthophosphate by *P. oceanica* decreased by 40% due to the presence of competition phenomenon between orthophosphate and sulphate. This competition phenomenon was highlighted by Wahab et al. [104] using batch system who indicated that chloride anions cause low interference with the adsorption of orthophosphate onto *P. oceanica.* However, for sulphate concentration over 150 mg/L, a significant drop of orthophosphate adsorption was recorded and interpreted by the fact that sulphate anions are divalent and expected to be effectively exchanged onto *P. oceanica* through electrostatic interaction or formation of outer-sphere complexes [104]. Under dynamic flow using CSTR, for sulphate/phosphorus ratio lower than 20 in synthetic wastewater, the average decrease of orthophosphate biosorption capacity was about 5% and attained 75% for sulphate/phosphorus ratio equal to 100 [105]. To overcome this competition effect, authors proposed the increase of *P. oceanica* dosage in order to increase the number of adsorption sites available to adsorb orthophosphate. Increasing *P. oceanica* dosage from 5 to 10 g/L led to the increase of orthophosphate removal efficiency by 16%.

The operating costs and energy costs of the CSTR using *P. oceanica* as biosorbent were estimated for the treatment of 30 m^3/day of hotel effluents characterized by high content of phosphorus due to the excessive use of detergents and low content of organic matter and pollutants [105]. The estimated total treatment cost was about 0.48 $/$m^3$.

In addition to the economic advantages, low metal leaching risk was demonstrated by Jellali et al. [53] who indicated low concentrations of Cr, Fe, Mn, Ni, and Zn allowing the reuse of ammonium-loaded *P. oceanica* after biosorption as compost without major concerns regarding heavy metals risks. Similarly, phosphorous-loaded *P. oceanica* was proposed as an alternative to compensate for global exhaustion of high-grade phosphate ores [104]. Thermally treated *P. oceanica* recovered after the adsorption of phosphate from anaerobic effluent or synthetic solutions was investigated as fertilizer substitutes and the results revealed a positive effect on plant growth with germination index values ranging between 96.7 and 111.14% [84].

P. oceanica have also shown great promise for heavy metal removal from water and wastewater with biosorption capacities higher than those of nutrients. *P. oceanica* contains two types of metallothioneins which are known for their capacity to bind heavy metals through the thiol groups, which are also known as sulfhydryl groups (R-SH), and amino groups (R-NH_2) of its cysteine residues [44].

Pennesi et al. [82] examined the removal of Vanadium (III) and Molybdenum (V) from wastewater using the *P. oceanica* biomass as an alternative and cost-efficient biomaterial. The initial characterization of this biomaterial revealed that the cuticle's carboxylic groups may be responsible for cation sorption and validated the bioaccumulation of hazardous metals. The combined effects of equilibrium pH and metal concentrations on biosorption performance were examined in both an ideal single-metal system and more practical multicomponent systems. For the single-metal solutions, at pH 3, high removal rate of 70% was obtained for vanadium against 40% for molybdenum. For the more real-life multicomponent systems, these adsorption performances were improved by high ionic strength and co-existence of both metals suggesting a synergistic effect on biosorption rather than competition. A maximum sorption ability estimated at 16 and 18 mg/g for vanadium and molybdenum, respectively, was explained by chemical-physical interaction (mainly based on ion exchange) with carboxylic and amine groups.

More recently, Boulaiche et al. [19] used *P. oceanica* to remove Pb^{2+}, Cd^{2+}, Ni^{2+}, Cu^{2+}, and Zn^{2+} using batch system. The biosorption capacities were 48.33, 43.9, 41.02, 37.90, and 30.22 mg/g for Pb^{2+}, Cu^{2+}, Ni^{2+}, Zn^{2+}, and Cd^{2+}, respectively, obtained for an equilibrium time of 80 min. Thermodynamic characteristics indicated that the metal biosorption is endothermic and spontaneous.

The removal of toxic hexavalent chromium from aqueous solutions was also investigated using *P. oceanica* under batch conditions. Low Cr(VI) adsorption capacity of 1.64 mg/g was recorded when raw *P. oceanica* was used [77]. However, when *P. oceanica* was activated via pyrolysis and chemical activation the estimated maximum Cr(VI) adsorption capacity reached 120 mg/g [8]. The increase of the adsorption capacity after thermal and chemical activation is attributed to the high specific surface area of 1563 m^2/g and a cumulative pore volume of 0.74 cm^3/g, allocated to 74% micro-pores and 26% to meso-macro- pores. According to the XPS measurements, the possible mechanism of Cr(VI) adsorption onto activated *P. oceanica* includes electrostatic attraction and the adsorption and complexion with the oxygen-containing groups.

Acid activation of *P. oceanica* was also effective to increase its capacity to adsorb Pb^{2+} from synthetic solution. Allouche et al. [7] have pre-treated *P. oceanica* with HCl solutions which have increased the Pb^{2+} adsorption up to 140 mg/g. While *P. oceanica* pretreatment with one mol/L of acetic acid was more effective for Pb^{2+} adsorption which reached 631.13 mg/g [35]. According to the adsorption isotherm fitting and thermodynamic studies, Pb^{2+} was chemically adsorbed onto pre-treated *P. oceanica.* In another study, the chemical modification of Posidonia with succinic anhydride produced materials with various succinyl groups contents (from 29.8 to 39.2%) exhibited high Pb^{2+} adsorption (218.4 mg/g) as compared to unmodified material [25]. It has been demonstrated that metal ion adsorption efficiency is proportional to the number of carboxylate functional groups introduced into the material.

Among the few studies investigating heavy metal adsorption onto Posidonia fiber under dynamic conditions, Izquierdo et al. [50] have investigated the biosorption of copper from aqueous solutions using fixed bed columns. The results indicated a maximum adsorption capacity in dynamic conditions about 56.70 mg/g for pH ranging between 5.0 and 5.5. However, this performance was affected by the presence of EDTA in the wastewater as *P. oceanica* does not retain the anionic copper(II)–EDTA complexes and the neutral species $CuSO_4$(aq) [51].

5 Conclusions and Future Prospects

P. oceanica certainly plays a significant ecological role in the Mediterranean sea as an efficient bioindicator of marine environment contamination with different organic and inorganic contaminants. Furthermore, the dead biomass generated by this aquatic plant can be used to create high-quality materials such as nanocomposite and adsorbent with outstanding performance to remove different contaminants from wastewater or serve as a suitable raw material for the production of cellulose derivatives due to its chemical and physical versatility. The application of *P. oceanica* fiber to treat wastewater ensures both the reduction of wastewater treatment costs and the improvement of the treated wastewater quality which will reduce the impact of eutrophication and trace elements on the growth and distribution of seagrass and minimize the risk of food web contamination with trace elements. Nevertheless, further investigations are required to study the treatment performance of fixed bed column packed with *P. oceanica* which could give more realistic information about the practical operation of full-scale adsorption processes in wastewater treatment plants. In addition, the pretreatment and modification of *P. oceanica* could represent a new research avenue which will improve the biosorption capacity and selectivity of *P. oceanica* towards the removal of inorganic pollutants. To increase the economic profitability of using *P. oceanica* as a biosorbant material, low-cost desorption processes should be developed to ensure the regeneration of the biosorbent material and to increase the cycle number of sorption–desorption. Finally, the agronomic reuse of *P. oceanica* loaded with nutrients represent a promising strategy for organic waste and nutrient recycling aiming to manage organic marine waste and reduce the application of chemical fertilizer and their negative impact on marine ecosystems.

References

1. Abdelfattah A, Ali S, Ramadan H et al (2022) Microalgae-based waste water treatment: mechanisms, challenges, recent advances, and future prospects. Environ Sci Ecotechnol 13:100205
2. Adeniyi AG, Ighalo JO (2019) Biosorption of pollutants by plant leaves: an empirical review. J Environ Chem Eng 7:103100
3. Afonne OJ, Ifediba EC (2020) Heavy metals risks in plant foods—need to step up precautionary measures. Curr Opin Toxicol 22:1–6
4. Agawin NSR, Sunyer-Caldú A, Díaz-Cruz MS et al (2022) Mediterranean seagrass Posidonia oceanica accumulates sunscreen UV filters. Mar Pollut Bull 176:113417
5. Ahmad T, Danish M (2022) A review of avocado waste-derived adsorbents: characterizations, adsorption characteristics, and surface mechanism. Chemosphere 296:134036
6. Al-Dhabi NA, Arasu MV (2022) Biosorption of hazardous waste from the municipal wastewater by marine algal biomass. Environ Res 204:112115
7. Allouche FN, Mameri N, Guibal E (2011) Pb(II) biosorption on *Posidonia oceanica* biomass. Chem Eng J 168:1174–1184
8. Asimakopoulos G, Baikousi M, Salmas C et al (2021) Advanced Cr(VI) sorption properties of activated carbon produced via pyrolysis of the "*Posidonia oceanica*" seagrass. J Hazard Mater 405:124274
9. Bartl I, Liskow I, Schulz K et al (2018) River plume and bottom boundary layer—hotspots for nitrification in a coastal bay Estuarine. Estuar Coast Shelf Sci 208:70–82
10. Bastami KD, Bagheri H, Kheirabadi V et al (2014) Distribution and ecological risk assessment of heavy metals in surface sediments along southeast coast of the Caspian Sea. Mar Pollut Bull 81:262–267
11. Beca-Carretero P, Azcárate-García T, Teichberg M et al (2022) Predicted warming intensifies the negative effects of nutrient increase on tropical seagrass: a physiological and fatty acid approach. Ecol Indic 142:109184
12. Bełdowska M, Bełdowski J, Kwasigroch U et al (2022) Coastal cliff erosion as a source of toxic, essential and nonessential metals in the marine environment. Oceanologia 64(2):553–566
13. Benettayeb A, Usman M, Tinashe CC et al (2022) A critical review with emphasis on recent pieces of evidence of Moringa oleifera biosorption in water and wastewater treatment. Environ Sci Pollut Res 29(32):48185–48209
14. Bettaieb F, Khiari R, Dufresne A et al (2015) Nanofibrillar cellulose from *Posidonia oceanica*: properties and morphological features. Ind Crops Prod 72:97–106
15. Blanco-Murillo F, Jimenez-Gutierrez S, Martínez-Vidal J et al (2022) Spatiotemporal trends observed in 20 years of *Posidonia oceanica* monitoring along the Alicante Coast, Spain. Water 14:274

16. Bonanno G, Borg JA (2018) Comparative analysis of trace element accumulation in seagrasses Posidonia oceanica and Cymodocea nodosa: biomonitoring applications and legislative issues. Mar Pollut Bull 128:24–31
17. Bonanno G, Di Martino V (2017) Trace element compartmentation in the seagrass Posidonia oceanica and biomonitoring applications. Mar Pollut Bull 116:196–203
18. Bonanno G, Veneziano V, Orlando-Bonaca M (2020) Comparative assessment of trace element accumulation and biomonitoring in seaweed Ulva lactuca and seagrass *Posidonia oceanica*. Sci Total Environ 718:137413
19. Boulaiche W, Belhamdi B, Hamdi B et al (2019) Kinetic and equilibrium studies of biosorption of M(II) (M = Cu, Pb, Ni, Zn and Cd) onto seaweed *Posidonia oceanica* fibers. Appl Water Sci 9(8):173
20. Burkholder JM, Tomasko DA, Touchette BW (2007) Seagrasses and eutrophication. J Exp Mar Biol Ecol 350:46–72
21. Campagne CS, Salles JM, Boissery P et al (2015) The seagrass *Posidonia oceanica*: ecosystem services identification and economic evaluation of goods and benefits. Mar Pollut Bull 97:391–400
22. Casella N, Careddu G, Calizza et al (2022) Increasing nutrient inputs over the last 500 years in an Italian low-impacted seagrass meadow. Mar Pollut Bull 174:113298
23. Castillo C, Mantecón AR, Sotillo J et al (2014) The use of banquettes of Posidonia oceanica as a source of fiber and minerals in ruminant nutrition. An observational study. Animal 8:1663–1666
24. Cengiz S, Tanrikulu F, Aksu S (2012) An alternative source of adsorbent for the removal of dyes from textile waters: *Posidonia oceanica* (L.). J Chem Eng 189–190:32–40
25. Chadlia A, Mohamed K, Najah L et al (2009) Preparation and characterization of new succinic anhydride grafted Posidonia for the removal of organic and inorganic pollutants. J Hazard Mater 172:1579–1590
26. Chai WS, Cheun JY, Kumar PS (2021) A review on conventional and novel materials towards heavy metal adsorption in wastewater treatment application. J Clean Prod 296:126589
27. Chan SS, Khoo KS, Chew et al (2022) Recent advances biodegradation and biosorption of organic compounds from waste water: microalgae-bacteria consortium-a review. Bioresour Technol 344:126159
28. Conte C, Rotini A, Winters G et al (2021) Elective affinities or random choice within the seagrass holobiont? The case of the native *Posidonia oceanica* (L.) Delile and the exotic Halophila stipulacea (Forssk.) Asch. from the same site (Limassol, Cyprus). Aquat Bot 174:103420
29. Cozza R, Rende F, Ferrari M et al (2019) Biomonitoring of *Posidonia oceanica* beds by a multiscale approach. Aquat Bot 156:14–24
30. Crini G, Lichtfouse E (2019) Advantages and disadvantages of techniques used for wastewater treatment. Environ Chem Lett 17:145–155
31. Cui L, Jiang Z, Huang X et al (2021) Decade changes of the food web structure in tropical seagrass meadow: implication of eutrophication effects. Mar Pollut Bull 173:113122
32. Delile H, Masson M, Miège C et al (2020) Hydro-climatic drivers of land-based organic and inorganic particulate micropollutant fluxes: the regime of the largest river water inflow of the Mediterranean Sea. Water Res 185:116067
33. Dural MU, Cavas L, Papageorgiou SK et al (2011) Methylene blue adsorption on activated carbon prepared from *Posidonia oceanica* (L.) dead leaves: kinetics and equilibrium studies. J Chem Eng 168:77–85
34. Demirak A, Dalman Ö, Tilkan E et al (2011) Biosorption of 2,4 dichlorophenol (2,4-DCP) onto *Posidonia oceanica* (L.) seagrass in a batch system: equilibrium and kinetic modeling. Microchem J 99:97–102
35. Elmorsi RR, El-Wakeel ST, Shehab El-Dein WA et al (2019) Adsorption of Methylene Blue and Pb^{2+} by using acid-activated Posidonia oceanica waste. Sci Rep 9(1):3356
36. Faragò M, Damgaard A, Madsen JA et al (2021) From wastewater treatment to water resource recovery: environmental and economic impacts of full-scale implementation. Water Res 204:117554

37. Farrant DN, Frank KL, Larsen AE (2021) Reuse and recycle: integrating aquaculture and agricultural systems to increase production and reduce nutrient pollution. Sci Total Environ 785:146859
38. Ferchichi K, Amdouni N, Chevalier Y et al (2022) Low-cost *Posidonia oceanica* bio-adsorbent for efficient removal of antibiotic oxytetracycline from water. Environ Sci Pollut Res 29(55):83112–83125
39. Fernandes C, Fontaínhas-Fernandes A, Cabral D et al (2008) Heavy metals in water, sediment and tissues of Liza saliens from Esmoriz-Paramos lagoon, Portugal. Environ Monit Assess 136:267–275
40. Foley M, Askin N, Belanger MP et al (2021) Essential and non-essential heavy metal levels in key organs of winter flounder (Pseudopleuronectes americanus) and their potential impact on body condition. Mar Pollut Bull 168:112378
41. García-García JD, Sánchez-Thomas R, Moreno-Sánchez R (2016) Bio-recovery of non-essential heavy metals by intra- and extracellular mechanisms in free-living microorganisms. Biotechnol Adv 34:859–873
42. Garcias-Bonet N, Arrieta JM, Duarte CM et al (2016) Nitrogen-fixing bacteria in Mediterranean seagrass (Posidonia oceanica) roots. Aquat Bot 131:57–60
43. Gayathiri M, Pulingam T, Lee KT et al (2022) Activated carbon from biomass waste precursors: factors affecting production and adsorption mechanism. Chemosphere 294:133764
44. Giordani T, Natali L, Maserti BE et al (2000) Characterization and expression of DNA sequences encoding putative type-II metallothioneins in the seagrass *Posidonia oceanica*. Plant Physiol 123:1571–1582
45. Guillén JE, Sánchez Lizaso JL, Jiménez S et al (2013) Evolution of Posidonia *oceanica* seagrass meadows and its implications for management. J Sea Res 83:65–71
46. Haripriyan U, Gopinath KP, Arun J et al (2022) Bioremediation of organic pollutants: a mini review on current and critical strategies for waste water treatment. Arch Microbiol 204(5):286
47. Hauxwell J, Valiela I (2004) Effects of nutrient loading on shallow seagrass-dominated coastal systems: patterns and processes. In: Nielsen SL, Banta GT, Pedersen MF (eds) Estuarine nutrient cycling: the influence of primary producers: the fate of nutrients and biomass, aquatic ecology book series. Springer, Netherlands, Dordrecht, pp 59–92
48. Hong J, Cho KH, Presser V et al (2022) Recent advances in waste water treatment using semiconductor photocatalysts. Curr Opin Green Sustain Chem 36:100644
49. Invers O, Kraemer GP, Pérez M et al (2004) Effects of nitrogen addition on nitrogen metabolism and carbon reserves in the temperate seagrass *Posidonia oceanica*. J Exp Mar Biol 303:97–114
50. Izquierdo M, Gabaldón C, Marzal P (2014) Modeling of the effect of EDTA on copper(II) biosorption onto Posidonia oceanica waste in batch and fixed-bed systems. J Taiwan Inst Chem Eng 45:665–673
51. Izquierdo M, Gabaldón C, Marzal P et al (2010) Modeling of copper fixed-bed biosorption from wastewater by *Posidonia oceanica*. Bioresour Technol 101:510–517
52. Jaouadi S, Wahab MA, Anane M et al (2014) Powdered marble wastes reuse as a low-cost material for phosphorus removal from aqueous solutions under dynamic conditions. Desalin Water Treat 52:1705–1715
53. Jellali S, Wahab MA, Hassine RB et al (2011) Adsorption characteristics of phosphorus from aqueous solutions onto phosphate mine wastes. Chem Eng J 169(1–3):157–165
54. Jiang N, Shang R, Heijman SGJ et al (2018) High-silica zeolites for adsorption of organic micro-pollutants in water treatment: a review. Water Res 144:145–161
55. Kabuba J, Lephallo J, Rutto H (2022) Comparison of various technologies used to eliminate nitrogen from wastewater: a review. JWPE 48:102885
56. Kadim MK, Risjani Y (2022) Biomarker for monitoring heavy metal pollution in aquatic environment: an overview toward molecular perspectives. Emerg Contam 8:195–205
57. Khaled A, Ahdy HHH, Hamed ESAE et al (2021) Spatial distribution and potential risk assessment of heavy metals in sediment along Alexandria Coast, Mediterranean Sea, Egypt. Egypt J Aquat Res 47:37–43

58. Khan H, Iram Gul K, Ara B et al (2020) Adsorptive removal of acrylic acid from the aqueous environment using raw and chemically modified alumina: batch adsorption, kinetic, equilibrium and thermodynamic studies. J Environ Chem Eng 8:103927
59. Khiari R, Mhenni MF, Belgacem MN et al (2010) Chemical composition and pulping of date palm rachis and *Posidonia oceanica*–a comparison with other wood and non-wood fibre sources. Bioresour Technol 101(2):775–780
60. Kraiem K, Wahab MA, Kallali H et al (2019) Effects of short- and long-term exposures of humic acid on the Anammox activity and microbial community. Environ Sci Pollut Res 26:19012–19024
61. Krika F, Krika A, Azizi A (2021) Impact of NaOH-surface treatment on emerging pollutant biosorption performance using marine algua, *Posidonia oceanica*. Glob Nest J 23(1):127–136
62. Kumar PS, Korving L, van Loosdrecht MCM et al (2019) Adsorption as a technology to achieve ultra-low concentrations of phosphate: research gaps and economic analysis. Water Res X 4:100029
63. Lafabrie C, Pergent G, Pergent-Martini C et al (2007) *Posidonia oceanica*: a tracer of past mercury contamination. Environ Pollut 148:688–692
64. Lee KS, Park SR, Kim YK (2007) Effects of irradiance, temperature, and nutrients on growth dynamics of seagrasses: a review. J Exp Mar Biol Ecol 350:144–175
65. Lovio-Fragoso JP, de Jesús-Campos D, López-Elías JA et al (2021) Biochemical and molecular aspects of phosphorus limitation in diatoms and their relationship with biomolecule accumulation. Biology (Basel) 10:565
66. Luy N, Gobert S, Sartoretto S et al (2012) Chemical contamination along the Mediterranean French coast using *Posidonia oceanica* (L.) Delile above-ground tissues: a multiple trace element study. Ecol Indic 18:269–277
67. Malagó A, Bouraoui F, Grizzetti B et al (2019) Modelling nutrient fluxes into the Mediterranean Sea. J Hydrol Reg Stud 22:100592
68. MacDonnell C, Bydalek F, Osborne TZ et al (2022) Use of a wastewater recovery product (struvite) to enhance subtropical seagrass restoration. Sci Total Environ 838:155717
69. Malea P, Mylona Z, Kevrekidis T (2019) Improving the utility of the seagrass *Posidonia oceanica* as a biological indicator of past trace element contamination. Ecol Indic 107:105596
70. Mao G, Han Y, Liu X et al (2022) Technology status and trends of industrial waste water treatment: a patent analysis. Chemosphere 288:132483
71. McLaughlin K, Nezlin NP, Howard MDA et al (2017) Rapid nitrification of wastewater ammonium near coastal ocean outfalls, Southern California, USA. Estuar Coast Shelf Sci 186:263–275. Tracking impacts of anthropogenic nutrient inputs in an urban coastal ecosystem during a wastewater diversion in the southern California bight
72. McSkimming C, Tanner JE, Russell BD et al (2015) Compensation of nutrient pollution by herbivores in seagrass meadows. J Exp Mar Biol 471:112–118
73. Montalvo S, Huiliñir C, Castillo A et al (2020) Carbon, nitrogen and phosphorus recovery from liquid swine wastes: a review. J Chem Technol Biotechnol 95:2335–2347
74. MozaffariMajd M, Kordzadeh-Kermani V, Ghalandari V et al (2022) Adsorption isotherm models: a comprehensive and systematic review (2010–2020). Sci Total Environ 812:151334
75. Myachina M, Gavrilova N, Nazarov V (2022) Adsorption of molybdenum blue nanoparticles on the alumina surface. Colloids Surf A Physicochem Eng Asp 644:128819
76. Nas B, Dolu T, Argun ME et al (2021) Comparison of advanced biological treatment and nature-based solutions for the treatment of pharmaceutically active compounds (PhACs): a comprehensive study for wastewater and sewage sludge. Sci Total Environ 779:146344
77. Ncibi MC, Mahjoub B, Seffen M (2007) Adsorptive removal of textile reactive dye using *Posidonia oceanica* (L.) fibrous biomass. IJEST 4:433–440
78. Novikau R, Lujaniene G (2022) Adsorption behaviour of pollutants: heavy metals, radionuclides, organic pollutants, on clays and their minerals (raw, modified and treated): a review. J Environ Manage 309:114685
79. Orth RJ, Carruthers TJB, Dennison WC et al (2006) A global crisis for seagrass ecosystems. Bioscience 56:987–996

80. Pandey AK, Kumar RR, Kalidasan et al (2021) Utilization of solar energy for waste water treatment: challenges and progressive research trends. J Environ Manage 297:113300
81. Pazzaglia J, Santillán-Sarmiento A, Ruocco M et al (2022) Local environment modulates whole-transcriptome expression in the seagrass *Posidonia oceanica* under warming and nutrients excess. Environ Pollut 303:119077
82. Pennesi C, Totti C, Beolchini F (2013) Removal of vanadium (III) and molybdenum (V) from waste water using *Posidonia oceanica* (Tracheophyta) biomass. PLoS ONE 8(10):e76870
83. Pergent G, Labbe C, Lafabrie C et al (2011) Organic and inorganic human-induced contamination of Posidonia oceanica meadows. Ecol Eng 37:999–1002
84. Photiou P, Vyrides I (2021) Recovery of phosphate from dewatered anaerobic sludge and wastewater by thermally treated *P. oceanica* residues and its potential application as a fertilizer. J Environ Manag 298:113441
85. Priya AK, Gnanasekaran L, Dutta K et al (2022) Biosorption of heavy metals by microorganisms: evaluation of different under lying mechanisms. Chemosphere 307:135957
86. Puccinelli E, Porri F, Altieri K et al (2022) Coastal ecosystem services in South Africa's largest natural bay: the role of marine benthic filter feeders in mitigating pollution. Ecol Indic 139:108899
87. Richir J, Gobert S (2014) A reassessment of the use of Posidonia oceanica and Mytilus galloprovincialis to biomonitor the coastal pollution of trace elements: new tools and tips. Mar Pollut Bull 89:390–406
88. Sadiq AC, Olasupo A, Ngah WSW et al (2021) A decade development in the application of chitosan-based materials for dye adsorption: a short review. Int J Biol Macromol 191:1151–1163
89. Saleh TA, Mustaqeem M, Khaled M (2022) Water treatment technologies in removing heavy metal ions from wastewater: a review. Environ Nanotechnol Monit Manag 17:100617
90. Salvo A, La Torre GL, Rotondo A et al (2020) Accumulation of PCBs, PAHs, plasticizers and inorganic elements in Hexanchusgriseus from the strait of Messina (Central Mediterranean sea). Nat Prod Res 34:172–176
91. Sampat AM, Hicks A, Ruiz-Mercado GJ et al (2021) Valuing economic impact reductions of nutrient pollution from livestock waste. RCR Adv 164:105199
92. Sanz-Lázaro C, Malea P, Apostolaki ET et al (2012) The role of the seagrass *Posidonia oceanica* in the cycling of trace elements. Biogeosciences 9:2497–2507
93. Shabaka SH, Khalil MK, El-Sikaily A (2021) Posidonia oceanica litter along the Mediterranean Coast of Egypt: status and a preliminary assessment of nutrients and trace elements contents. Estuar Coast Shelf Sci 255:107342
94. Shindhal T, Rakholiya P, Varjani S, Pandey et al (2021) A critical review on advances in the practices and perspectives for the treatment of dye industry wastewater. Bioengineered 12(1):70–87
95. Sikosana ML, Sikhwivhilu K, Moutloali R et al (2019) Municipal wastewater treatment technologies: a review. Procedia Manuf 35:1018–1024. The 2nd International conference on sustainable materials processing and manufacturing (SMPM 2019), 8–10 March 2019, Sun City, South Africa
96. Smulders FOH, Becker ST, Campbell JE et al (2022) Fish grazing enhanced by nutrient enrichment may limit invasive seagrass expansion. Aquat Bot 176:103464
97. Stoner EW, Yeager LA, Sweatman JL et al (2014) Modification of a seagrass community by benthic jellyfish blooms and nutrient enrichment. J Exp Mar Biol 461:185–192
98. Tarchoun AF, Trache D, Klapötke TM (2019) Microcrystalline cellulose from *Posidonia oceanica* brown algae: Extraction and characterization. Int J Biol Macromol 138:837–845
99. Taslima K, Al-Emran M, Rahman MS et al (2022) Impacts of heavy metals on early development, growth and reproduction of fish—a review. Toxicol Rep 9:858–868
100. Tee GT, Gok XY, Yong WF (2022) Adsorption of pollutants in wastewater via biosorbents, nanoparticles and magnetic biosorbents: a review. Environ Res 212:113248
101. Telesca L, Belluscio A, Criscoli A et al (2015) Seagrass meadows (*Posidonia oceanica*) distribution and trajectories of change. Sci Rep 5:12505

102. Temino-Boes R, Romero I, Pachés M et al (2019) Anthropogenic impact on nitrification dynamics in coastal waters of the Mediterranean Sea. Mar Pollut Bull 145:14–22
103. Wahab MA, Boubakri H, Jellali S et al (2012) Characterization of ammonium retention processes onto Cactus leaves fibers using FTIR, EDX and SEM analysis. J Hazard Mater 241–242
104. Wahab MA, Hassine RB, Jellali S (2011) *Posidonia oceanica* (L.) fibers as a potential low-cost adsorbent for the removal and recovery of orthophosphate. J Hazard Mater 191:333–341
105. Wahab MA, Hassine RB, Jellali S (2011) Removal of phosphorus from aqueous solution by *Posidonia oceanica* fibers using continuous stirring tank reactor. J Hazard Mater 189:577–585
106. Wahab MA, Jellali S, Jedidi N (2010) Effect of temperature and pH on the biosorption of ammonium onto *Posidonia oceanica* fibers: Equilibrium, and kinetic modeling studies. Bioresour Technol 101:8606–8615
107. Wahab MA, Jellali S, Jedidi N (2010) Ammonium biosorption onto sawdust: FTIR analysis, kinetics and adsorption isotherms modeling. Bioresour Technol 101:5070–5075
108. Waycott M, Duarte CM, Carruthers TJB et al (2009) Accelerating loss of seagrasses across the globe threatens coastal ecosystems. PNAS 106:12377–12381
109. Yan T, Ye Y, Ma H et al (2018) A critical review on membrane hybrid system for nutrient recovery from wastewater. J Chem Eng 348:143–156
110. Zareh MM, El-Sayed AS, El-Hady DM (2022) Biosorption removal of iron from water by Aspergillus niger. npj Clean Water 5(1):58
111. Zhang Y, Zhao M, Cheng et al (2021) Research progress of adsorption and removal of heavy metals by chitosan and its derivatives: a review. Chemosphere 279:130927
112. Zhao H, Yuan M, Strokal M et al (2021) Impacts of nitrogen pollution on corals in the context of global climate change and potential strategies to conserve coral reefs. Sci Total Environ 774:145017

The Importance of the Adsorption for the Removal of Textile Dyes Using Organic and Inorganic Materials

Meriem Bembli, Ramzi Khiari, and Khaled Boughzala

Abstract There is a large amount of dye-containing waste water discharged by the textile industry, which pollutes water and soil, as well as harming human health, animals, and plants. The textile dyes can cause acute environmental issues for aquatic and terrestrial ecosystems. Moreover, they cause surface water contamination and accumulation of toxic and carcinogenic substances. Several physicochemical or biological methods can be used to remove dyes from dye-containing waste water. These methods should be efficient, cheap, user-friendly, and using low-cost adsorbents. As inorganic adsorbents, apatites are widely used thanks to their high adsorption capacity and their availability. And for organic adsorbent, Posidonia is considered as one of the most common used adsorbents due to its local availability in the Mediterranean, and it is used as renewable biological adsorbent.

Keywords Adsorption · Textile dye · Waste water · Apatite · *Posidonia oceanica* · Dead leaves

M. Bembli · K. Boughzala (✉)
Laboratory of Physical-Chemistry of Materials, Faculty of Sciences of Monastir, University of Monastir, 5019 Monastir, Tunisia
e-mail: khaledboughzala@gmail.com

R. Khiari
Department of Textile, Higher Institute of Technological Studies of Ksar Hellal, 5070 Ksar Hellal, Tunisia

CNRS, Grenoble INP, LGP2, Université Grenoble Alpes, F-3800 Grenoble, France

K. Boughzala
Department of Textile, Higher Institute of Technological Studies of Ksar Hellal, Monastir, Tunisia

R. Khiari and M. Jawaid (eds.), *Current Status and Opportunity in Fibre and Composites*, Sustainable Materials and Technology,
https://doi.org/10.1007/978-981-96-4546-6_12

1 Introduction

Textile dyes are considered as the major source of environmental damage. They upset the biological effect in aquatic ecosystem and also the photosynthesis process of algae. Their long-term presence in the environment and their accumulation in sediments, particularly in fish and other aquatic life forms, can cause many toxic effects [1]. Under environmental conditions, chemical changes occur in dyes and the transformation products are harmful and carcinogenic even more than the original molecules [2]. In addition, many studies showed that textile effluents decrease light penetration in water resources and causing oxygen deficiency which is a grave threat to aquatic ecosystem [3, 4]. Another problem is that the textile industry uses a huge amount of water. As a consequence of the big level of contamination that can occur during dyeing and finishing processes, treated waste water should be recycled. Flowing textile dye effluents into drains, rivers, ponds, and lagoons can affect drinking water quality and make it unsafe to consume. Water bodies should be cleaned of these pollutants before final disposal [5]. Textile dye removal methods are extensive and varied [6–10], such as ozonation, flocculation, membrane filtration, oxidation, coagulation/flocculation, photochemical, electrochemical methods, biological or photocatalytic degradation, microbial degradation, chemical or advanced oxidation, membrane separation process, precipitation, solvent extraction, and adsorption [11–14]. But all the current methods have some limitations in effectiveness, applicability, and costs, and there have been no successful attempts to completely remove color from wastewater [15]. In terms of textile dyes removal from water at low costs, adsorption is probably the simplest and most powerful treatment method. It is proven that colored organics have been successfully removed with adsorption techniques. [16]. Scientists have been concerned with adsorption at interfaces since the beginning of the twentieth century. This key method underlies many highly important processes of utility. Adsorption's biological, environmental, and technological importance can never be doubted. There is a great deal of value in its practical application in industry and in environmental protection [17]. However, sometimes the high cost and limited availability of adsorbent materials represent a constraint [14, 18]. There are some alternative solutions using cheap adsorbents which are extracted from agricultural residues, such as corn cob, palm fruit parts, waste material from the chestnut peel, almond shell waste, rice husk, waste orange and lime peels, pine fruit shells, rice hulls, maize cob and bagasse pith, and others [19, 20]. We can use also some materials as adsorbents include activated carbon, peat, fly ash, and coal [21]. In the following paragraphs, we will focus on the organic and inorganic adsorbents. Generally, cheap adsorbents are used, and they can be natural or synthetic. For the organic ones, various materials can be used to adsorb dyes, such as simple and modified polyurethane which is used as powerful, effective, and inexpensive adsorbent. There has been extensive research into the bioremediation of dyes, and polyurethane has been shown to be an excellent inert support material [22]. Metal organic frameworks (MOFs) also have currently attracted a great deal of scientific concern because of their unique characteristics, such as tunable pore properties, large porosity, volumes, hierarchical structures, and great adsorption and regeneration performances. MOFs can be used to remove harmful organic pollutants from aqueous solutions as an eco-friendly alternative to conventional adsorbents [23]. Posidonia or *Posidonia Oceanica*, marine flowering plant, is a miracle in the Mediterranean. It is usually used a low-cost eco-adsorbent and bio-adsorbent for the elimination of dyes from aqueous environments [24, 25]. For the inorganic adsorbents, phosphates are classified among those most frequently used for treatment of water. And Tunisia is ranked as one of the most reproducing countries of phosphate in the world [26]. So, these materials are considered as low-cost adsorbents and even we can use the waste of the phosphate industry [6, 7, 27, 28]. Apatites, synthetic phosphate, are also widely used thanks to their simple methods of synthesis, their low costs, and their well-defined structure comparing to natural phosphate with non-defined structure [29, 30]. Besides these, many other inorganic adsorbents can be used [32, 33]. Moreover, dyes can be classified in several ways but their classification according to their applications is the best appropriate (Table 1).

Table1 summarizes the classification of different dyes according to their methods of use. It is classified corresponding to C.I. applications. For each application type, it shows the principal substrate and the representative chemicals [35]. In the following paragraphs, a simple comparison between the adsorption using organic material (Posidonia) and inorganic material (apatite) will be discussed.

Table 1 Classification of dyes and types of dyes used in different substrate [35]

Class	Principle substrate to be used	Chemicals type used
Acid	Nylon, wool, silk, paper,	Azo (including premetallized), anthraquinone, triphenylmethane, azine, xanthene, nitro, and nitroso
Azoic components	Cotton, rayon, cellulose acetate, and polyester	Azo
Basic	Paper, polyacrylonitrile, modified nylon, polyester, and inks	Cyanine, hemicyanine, diazhemicyanine, diphenylmethane, triarylmethane, azo, azine, xanthenes, acridine, oxazine, and anthraquinone
Direct	Cotton, rayon, paper, leather, and nylon	Azo, anthraquinone, styryl, nitro, and benzodifuranone
Disperse	Polyester, polyamide, and acetate	Azo, anthraquinone, and styryl
Reactive	Cotton, wool, silk, and nylon	Azo, anthraquinone, and phthalocyanine
Vat	Cotton and rayon	Anthraquinone and indigoids
Solvent	Plastics, gasoline, varnishes, lacquers, stains, fats, oil, and waxes	Azo, triphenylmethane, anthraquinone, and phthalocyanine

2 Adsorption

Even a little quantity of dye in water can be noxious and clearly visible, making it environmentally important to remove it from waste effluents [36]. Dye removal from wastewater is considered an environmental challenge, government regulations require that wastewater from textile industries to be handled, and an efficient process is needed to remove dyes efficiently [37]. For highly developed sorption processes, adsorption is the preferred method thanks to its low price and the excellent quality of treated effluents [38]. It is essential to have a thorough understanding of adsorption equilibrium in the aim to well understand an adsorption process [39]. The effect of the different process parameters, such as time of contact, adsorbate concentration, adsorbent concentration, temperature, pH, and $pH_{0,}$ should be investigated [40]. pH, initial dye concentration, adsorbent concentration, and process temperature all affect dye adsorption capacity of adsorbents. The pH of solution is strongly linked to the dye-classified adsorption, where it affects several parameters including the surface charge of the adsorbent and the degree of ionization of the adsorbate [41]. Modeling experimental data from adsorption methods is a significant method to estimate the mechanisms of the different adsorption systems. Thus, this paragraph presents an overview of the most commonly used adsorption isotherms.

2.1 *Langmuir Isotherm*

As well as being used to define gas–solid adsorption, Langmuir adsorption is used to measure and compare the capacities of adsorption of diverse adsorbents [42]. By equilibrating the proportional rates of adsorption and desorption (dynamic equilibrium), this isotherm calculates the surface coverage. Desorption is proportional to the fraction of open adsorbent surface, whereas adsorption is proportional to covered adsorbent surface [43].

Langmuir's equation is written as follows [44]:

$$q_e = \frac{q_m K_L.C_e}{1 + K_L.C_e} \tag{1}$$

Where *Ce* is the equilibrium concentration of adsorbate ($mg.g^{-1}$).

K_L is Langmuir constant linked to the capacity of adsorption ($mg.g^{-1}$), and its variation is related to the porosity and appropriate area of the adsorbent. Therefore, a larger surface area and volume of pores will result in a greater adsorption capacity.

2.2 Freundlich Isotherm

Adsorption on heterogonous surfaces are described by Freundlich isotherms [45]. As a result of this isotherm, active sites and their energies can be defined in terms of surface heterogeneity and exponential distribution [46].

The Freundlich equation is as follows:

$$q_e = K_F\, C_e^{\frac{1}{n}} \tag{2}$$

where q_e (mg/g) is the equilibrium removed amount. C_e is the equilibrium concentration, and K_F and 1/n are the Freundlich constants. K_F (L/mg) is a set of measures of adsorption capacity and affinity, and 1/n indicates the level of intensity of the removed dye.

2.3 Temkin Isotherm

Temkin isotherm model considers indirect interactions between adsorbate and adsorbate during adsorption. In a linear manner with increasing coverage of the adsorbent surface, the heat of adsorption (ΔH_{ads}) decreases for all molecules. Until obtaining maximum binding energy, adsorption is characterized by an identical distribution of binding energies [47]. Eq. (3) describes the Temkin isotherm:

$$q_e = \frac{\text{RT}}{b}\,\text{Ln}K_T + \frac{\text{RT}}{b}\,\text{Ln}\ C_e \tag{3}$$

where K_T represents the equilibrium binding constant or the maximum binding energy (L.mol^{-1}), b is factor bound up with to the heat of adsorption, R is the universal gas constant ($8.314\ \text{J.K}^{-1}.\text{mol}^{-1}$), and T is the temperature (K) [47].

2.4 Dubinin–Radushkevich Isotherm

This isotherm model estimates that adsorbent size is comparable to the micropore size and the adsorption equilibrium relation for a well-given adsorbate-adsorbent might be demonstrated separately of temperature while using the adsorption potential (ε):

$$\varepsilon = \text{RT}\,\text{Ln}\,(1 + \frac{1}{C_e}) \tag{4}$$

Dubinin–Radushkevich isotherm equation is demonstrated in this manner [46]:

$$\text{Ln } q_e = \text{Ln } q_m - \beta E^2 \tag{5}$$

where q_m is the *D-R* constant (mol.g^{-1}) and *B* represents the free energy of sorption, and these values can be calculated from the linearization of the *D-R* isotherm curve. *E* (kJ mol^{-1}) describes the energy per molecule of sorbate when it transfers to the solid surface of the bulk solution. This parameter can be calculated according to the following expression (6) [47]:

$$E = \frac{1}{\sqrt{2B}} \tag{6}$$

2.5 *Harkins–jura Isotherm*

The Harkins–Jura isotherm model anticipates that multilayer adsorption on surfaces of absorbents with the distribution of heterogeneous pores may occur [39]. This model has as equation:

$$\frac{1}{q_e^2} = \frac{B}{A} - \frac{1}{(A)}\log C_e \tag{7}$$

where *A* and *B* are Harkins–Jura constants. They can be determined through plotting $\frac{1}{q_e^2}$ versus $\log C_e$.

3 Adsorption of Textile Dyes Using Inorganic Materials (Apatites)

Apatites are classified among the largest families in inorganic chemistry, and they are phosphate-based materials. They are identified by their unique structure, thermal stability, and insolubility. These materials have fascinating physicochemical and thermodynamic characteristics. Apatites can be synthesized by several methods and by both solid and wet ways. Their fields of application are vast and numerous [48].

3.1 *Structure of Apatites*

Because of their chemical and thermal stability, apatites have attracted wide attention over the past two decades. Apatite is generally characterized by its hexagonal structure (Space Group P63/m) and by its chemical formula: $Me_{10}(XO_4)_6Y_2$ where Me is a divalent cation (Sr^{2+}, Ca^{2+}, Br^{2+}, Pb^{2+}…), the XO_4 group is a trivalent anion (PO_4^{3-}, VO_4^{3-}, AsO_4^{3-}…), and Y is a monovalent anion (F^-, Cl^-, OH^-…) [49]. The flexibility of the apatitic structure allows several substitutions: cationic, anionic, or both types at the same time [50–52]. A coupled substitution is performed when a trivalent rare earth, for example, Ln^{3+} replaces Me^{2+}, the electric neutrality is then ensured by the substitution of the group PO_4^{3-} by SiO_4^{4-}, and the resulting compounds are called britholites. Britholites are then silicate-apatites which contain a large quantities of the elements of bloc f of the periodic table [48].

3.2 Synthesis Methods of Apatites

Apatites can be synthesized by solid and wet methods, and several methods lead to the formation of these materials. Solid-state synthesis methods include solid-state reaction, mechanosynthesis, fusion, etc. And wet synthesis methods include sol–gel process, hydrothermal method, double decomposition synthesis, etc. [48].

3.3 Applications of Apatites

Due to their chemical and thermal stabilities and their ability to adapt a wide range of substitutions, apatites have shown a keen interest and they have attracted a great deal of attention. Nowadays, they are used in a wide variety of applications [53–56]. They are used as catalysts [57], laser materials [58], gas sensors [59], and oxide ion conductors [60]. Their chemical composition and crystallographic structure, similar to those of human tissue, offer excellent biocompatibility for bone and dental implantations [61]. They are widely used as biomaterials for example as bone grafting materials in the treatment of bone defects because of osteoporosis or injury. Ceramics based on calcium phosphate are applied as bone grafts in hard tissue surgery thanks to their chemical composition which is so similar to human bone composition, bioresorbable, and have a strong bio-activity [62]. There are some examples of use as cement used in facial surgery to replace autogenous bone grafts [63]. And they are presented in many applications as a drug carrier through their unique structure with interconnected micropores [64]. One of the most common areas of application for the apatites is adsorption, which occurs on the surface between this material which is called adsorbent and an adsorbate which can be a dye [65], an organic compound [66], a polymer [67], DNA [68], proteins [69], drugs [70], etc. This paragraph focuses only on the adsorption of textile dyes. These dyes cause pollution of surface water and groundwater. Thanks to their effectivity and low cost, apatites are used as adsorbents for the elimination of textile dyes [71]. They are even better than natural phosphates because apatites have a well-defined structure while natural phosphates have a complex structure that cannot be easily defined [72].

3.4 Adsorption of Basic Blue 41 by Strontium Oxybritholites

In the aim to removal textile dyes by britholites, three compounds of strontium oxybritholites with the general formula $Sr_{10-x}La_x(PO_4)_{6-x}(SiO_4)_xO$, where $x = 1$, 3, and 5 were prepared by mechanochemical method using a Fritsch planetary micro mill PULVERISETTE 7. The synthesis reaction occurs as follows:

$$4SrCO_3 + (6-x)/2\,Sr_2P_2O_7 + xSiO_2 + x/2\,La_2O_3Sr_{10-x}La_x(PO_4)_{6-x}(SiO_4)_xO + 4CO_2$$

A part of each sample have been calcined at 1000 °C /24 h. These powders are used to remove a textile dye which is Basic Bleu 41 (BB41). A series of experiments are done while varying some parameters like concentration of the aqueous solution, temperature, mass of the apatites, pH, pH_0, and kinetics of adsorption. According to the experimental results, these strontium oxybritholites removed effectively the dye. The adsorption of the Basic Bleu 41 by the oxybritholites materials is well described by the pseudo-second-order model, whereas the isotherm of adsorption is determined by the Langmuir model. Additionally, the thermodynamic study revealed that the adsorption process is spontaneous and endothermic.

4 Adsorption of Textile Dyes Using Organic Material (Posidonia)

4.1 Posidonia

Posidonia *(P. oceanica),* the miracle of the Mediterranean, is a marine flowering plant present in most of the Mediterranean and only in the Mediterranean. Posidonia is considered one of the extremely important and best-studied seagrass species in the Mediterranean Sea [73]. Like terrestrial plants, *P. oceanica* which is an endemic species in the Mediterranean. It loses its leaves seasonally. These dead leaves called "mattes" are accrued on the beaches in enormous quantities, provoking serious environmental and economical issues [74]. Mattes act as significant carbon sinks, storing up to around 90 kg C_{org} m^{-2} in their top meter of sediment [75–78] which ranks *P. oceanica* among the top blue carbon sinks compared to other seagrass species [79, 80]. In order to take advantage, these dead leaves are used as an alternative and low-cost adsorbent for dyes removal [81].

4.2 Removal of Textile Dyes by Posidonia

Aqueous solutions were treated with *P. oceanica* (L.) leaf sheaths as inexpensive, renewable biological, and available adsorbents to remove reactive textile dyes [82]. Several studies have shown that many dyes, including methylene blue, alpacide gray, and alpacide yellow, have been shown to be biosorbed by *P. oceanica* fibers [83–85]. The pseudo-second-order model give adequate description of the biosorption kinetics of all these dyes and with a high maximum adsorption capacities of *P. oceanica* fibers. Many other reports [86–89] showed that this cellulosic support was so efficient for the removal of many dyes. The relevant results have proven the high adsorption capacity of this seagrass. A series of experiments were generally made to evaluate the effect of the different parameters including contact time, solution pH, biosorbent dosage and initial dye concentration, temperature, etc. [83]. Comparing the biosorption behavior of this marine lignocellulosic plant with other adsorbent materials, it is shown that it has a high and a considerable biosorption capacity.

This marine waste (*P. oceanica)* is an eco-friendly material and sustainable natural resource, as it is used for a large variety of applications, including the treatment of waste water. It is also possible to prepare activated carbon from *P. oceanica* at low cost [90].

5 Conclusion

A huge amount of synthetic dyes, almost around 1 million tons, are produced annually in the world. They are used as colorants in various industries, such as textile, food, cosmetic, and paper. In addition, wastewaters of textiles industries present an ecological and environmental problem. These dyes release chemical compounds which are highly hazardous to human health. In order to reduce this phenomenon, many methods are proposed but most of them are very expensive. The adsorption is considered as an efficient technique of removal which is simple, cheap, and widely used. The adsorbents can be organic or inorganic materials. Among the most commonly used inorganic adsorbent, apatites are low-cost compounds, their synthesis methods are simple, and there are known for their high adsorption capacity. As organic adsorbent, *P. oceanica*, this Mediterranean endemic species is considered as an alternative and low-priced biomaterial for the elimination of dyes. Several research studies have proven that the capacity of adsorption of this seagrass is important.

References

1. Gita S, Hussan A, Choudhury TG (2017) Impact of textile dyes waste on aquatic environments and its treatment. 35:2349–2353
2. Ratna, Padhi BS (2012) Pollution due to synthetic dyes toxicity and carcinogenicity studies and remediation. 3:940–955. https://doi.org/10.6088/ijes.2012030133002

3. Phugare SS, Kalyani DC, Surwase SN, Jadhav JP (2011) Ecofriendly degradation, decolorization and detoxification of textile effluent by a developed bacterial consortium. 74:1288–1296. https://doi.org/10.1016/j.ecoenv.2011.03.003
4. Bensalah H, Bekheet MF, Younssi SA, Ouammou M, Gurlo A (2017) Removal of cationic and anionic textile dyes with moroccan natural phosphate. 5:2189–2199. https://doi.org/10.1016/j.jece.2017.04.021
5. Saini RD (2017) Textile organic dyes: polluting effects and elimination methods from textile waste water. 9:121–136
6. Bembli M, Kooli F, Khiari R, Boughzala K (2022) Removal of basic blue 41 by waste product from the phosphate industry: batch design and regeneration. 246:291–303. https://doi.org/10.5004/dwt.2022.28013
7. Bembli M, Zine Elabidine M, Bechrifa A, Boughzala K (2022) Etude de l'adsorption de l'acide rouge 52 par de sous-produits de l'industrie phosphatière. 15:51–65. https://doi.org/10.48419/IMIST.PRSM/rhazes-v15.31707
8. Bhatia D, Sharma NR, Singh J, Kanwar RS (2017) Biological methods for textile dye removal from wastewater: a review. 47:1836–1876. https://doi.org/10.1080/10643389.2017.1393263
9. Lahkimi A, Oturan MA, Oturan N, Chaouch M (2007) Removal of textile dyes from water by the electro-Fenton process. 5:35–39. https://doi.org/10.1007/s10311-006-0058-x
10. Chu W (2001) Dye removal from textile dye wastewater using recycled alum sludge. 35:3147-3152. https://doi.org/10.1016/s0043-1354(01)00015-x
11. Wang S, Zhu ZH (2006) Characterisation and environmental application of an Australian natural zeolite for basic dye removal from aqueous solution. 136:946–952. https://doi.org/10.1016/j.jhazmat.2006.01.038
12. Boughzala K, Kooli F, Meksi N, Bechrifa A, Bouzouita K (2020) Waste products from the phosphate industry as efficient removal of Acid Red 88 dye from aqueous solution: their regeneration uses and batch design adsorber. 202:410–419. https://doi.org/10.5004/dwt.2020.26149
13. Robinson T, McMullan G, Marchant R, Nigam P (2001) Remediation of dyes in textile effluent: a critical review on current treatment technologies with a proposed alternative. 77:247–255. https://doi.org/10.1016/S0960-8524(00)00080-8
14. Beyene H.D (2014) The potential of dyes removal from textile wastewater by using different treatment technology: a review. 347–353. https://doi.org/10.11648/J.IJEMA.20140206.18
15. Edokpayi O, Osemwenkhae O, Ayodele BV, Ossai J, Fadilat SA, Ogbeide SE (2018) Batch adsorption study of methylene blue in aqueous solution using activated carbons from rice husk and coconut shell. 22:631–635. https://doi.org/10.4314/jasem.v22i5.4
16. Erdem E, Çölgeçen G, Donat R (2005) The removal of textile dyes by diatomite earth. 282:314–319. https://doi.org/10.1016/j.jcis.2004.08.166
17. Dabrowski A (2001) Adsorption- from theory to practice. 93:135–224. https://doi.org/10.1016/S0001-8686(00)00082-8
18. Yagub MT, Sen TK, Afroze S, Ang HM (2014) Dye and its removal from aqueous solution by adsorption: a review. 209:172–184. https://doi.org/10.1016/j.cis.2014.04.002
19. Kyzas GZ, Fu J, Matis KA (2013) The change from past to future for adsorbent materials in treatment of dyeing wastewaters. 6:5131–5158. https://doi.org/10.3390/ma6115131
20. Salleh MAM, Mahmoud DK, Abdul Karim WAW, Idris A (2011) Cationic and anionic dye adsorption by agricultural solid wastes: a comprehensive review. 280:1–13. https://doi.org/10.1016/j.desal.2011.07.019
21. Nigam P, Armour G, Banat IM, Singh D, Marchant R (2000) Physical removal of textile dyes from effluents and solid-state fermentation of dye-adsorbed agricultural residues. 72:219–226. https://doi.org/10.1016/S0960-8524(99)00123-6
22. Sultan M (2017) Polyurethane for removal of organic dyes from textile wasewater. 15:347–366. https://doi.org/10.1007/s10311-016-0597-8
23. Tchinsa A, Faysal Hossain Md, Wang T, Zhou Y (2021) Removal of organic pollutants from aqueous solution using metal organic frameworks (MOFs)-based adsorbents: a review. 284:131393. https://doi.org/10.1016/j.chemosphere.2021.131393

24. Elmorsi RR, El-Wakeel ST, Shehab El-Dein WA, Lotfy HR, Rashwan WE, Nagah M, Shaaban SA, Sayed Ahmed SA, El-Sherif IY , Abou-El-Sherbini KS (2019) Adsorption of Methylene Blue and Pb2+ by using acid-activated *Posidonia Oceanica* waste. 9:3356. https://doi.org/10.1038/s41598-019-39945-1
25. Ferchichi K, Amdouni N, Chevalier Y, Hbaieb S (2022) Low-cost *Posidonia oceanica* bio-adsorbent for efficient removal of antibiotic oxytetracycline from water. https://doi.org/10.1007/s11356-022-21647-3
26. Boughzala K, Fattah N, Bouzouita K, Ben Hassine H (2015) Etude minéralogique et chimique du phosphate naturel d'Oum El Khecheb (Gafsa, Tunisie), Science des matériaux (Laboratoire LARHYSS). 6
27. Nasr S, Jeahi S, Hidouri M, Boughzala K (2021) Adsorption of Acid Red 52 contained in waste water by co-products pf the phosphate industry: kinetics and thermodynamic. preprints 2021120319. https://doi.org/10.20944/preprints202112.0319.v1
28. Jebahi S, Hidouri M, Boughzala K (2021) Kinetic and thermodynamic study of the retention of textile Effluent bu co-products from the phosphate industry. 14:76–94. https://doi.org/10.48419/IMIST.PRSM/rhazes-v14.31154
29. Sboui N, Agougui H, Jabli M, Boughzala K (2022) Synthesis, Physico-chemical and structural properties of Silicate Apatites: effect of synthetic methods on apatite structure and dye removal. 142:109628. https://doi.org/10.1016/j.inoche.2022.109628
30. Bahdod A, El Asri S, Saoiabi A, Coradin T, Laghzizil A (2009) Adsorption of phenol from an aqueous solution by selected apatite adsorbents: kinetic process and impact of the surface properties. 43:313–318. https://doi.org/10.1016/j.watres.2008.10.023
31. El-Zahhar AA, Awwad NS (2016) Removal of malachite green dye from aqueous solutions using organically modified hydroxyapatite. 4:633–638. https://doi.org/10.1016/j.jece.2015.12.014
32. Asl MN, Mahmodi NM, Teymouri P, Shahmoradi B, Rezaee R, Maleki A (2016) Adsorption of organic dyes using copper oxide nanoparticles: isotherm and kinetic studies. 57:25278–25287. https://doi.org/10.1080/19443994.2016.1151832
33. Zhu HY, Jiang R, Fu YQ, Li RR, Yao J, Jiang ST (2016) Novel multifunctional NiF_2O_4/ZnO hybrids for dye removal by adsorption, photocatalysis and magnetic separation. 369:1–10. https://doi.org/10.1016/j.apsusc.2016.02.025
34. Zhang Z, Kong J (2011) Novel magnetic Fe_3O_4/C nanoparticles as adsorbents for removal of organic dyes from aqueous solution. 193:325–329. https://doi.org/10.1016/j.jhazmat.2011.07.033
35. Singh K, Kumar P, Srivastava R (2017) An overview of textile dyes and their removal techniques: Indian perspective. 36:790-797
36. Chiou MS, Ho PY, HY Li (2004) Adsorption of anionic dyes in acid solutions using chemically cross-linked chitosan beads. 60:69–84. https://doi.org/10.1016/S0143-7208(03)00140-2
37. Lee JW, Choi SP, Thiruvenkatachari R, Shim WG, Moon H (2006) Evaluation of the performance of adsorption and coagulation processes for the maximum removal of reactive dyes. 69:196–203. https://doi.org/10.1016/j.dyepig.2005.03.008
38. Qadeer R (2007) Adsorption behavior of ruthenium ions on activated charcoal from nitric acid medium. 293:217–223. https://doi.org/10.1016/j.colsurfa.2006.07.035
39. Ayawei N, Ebelegi AN, Wankasi D (2017) Modelling and interpretation of adsorption isotherms. Article ID 3039817 https://doi.org/10.1155/2017/3039817
40. Khenifi A, Bouberka Z, Sekrane F, Kameche M, Derriche Z (2007) Adsorption study of an industrial dye by an organic clay. 13:149–158. https://doi.org/10.1007/s10450-007-9016-6
41. Mohd Salleh MA, Mahmoud DK, Abdul Karim WAW, Idris A (2011) Cationic and Anionic dye adsorption by agricultural solid wastes: a comprehensive review. 280:1–13
42. Elmorsi TM (2011) Equilibrium isotherms and kinetic studies of removal of methylene blue dye by adsorption onto miswak leaves as anatural adsorbent. 2:817–827. https://doi.org/10.4236/jep.2011.26093
43. Günay A, Arslankaya E, Tosun I (2007) Lead removal from aqueous solution by natural and pretreated clinoptilolite: adsorption equilibrium and kinetics. 146:362–371. https://doi.org/10.1016/j.jhazmat.2006.12.034

44. Vasanth Kumar K, Sivanesan S (2005) Prediction of optimum sorption isotherm: comparison of linear and non-linear method. 126:198–201. https://doi.org/10.1016/j.jhazmat.2005.06.007
45. Ayawei N, Angaye SS, Wankasi D, Dikio ED (2015) Synthesis, characterization andapplication of Mg/Al layered double hydroxide for the degradation of congo red in aqueous solution. 5:56–70. https://doi.org/10.4236/ojpc.2015.53007
46. Ayawei N, Ekubo AT, Wankasi D, Dikio ED (2015) Adsorption of congo red by Ni/Al-CO_3: equilibrium, thermodynamic and kinetic studies. 31:1307–1318. https://doi.org/10.13005/ojc/310307
47. Piccin JS, Dotto GL, Pinto LAA (2011) Adsorption isotherms and thermochemical data of FD&C Red N° 40 Binding by Chitosan. 28:295–304. https://doi.org/10.1590/S0104-66322011000200014
48. Bembli M, Boughzala K (2022) Structure des apatites, méthodes de synthèse et domaines d'application. 15:66–70. https://doi.org/10.48419/IMIST.PRSM/rhazes-v15.31705
49. Boughzala K, Bouzouita K (2015) Synthesis and characterization of strontium–calcium–lanthanum apatites $Sr_{7-x}Ca_xLa_3(PO_4)_3(SiO_4)_3F_2$ $0 \leq x \leq 2$. 18:858–866, https://doi.org/10.1016/j.crci.2014.11.011.
50. Veiderma M, Kaia T, Knubovets R, Peld M (2005) Impact of anionic substitutions on apatite structure and properties. 690:2638–2643. https://doi.org/10.1016/j.jorganchem.2004.11.022
51. Tamm T, Peld M (2006) Computational study of cation substitutions in apatites. 179:1581–1587 https://doi.org/10.1016/j.jssc.2006.02.012
52. Carpéna J, Boyer L, Fialin M, Kiénast JR, Lacout JL (2001) Ca^{2+}, $PO_4{}^{3-} \rightleftharpoons Ln^{3+}$, $SiO_4{}^{4-}$ coupled substitution in the apatitic structure: stability of the mono-silicated fluor-britholite. 333:379–379. https://doi.org/10.1016/S1251-8050(01)01656-1
53. Wang H, Cai K, Sun M, Xia XP, Lai CK, Li P, Wan B, Zhang Z (2022) Apatite as a magma redox indicator and its application in metallogenic research. 422–423
54. Choimet M, Tourrette A, Marsan O, Rassu G, Drouet C (2020) Bio-inspired apatite particles limit skin penetration of drugs for dermatology applications. 111:418–428. https://doi.org/10.1016/j.actbio.2020.05.010
55. Youness RA, Ibrahim MA, Taha MA (2022) Evaluation of the electrical and dielectric behavior of the apatite layer formed on the surface of hydroxyapatite /hardystonite/ copper oxide hybrid nanocomposites for bone repair applications. 48:19837–19850. https://doi.org/10.1016/j.ceramint.2022.03.259
56. Lemlikchi W, Drouiche N, Belaicha N, Oubagha N, Baaziz B, Mecherri MO (2015) Kinetic study of the adsorption of textile dyes on synthetic hydroxyapatite in aqueous solution. 32:233–237. https://doi.org/10.1016/j.jiec.2015.08.023
57. Gruselle M (2015) Apatites: A new family of catalysts in organic synthesis. 793:93–101. https://doi.org/10.1016/j.jorganchem.2015.01.018
58. Stephen A. Payne, Laura D. DeLoach, Larry K. Smith, Wayne L. Kway, John B. Tassano, William F. Krupke (1994) Ytterbium-doped apatite-structure crystals: a new class of laser materials. 76:497. https://doi.org/10.1063/1.357101
59. Takeda N, Itagaki Y, Aono H, Sadaoka Y (2006) Preparation and characterization of $Ln_{9.33+x/3}Si_{6-x}Al_xO_{26}$ (Ln = La, Nd and Sm) with apatite-type structure and its application to a potentiometric O_2 gas sensor. 115:455–459. https://doi.org/10.1016/j.snb.2005.10.009
60. Slater PR, Sansom JEH, Tolchard JR(2004) Development of apatite-type oxide ion conductors. 4:373–384. https://doi.org/10.1002/tcr.20028
61. Okada M, Matsumoto T (2015) Synthesis and modification of apatite nanoparticles for use in dental and medical applications. 51:85–95. https://doi.org/10.1016/j.jdsr.2015.03.004
62. Chaair H, Labjar H, Britel O (2017) Synthesis of β-tricalcium phosphate. 101:120–124. https://doi.org/10.1016/j.morpho.2017.06.002
63. Lee DW, Kim JY, Lew DH (2010) Use of rapidly hardening hydroxyapatite cement for facial contouring surgery. 21:1084–1088. https://doi.org/10.1097/SCS.0b013e3181e1b64b
64. Hong MH, Oh DS, Lee YK (2011) Drug release from porous hydroxyapatite spheres as a synthetic bone-graft material. 48:S186. https://doi.org/10.1016/j.bone.2011.03.434

65. Vasugi G, Kumar G Suresh, Thamizhavel A, Girjia EK (2016) Removal of reactive textile dyes using carbonate substituted Nanocrystalline Hydroxyapatite. 10:38–46. https://doi.org/10.1166/jbns.2016.1357
66. Yang K, Xing B (2010) Adsorption of organic compounds by carbon nanomaterials in aqueous phase: polanyi theory and its application. 110:5989–6008. https://doi.org/10.1021/cr100059s
67. Veloso CH, Filippov LO, Filippova IV, Ouvrad S, Araujo AC (2020) Adsorption of polymers onto iron oxides: equilibrium isotherms. 9:779–788. https://doi.org/10.1016/j.jmrt.2019.11.018
68. Grunenwald A, Keyser C, Sautereau AM, Crubézy E, Ludes B, Drouet C (2014) Adsorption of DNA on biomimetic apatites: toward the understanding of the role of bone and tooth mineral on the preservation of ancient DNA. 292:867–875. https://doi.org/10.1016/j.apsusc.2013.12.063
69. Barberi J, Spriano S (2021) Titanium and protein adsorption: an overview of mechanisms and effects of surface features. 14:1590. https://doi.org/10.3390/ma14071590
70. Errassifi F, Menbaoui A, Autefage H, Benaziz L, Ouizat S, Santran V, Sarda S, Lebugle A, Combes C, Barroug A, Sfihi H, Rey C (2010) Adsorption on apatitic calcium phosphates: applications to drug delivery. 218:159–174 https://doi.org/10.1002/9780470909898.ch16
71. El Boujaady H, El Rhilassi A, Bennani-Ziatni M, El Hamri R, Taiti A, Lacout JL (2011) Removal of a textile dye by adsorption on synthetic calcium phosphates.275:10–16.https://doi.org/10.1016/j.desal.2011.03.036
72. Leventouri Th, Chakoumakos BC, Papanearchou N, Perdikatsis V (2001) Comparison of crystal structure parameters of natural and synthetic apatites from neutron powder diffraction. 16:2600–2606. https://doi.org/10.1557/JMR.2001.0357
73. Telesca L, Belluscio A, Criscoli A, Ardizzone G, Apostolaki ET, Fraschetti S, Gristina M, Knittweis L, Martin CS, Pergent G, Alagna A, Badalamenti F, Garofalo G, Gerakaris V, Pace ML, Pergent-Martini C, Salomidi M (2015) Seagrass meadows (*Posidonia oceanica*) distribution and trajectories of change. 5:12505. https://doi.org/10.1038/srep12505
74. Apostolaki ET, Caviglia L, Santinelli V, Cundy AB, Tramati CD, Mazzola A, Vizzini S (2022) The Importance of Dead Seagrass (*Posidonia oceanica*) Matte as a Biogeochemical Sink. https://doi.org/10.3389/fmars.2022.861998
75. Serrano O, Davis G, Lavery PS, Duarte CM, Martinez-Cortizas A, Mateo MA, Masqué P, Arias-Ortiz A, Rozaimi M, Kendrick GA (2016) Reconstruction of centennial-scale fluxes of chemical elements in the Australian coastal environment using seagrass archives. 541:883–894. https://doi.org/10.1016/j.scitotenv.2015.09.017
76. Mazarrasa I., Marbà N., Garcia-Orellana J., Masqué P., Arias-Ortiz A., Duarte C. M. (2017b). Effect of Environmental Factors (Wave Exposure and Depth) and anthropogenic pressure in the c sink capacity of *Posidonia Oceanica* meadows. 62:1436–1450. https://doi.org/10.1073/pnas.0905620106
77. Apostolaki ET., Vizzini S., Santinelli V., Kaberi H., Andolina C., Papathanassiou E (2019) Exotic *Halophila Stipulacea* is an introduced carbon sink for the Eastern Mediterranean Sea. 9:9643. https://doi.org/10.1038/s41598-019-45046-w
78. Monnier B., Pergent G., Mateo M.Á., Carbonell R., Clabaut P., Pergent-Martini C (2021) Sizing the carbon sink associated with *Posidonia Oceanica* Seagrass meadows using very high-resolution seismic reflection imaging. 170:105415. https://doi.org/10.1016/j.marenvres.2021.105415
79. Sousa AI, da Silva JF, Azevedo A, Lillebø AI (2019) Blue Carbon stock in *Zostera Noltei* Meadows at Ria De Aveiro Coastal Lagoon (Portugal) over a decade. 9:14387. https://doi.org/10.1038/s41598-019-50425-4
80. Serrano O, Almahasheer H, Duarte CM, Irigoien X (2018) Carbon stocks and accumulation rates in red sea seagrass meadows. 8:113. https://doi.org/10.1038/s41598-018-33182-8
81. Cengiz S, Tanrikulu F, Aksu S (2012) An alternative source of adsorbent for the removal of dyes from textile waters: *Posidonia oceanica* (L.). 189-190:32–40. https://doi.org/10.1016/j.cej.2012.02.015
82. Ncibi MC, Mahjoub B, Seffen M (2007) Adsorptive removal of textile reactive dye using *Posidonia oceanica* (L.) fibrous biomass. 4:433–440. https://doi.org/10.1007/BF03325978

83. Ncibi MC , Mahjoub B, Seffen M (2007) Kinetic and equilibrium studies of methylene blue biosorption by *Posidonia oceanica* (L.) fibres. 139:280–285. https://doi.org/10.1016/j.jhazmat.2006.06.029
84. Ncibi M.C , Mahjoub B, Seffen M (2008) Investigation of the sorption mechanisms of metal—complexed dye onto *Posidonia oceanica* (L.) fibres through kinetic modelling analysis. 99:5582–5589. https://doi.org/10.1016/j.biortech.2007.10.040
85. Ncibi MC, Mahjoub B, Ben Hamissa AM , Ben Mansour R, Seffen M (2009) Biosorption of textile metal-complexed dye from aqueous medium using *Posidonia oceanica* (L.) leaf sheaths: mathematical modelling. 243:109–121. https://doi.org/10.1016/j.desal.2008.04.018
86. Guezguez I, Dridi-Dhaouadi S, Mhenni F (2009) Sorption of Yellow 59 on *Posidonia oceanica*, a non-conventional biosorbent: comparison with activated carbons. 29:197–204. https://doi.org/10.1016/j.indcrop.2008.05.002
87. Cengiz S, Cavas L (2010) A promising evaluation method for dead leaves of *Posidonia oceanica* (L.) in the adsorption of methyl violet. 12:728–736. https://doi.org/10.1007/s10126-010-9260-8
88. Cavas L, Karabay Z, Alyuruk H, Doğan H, Demir G.K (2011) Thomas and artificial neural network models for the fixed- bed adsorption of methylene blue by a beach waste *Posidonia oceanica* (L.) dead leaves. 71:557–562. https://doi.org/10.1016/j.cej.2011.04.030
89. Dural U, Cavas L, Katsaros F, Papageorgiou S (2011) Methylene blue adsorption on the activated carbon prepared from *Posidonia oceanica* (L.) dead leaves: kinetics and equilibrium studies. 168:77–85. https://doi.org/10.1016/j.cej.2010.12.038
90. Saad MEK, Khiari Ramzi, Elaloui E, Moussaoui Y (2014) Adsorption of anthracene using activated carbon and *Posidonia Oceanica*. 7:109-113. https://doi.org/10.1016/j.arabjc.2013.11.002

The manufacturer's authorised representative in the EU is Springer Nature Customer Service Centre GmbH, Europaplatz 3, 69115 Heidelberg, Germany. If you have any concerns regarding our products, please contact ProductSafety@springernature.com

Printed and bound by CPI Group (UK) Ltd, Croydon, CR0 4YY
01/10/2026
02225850-0007